The Ecology of Agroecosystems

The Ecology of Agroecosystems

Editor: Harvey Richardson

R CALLISTO REFERENCE

www.callistoreference.com

Callisto Reference,
118-35 Queens Blvd., Suite 400,
Forest Hills, NY 11375, USA

Visit us on the World Wide Web at:
www.callistoreference.com

ISBN: 978-1-64116-269-2 (Hardback)

Cataloging-in-Publication Data

The ecology of agroecosystems / edited by Harvey Richardson.
    p. cm.
Includes bibliographical references and index.
ISBN 978-1-64116-269-2
1. Agricultural ecology. 2. Ecology. 3. Ecosystem management.
4. Agriculture--Environmental aspects. I. Richardson, Harvey.
S589.7 .E26 2020
577.55--dc23

Table of Contents

Permissions

List of Contributors

Index

Preface

The study of ecological processes that are important to agricultural production systems is called agroecology. The basic unit of study in agroecology is called an agroecosystem. It can be viewed as a subset of a conventional ecosystem and may be defined as a unit of agricultural activity that includes the living and abiotic components as well as their interactions. It embraces not only the immediate site of agricultural activity but also the region that is impacted by agriculture. Current farming practices have led to over-exploitation of water resources, reduced soil fertility and high levels of erosion. Therefore there has been a burgeoning interest in farming with agroecosystems. A major initiative in agroecology is to promote practices that decrease the distinction between natural ecosystems and agroecosystems by reducing the impact of agriculture. The book aims to shed light on some of the unexplored aspects of agroecosystems and the recent researches in agroecology. A number of latest researches have been included to keep the readers up-to-date with the global concepts in these areas of study. The extensive content of this book provides the readers with a thorough understanding of the subject.

This book has been the outcome of endless efforts put in by authors and researchers on various issues and topics within the field. The book is a comprehensive collection of significant researches that are addressed in a variety of chapters. It will surely enhance the knowledge of the field among readers across the globe.

It gives us an immense pleasure to thank our researchers and authors for their efforts to submit their piece of writing before the deadlines. Finally in the end, I would like to thank my family and colleagues who have been a great source of inspiration and support.

Editor

Prevalence of *Trypanosoma Cruzi* in Backyard Mammals from a Rural Community of Yucatan, Mexico

Hugo Antonio Ruiz-Piña[1*], Edwin Gutierrez-Ruiz[2],
Francisco Javier Escobedo-Ortegon[1], Roger Ivan Rodriguez-Vivas[2],
Manuel Bolio-Gonzalez[2] and Dianelly Ucan-Leal[2]

[1]*Laboratorio de Zoonosis y otras Enfermedades Transmitidas por Vector,
Centro de Investigaciones Regionales "Dr. Hideyo Noguchi", Universidad
Autónoma de Yucatán. email: rpina@correo.uady.mx*
[2]*Departamento de Salud Animal y Medicina Preventiva, Facultad de Medicina
Veterinaria y Zootecnia, Universidad Autónoma de Yucatán.*
Corresponding author

SUMMARY

Trypanosoma cruzi the causative agent of Chagas disease, can be found in more than 150 different species of domestic and wild mammals in the American continent. Domestic mammals like dogs, and commensal rodents, have been frequently implicated as hosts and/or reservoirs in the domestic and peridomestic cycle of transmission, however, little attention have been paid to backyard mammals as potential maintaining hosts of *T. cruzi* in the peridomicile. In the present work, we reported the first data of prevalence of infection with *T. cruzi* in the backyard mammals of households in a rural community of Yucatan, Mexico. A total of 84 animals, including swine, sheep, horses, cattle and rabbits, were sampled. Blood samples were collected and processed for DNA extraction of parasite and analyzed by PCR technique. From all animals tested, 75% of the pigs (21/28), 85.71% of the sheep (6/7) and 100% of the horses (8/8), were positive for *T. cruzi*. Cattle (28) and rabbits (13) were negative. The results obtained constitute an important preliminary evidence to implicate backyard mammals as potential maintaining hosts of *T. cruzi* in the peridomestic habitat of rural communities in Yucatan.

Key words: *Trypanosoma cruzi;* prevalence; backyard mammals; Yucatan.

INTRODUCTION

American Trypanosomiasis, also known as Chagas disease, is a parasitic disease caused by *Trypanosoma cruzi* and is endemic to the American continent, from the south of the United States of America to Argentina and Chile. It is considered a serious health problem with an estimate of 15 million people infected (OMS, 2007). The Chagas disease is considered endemic in several regions of Mexico (Ramsey et al., 2003). The infection is maintained by more than 150 domestic and wild animals and transmitted from infected to susceptible animals through the bite of triatomine insects (Días, 2000). The backyard of the houses, especially in rural communities of Mexico, represents a very complex system where plants and animals are kept for consumption of families (Ruiz-Piña y Reyes-Novelo, 2012). Many different animal species can be found in the system but domestic fowl and pigs are the commonest, with cattle, sheep, horses and other minor species like rabbits are host as well (Acosta-Casanova, 2004). The role that backyard mammals of rural households could play in the maintaining of peridomestic transmission of *T. cruzi*, has not been investigated in Yucatan; the habitat conditions found in the rural backyard system clearly favor the presence of the triatomine vector, probably representing an important source of blood, and consequently a risk for people living in the same environment (Ruiz-Piña et al., 2013). Previous studies in the Yucatan peninsula have reported the circulation of *T. cruzi* in rural peridomicile, both in synanthropic mammals as opossums, and in livestock kept in backyard (Ruiz-Piña and Cruz-Reyes, 2002; Duarte-Ubaldo, 2005; Jiménez-Coello et al. 2012).

The objective of this study was to determine the prevalence of infection in mammals kept in the backyard system, in order to go further into the knowledge of dynamics of *T. cruzi* transmission in households of rural communities of Yucatan, Mexico.

MATERIALS AND METHODS

Area of study

The study was carried out in the community of Molas, Yucatan, Mexico, located 16 km South of Merida capital city of the state of Yucatan, at 20° 48' 58.54" LN and 89° 37' 45.43" LW.

Diagnostic tests were carried out in the "Laboratorio de Zoonosis y otras Enfermedades Transmitidas por Vector" of the Centro de Investigaciones Regionales (CIR) "Dr. Hideyo Noguchi", of Universidad Autonoma de Yucatan.

Study population

As part of a larger study on zoonosis diseases in a rural community of Yucatan, Mexico, 156 out of 300 houses in the community were visited, 33 of them had backyard animals but only in 16 of them it was possible to obtain blood samples from animals. In some cases, owners did not authorize to include their animals in the study.

A total of 84 backyard animals distributed in 16 houses were sampled. Five mililiters of blood were obtained from cattle (28), sheep (7), horses (8) and pigs (28) from the jugular vein, but in the case of rabbits (13)

one mililiter was obtained from the auricular vessels. Blood was placed in tubes with EDTA and kept refrigerated (4 °C) until the tests were performed. All procedures were followed accordingly with NOM-062-ZOO-1999 for animal care.

Diagnostic test

To detect the presence of *T. cruzi* in blood samples a polymerase chain reaction test (PCR) was carried using the methodology described by Monteón et al. (1994). This methodology uses primers KNS1 and KNS2 designed from kinetoplast sequence, this reaction has proved higher sensitivity than other methods, with 100% sensitivity and 86% specificity.

DNA was extracted from the samples using common method previously described. Briefly, 80 µl of sterile water (Baxter) were mixed with 20 µl of sample (dilution 1:5), then the mixture was heated at 95 °C for 10 minutes and then centrifuged to maximum (approximately 15000 x g) in an microfuge for 5 min; afterwards, 8 µl of supernatant were placed in a new PCR tube with 10 µl of the master mix (Gotaq Green Master Mix 2x, Promega, Madison, WI, USA)) and 1 µl (10 pmol) of each of the primers to give a final volume of 20 µl. After that, the amplification program, consisting of 35 cycles of 92 °C for 1 minute, 56 °C for 2 minutes and 72 °C for 1 minute was run in a Bio-RadiCycler thermal cycler (Bio-Rad Laboratories, Hercules, CA, USA). This program also includes a denaturing step before de cycles at 95 °C for 3 min and an additional step after, at 72 °C for 10 min.

PCR products were electrophoresed in a molecular biology grade agarose gel (1.2 %) and stained with ethidium bromide (10 µg/ml). A DNA weight marker Sigma ΦX174 DNA/Hae III Marker (Sigma-Aldrich Mexico, DF, Mexico) was included on each electrophoresis run. Electric current was applied at 100 volts for 25 minutes and fragments visualized in a UV light transilluminator (UVP, Upland, CA, USA) at 330 nm and the image documented with EDAS 290 1D Gel documentation system v. 3.5 from Kodak (Scientific Imaging Systems, Rochester, NY, USA).

Data Analysis

The prevalence of infection for each animal species was obtained using formula described by Thrusfield (2007).

RESULTS AND DISCUSSION

Thirty five out of 84 samples were positive (42 %) for the amplification of PCR products corresponding to *T. cruzi* (Figure 1). Total results for *T. cruzi* infection prevalence in all mammal species analyzed are presented in Table 1. Cattle and rabbit samples were negative in the PCR test for *T. cruzi*.

Novel and preliminary findings were obtained in the present study, especially regarding sheep and horses with the highest *T. cruzi* infection prevalence (85.7% and 100% respectively). According to Noireau (1999) these animal species generally present low indices of infection and do not play an important role as hosts; however, despite the small number of animals tested in this study, our results obligate us to continue the research in order to determine if that these species could play a part in the maintenance of *T. cruzi* in the backyard system. Pigs also shown a high prevalence of *T. cruzi* infection, the only previous report of pigs naturally infected in Mexico was published by Salazar-Schettino *et al.* (1997) and Jiménez-Coello et al. (2012), Nonetheless, the later study showed low seroprevalence in pig farms, making the finding of this study relevant, since backyard pigs apparently had higher prevalence, probably because of *T. cruzi* cycle stablished in the peridomicile, including other domesticated and synanthropic mammals and infected *T. dimidiata* vectors in the locality (Koyoc-Cardeña et al., 2015). The potential role of horses, pigs and sheep in the cycle of *T. cruzi* in the backyard system of rural communities in Mexico, needs further investigation.

In relation to the relative importance of larger mammals as hosts for *T. cruzi*, Noireau (1999) argued that there is a negative correlation between body size and infection index with *T. cruzi*, suggesting short infection rates and duration in large compared to small animals. Other researchers claim that cattle could be a host for the parasite (Fujita *et al.*, 1994). The different prevalence data found for different animal species could also have been influenced by the feeding preferences of the insect vector. Many animal species have been found to be a source of food for *Triatoma dimidiata*, the main vector for *T. cruzi* in Yucatan, Mexico (Reyes-Novelo *et al.*, 2011). Chickens, turkeys, cattle, horses, pigs, dogs, rats, opossums and humans have been reported as source of food for the insect (González-Angulo and Ryckman, 1967; Christiansen *et al.*, 1988). Considering the generalist feeding behavior of *T. dimidiata* (Zeledón, 1981), backyard animals support peridomicile populations acting as feeding hosts, and at the same time, potential hosts for the maintenance of *T. cruzi* peridomicile infection (Reyes-Novelo et al., 2013; Koyoc-Cardeña et al., 2015).

Table 1. Prevalence of infection by *Trypanosoma cruzi* in backyard mammals of rural households from Molas, Yucatan, Mexico.

Backyard Mammal species	Tested animals (number)	Positive (number)	Prevalence
Pigs	28	21	75%
Sheep	7	6	85.71%
Horses	8	8	100%
Cattle	28	0	0%
Rabbits	13	0	0%
Total	84	35	42%

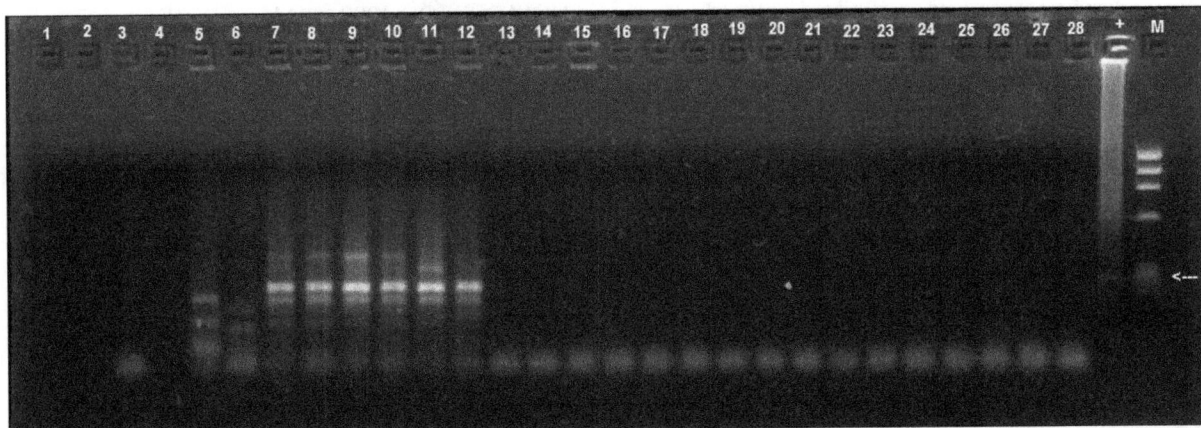

Figure 1. PCR results for *Trypanosoma cruzi* from backyard animal species of the community of Molas, Yucatan, Mexico. DNA bands of 188 bp were considered positive. Columns 1-4 are from negative rabbits, 5 is from a negative sheep, 6 from negative cattle, 7-12 from positive pigs and 18-28 from negative cattle, (+) positive control, (M) molecular weight marker (arrow).

All animal species analyzed and others like sheep, cats and rabbits are present in the backyard system of Molas, Yucatan, and are raised in close proximity to human beings. More studies on the specific role of infected animal species regarding the cycle of *T. cruzi*, are necessary to understand the rural epidemiology of Chagas disease in the backyard system. Special attention must be paid for future research in the potential role of pigs and horses because their economic importance and abundance in the backyard of rural houses in Yucatan (Gutiérrez-Ruiz et al., 2013).

CONCLUSION

The results obtained constitute an important preliminary evidence to implicate backyard mammals as potential maintaining hosts of *T. cruzi* in the peridomestic habitat of rural communities in Yucatan.

REFERENCES

Acosta-Casanova, M.I. 2004. Caracterización del sub-sistema de ganadería de traspatio en 33 comunidades rurales del estado de Yucatán. Tesis de Licenciatura. Facultad de Medicina Veterinaria y Zootecnia, Universidad Autónoma de Yucatán, Mérida, México.

Christensen, H.A., Sousa, O.E., Vázquez, A.M. 1988. Host feeding profiles of *Triatoma dimidiata* in peridomestic habitats of Western Panamá. American Journal of Tropical Medicine and Hygiene. 38(3): 477-479.

Días, J.C.P. 2000. Epidemiology. *Trypanosoma cruzi* and Chagas disease, 2ª ed. Rio de Janeiro. pp 55-58.

Duarte-Ubaldo, I.E. 2005. Aspectos zoosanitarios de la interacción entre fauna silvestre y animales domésticos en una comunidad maya de la región de Calakmul. Tesis de Maestría. El Colegio de la Frontera Sur. Campeche, México. pp 87-88.

Fujita, O., Sanabria, L., Inchaustti, A., De Arias, A.R., Tomizawa, Y., Oku, Y. 1994. Animal reservoirs for *Trypanosoma cruzi* infection in an endemic area in Paraguay. Journal of Veterinary Medical Science. 56(2): 305-308.

González-Angulo, W. Ryckman, R.E. 1967. Epizootiology of *Trypanosoma cruzi* in Southwestern North America. Part IX. An investigation to determine the incidence of *Trypanosoma cruzi* infections in triatomine and man on the Yucatán peninsula of Mexico. Journal of Medical Entomology. 4 (1): 44-47.

Gutiérrez-Ruiz, E., Aranda-Cirerol, F.J., Rodríguez-Vivas, R. I., Bolio-González, M., Cámara-Gamboa, E., Ramírez-González, S., Estrella-Tec, J., Acosta-Casanova, M. 2013. Características del subsistema de producción

animal de traspatio en Yucatán, México. In: Pacheco Castro J, Lugo Pérez JA, Tzuc Canché L, Ruiz Piña H, editores. Estudios multidisciplinarios de las enfermedades zoonóticas y ETVs en Yucatán. Mérida: Universidad Autónoma de Yucatán. p. 184-94.

Jiménez-Coello, M., Acosta-Viana, K.Y., Guzmán-Marín, E., Ortega-Pacheco, A. 2012. American Trypanosomiasis infection in fattening pigs from the South-east of Mexico. Zoonoses and Public Health. 59(S2): 166-169.

Koyoc-Cardeña E, Medina-Barreiro A, Escobedo-Ortegón FJ, et al (2015) Chicken coops, Triatoma dimidiata infestation and its infection with Trypanosoma cruzi in a rural village of Yucatan, Mexico. Rev Inst Med Trop Sao Paulo 57:269–272 .

Monteón-Padilla, V.M., Reyes-López, P.A., Rosales-Encina, J.L. 1994. Detección de *Trypanosoma cruzi* en muestras experimentales por el método de reacción en cadena de la ADN polimerasa. Archivos del Instituto de Cardiología México. 64:135-143.

Noireau, F. 1999. La enfermedad de Chagas y sus particularidades epidemiológicas en Bolivia In: La enfermedad de Chagas en Bolivia-Conocimientos científicos al inicio del programa control 1998-2002, OMS/OPS, La Paz. pp 33,38 y 39. En: http:www.ops.org.bo/textocompleto/ncha293 50.pdf.

NOM-062-ZOO-1999. México. Secretaria de Agricultura, Ganadería, Desarrollo Rural, Pesca y Alimentación. Especificaciones técnicas para el cuidado y uso de los animales de laboratorio. Ciudad de México. Diario Oficial de la Federación.

Organización Mundial de la Salud. 2007. Reporte sobre la enfermedad de Chagas. Grupo de trabajo científico, Buenos Aires, Argentina. pp 1-5.

Ramsey J M, Ordonez R, Tello L A, Pohls J L, Sánchez V, Peterson A T (2003) Actualidades sobre la epidemiología de la Enfermedad de Chagas en México; iniciativa para la vigilancia y el control de la Enfermedad de Chagas, en la República Mexicana. Instituto Nacional de Salud Pública de México, Salud Pública de México, 142p

Rey, J., Kobylinski, K., Rutledge Connelly R. 2006. La Tripanosomiasis Americana – Mal de Chagas, UF University of Florida. pp 1-4

Reyes-Novelo, E., Ruíz-Piña, H., Escobedo-Ortegón, J., Rodríguez-Vivas, R.I., Bolio-González, M., Polanco-Rodríguez, A., Manrique-Saide, P. 2011. Situación actual y perspectiva para el estudio de las enfermedades zoonóticas emergentes, reemergentes y olvidadas en la península de Yucatán, México. Tropical and Subtropical Agroecosystems. 14 (1): 35-54.

Ruiz-Piña, H.A., Cruz-Reyes, A. 2002. The opossum *Didelphis virginiana* as synanthropic reservoir of *Trypanosoma cruzi* in Dzidzilché, Yucatán, Mexico. Memorias del Instituto Oswaldo Cruz, 97(5): 613-620.

Ruiz-Piña H, Reyes-Novelo E. El huerto familiar yucateco y las zoonosis. In: Flores JS, editor. Huertos familiares de la Península de Yucatán. Mérida: Universidad Autónoma de Yucatán; 2012. p. 359-74.

Ruiz-Piña HA, Reyes-Novelo E, Escobedo-Ortegón F, Barrera-Perez M. Mamíferos sinantrópicos y la transmisión de enfermedades zoonóticas en el área rural de Yucatán. In: Pacheco Castro J, Lugo Pérez JA, Tzuc Canché L, Ruiz Piña H, editores. Estudios multidisciplinarios de las enfermedades zoonóticas y ETVs en Yucatán.

Mérida: Universidad Autónoma de Yucatán. p. 184-94.

Sabino, E. C., Ribeiro, A. L., Lee, T. ., Oliveira, C. L., Carneiro-Proietti, A. B., Antunes, A. P., ... Busch, M. P. (2015). Detection of T. cruzi DNA in Blood by PCR is associated with Chagas cardiomyopathy and disease severity. European Journal of Heart Failure, 17(4), 416–423. http://doi.org/10.1002/ejhf.220

Salazar-Schettino, P.M., Bucio, M.I., Cabrera, M., Bautista, J. 1997. First case of natural reservoirs in México. Memorias del Instituto Oswaldo Cruz. 92(4): 499-502.

Tay, J., Schenone, H., Sanchez, J.T., Robert, L. 1992. Estado actual de los conocimientos sobre la enfermedad de Chagas en la República Mexicana. Boletín Chileno de Parasitología. 47: 43-53.

Thrusfield, M. 2007. Veterinary Epidemiology, 3ª ed. Blackwell Science Ltd. Edinburgh. pp 53.

Zeledón, R. 1981. El *Triatoma dimidiata* (Latreille, 1811) y su relación con la Enfermedad de Chagas, Editorial Universidad Estatal a Distancia, San José, Costa Rica, 146 pp.

Evaluation of Bacterial Isolates in the Control of *Plutella Xylostella* L. in Broccoli

J. C. Mazetto Júnior[1]; R. C. A. Sene[2]; F. H. Iost Filho[3]; R. T. Thuler[2];
J. L. R. Torres[2*] and E. F. A. Moreira[2]

*[1]Institute of Agricultural Sciences at the Federal University of Uberlandia.
Amazonas Av., Umuarama. Uberlândia-MG, Brazil. Zip code: 38400-902.
E-mail: jcmazettojr@hotmail.com
[2]Federal Institute of Triangulo Mineiro (IFTM-Uberaba), 4000 João Batista
Ribeiro St., Uberaba-MG, Brazil. Zip code: 38064-790.
E-mail: ruancairo@hotmail.com, rthuler@iftm.edu.br; jlrtorres@iftm.edu.br;
edimo@iftm.edu.br
[3]University of Sao Paulo, College of Agriculture "Luiz de Queiroz" (ESALQ-
USP), 11 Padua Dias Av., Piracicaba-SP, Brazil. Zip code: 13418-900.
E-mail: fernandohiost@gmail.com
Corresponding author

SUMMARY

The diamondback moth (*Plutella xylostella* L.) is an insect that causes great losses in Brassica plantations, which may reach 100% loss in some cases. The most common method to control this pest is still the use of insecticide. However, its successive and arbitrary use has contributed to the development of insecticide resistance, and the study of alternative methods has become essential to successfully control the diamondback moth in broccoli. The objective of this study was to select isolates of rhizobacteria that act to control diamondback moth in broccoli. The experimental design was in randomized blocks, with three bacterial isolates [1 - *Kluyvera ascorbata* (EN4); 2 - *Bacillus subtllis* (R14); 3 - *Bacillus cereus* (C210) and 4 - Control (distilled water + spreader sticker)] tested with 5 repetitions in 20 m² plots. Larval viability (LV), pupal viability (PV), larval stage duration (LD), pupal stage duration (PD), pupal weight (PW) and total mortality (TM) were evaluated. The rhizobacteria isolates performed better to reduce larval and pupal viability, larval and pupal duration, and increase total mortality, and can be used as a management option to complement the control strategies of the diamondback moth in broccoli.
Keywords: *Brassica oleraceae* var. *italica*; entomopathogens; biological control.

INTRODUCTION

The Brassicacea plant family encloses about 350 genera and 3000 species, distributed throughout the world, majorly in temperate and tropical climates (Mendes, 2016). Within this family, the *Brassica* genus stands out by their socioeconomic importance. In tropical countries, *Brassica* species can be produced throughout the year, and have fast growth, high commercial and nutritional value (Kano *et al.*, 2010).

Cabbage, cauliflower and broccoli are among the most cultivated and consumed *Brassica* species in Brazil (Bernardes *et al.*, 2016). Broccoli (*Brassica oleracea* var. *italica*) is mainly grown in regions with mild temperatures (South and Southeast of Brazil), having most of its production destined for *in natura* consumption (Lalla *et al.*, 2010; Filgueira, 2013; Carvalho *et al.*, 2016).

The main factor of reduction of the production of these brassicas in Brazil and over the world has been the occurrence of the diamondback moth, *Plutella xylostella* (Linnaeus, 1758) (Lepidoptera: Plutellidae), that can cause losses of up to 100% in the crop. The importance of this pest is due to its high reproductive potential and short life cycle, with a high number of generations per year (Thuler *et al.*, 2009). In addition, it has a cosmopolitan distribution, a voracious feeding behaviour and is hosted by many commercial plants (Renwick *et al.*, 2006). This pest causes serious damage, depreciates the product, interferes with the growth of the plant and may lead to the death or total loss of the crop (Carvalho, 2008).

Its presence in Brazil is regularly reported and the chemical control is often the option chosen for managing it. According to Delen and Tosun (2004), chemical control has been the most widely used method in pest control because it is easy, provides quick and, most of the times, effective results, even though it increases production cost, changes the environmental balance and can be harmful to human health.

The successive use of insecticides to control diamondback moth has contributed to the evolution of resistance to most insecticides commercially available (De Bortoli *et al.*, 2013). In 1989, this insect had proven resistance to 50 products insecticides (Georghiou and Lagunes-Tejada, 1991), while Maia (2005) reported that 504 species of insects presented resistance to pyrethroids, carbamates and phosphorous insecticides. Situations like that rise the need for alternative methods to control the diamondback moth.

The use of alternative methods to control this pest involves practices of integrated pest management, such as biological control, which can be performed in a straightforward way of releasing entomopathogenic microorganisms in the environment to control the pest; and in an indirect way using biopesticides containing parts or secretions of organisms (Federici *et al.* 2010).

According to Cardoso *et al.* (2010) and Filgueira (2013) to control Lepidoptera insect species, one should preferably use the *Bacillus thuringienses* biological control, which is a bacterium with exclusive action on caterpillars. The active ingredient spinosad is a secondary metabolite of aerobic fermentation of the actinomycete *Sacchapolyspora spiniosa*, which has moderate selectivity to the cruciferous moth parasitoids (Villas Boas *et al.*, 2004). Also, among these microorganisms, highlight plant growth promoting rhizobacteria, which comprise a group with great potential as biological control agents (Thuler *et al.*, 2006). In this context, the objective of this study was to select isolates of rhizobacterias that act in effective control of the diamondback moth in broccoli.

MATERIAL AND METHODS

Experimental area

The study was conducted in an experimental area from the Federal Institute of Triangulo Mineiro (IFTM), campus Uberaba, Minas Gerais, Brazil, between August 2010 and July 2012. The area is located at the latitude 19°39'44" South and longitude 47°57'58" West were the average altitude is 795 m.

Type of soil

The soil of the experimental area was characterized as a Red Latosol Dystrophic (Embrapa, 2013), medium texture, relief soft wavy. The layer of 0-0.2 m presents: 210 g kg^{-1} of clay, 710 g kg^{-1} of sand and 80 g kg^{-1} of silt, pH $CaCl_2$ 5.5; 76 mg dm^{-3} of P (resin); 2 $mmol_c$ dm^{-3} of K^+; 22 $mmol_c$ dm^{-3} of Ca^{2+}; 10 $mmol_c$ dm^{-3} of Mg^{2+}; 17 $mmol_c$ dm^{-3} of H+Al and 19 g dm^{-3} of soil organic matter.

Region's climate

The climate of the region is classified as Aw, tropical, hot, according to the classification of Köppen, having hot and rainy summer and cold and dry winter. In the region, the annual averages of rainfall, temperature and relative humidity of the air are 1.600 mm, 22.6 °C and 68%, respectively (Uberaba, 2009).

Experimental Design

The experimental design was in randomized blocks, with three bacterial isolates: 1 - *Kluyvera ascorbata*

(EN4); 2 - *Bacillus subtllis* (R14); 3 - *Bacillus cereus* (C210) and 4 - Control (distilled water + spreader sticker), all conceded by the Laboratory of Phytobacteriology (Federal Rural University of Pernambuco), in broccoli plants (BR-068® Syngenta), with 5 repetitions in 20 m² plots (4 x 5 m). This broccoli hybrid features open canopy plants, with few leaves, uniformity of maturity and medium cycle (85 to 90 days in winter).

Soil Preparing and Fertilization

The soil was conventionally prepared with deep plowing, harrowing, then mechanically entourage and fertilized five days before transplanting the seedlings. The planting beds were mechanically prepared. During the preparation of the furrows, before transplanting the seedlings, manure was applied at an equivalent dose of 20 t ha^{-1}. Mineral fertilizer was applied during planting according to the Committee of Soil Fertility of the State of Minas Gerais (1999) at an equivalent dose of 100 kg ha^{-1} of P_2O_5, 100 kg ha^{-1} of K_2O and 50 kg ha^{-1} of N, 2 kg ha^{-1} of boric acid (17.5% of B). Nitrogen (50 kg ha^{-1} application^{-1}) was also applied at 30 and 45 days after planting.

Seed Inoculation

The inoculation was done by seed immersion in a suspension of each isolate [9x10^8 CFU mL^{-1}, in distilled water + spreader sticker (0.05%)]. The concentration of suspensions was calculated by the equation: $y = e^{(6.702 - 9,041x + 11,159x^2)}$, using the values of absorbance of 0.77 nm (spectrophotometer). A control containing distilled water + spreader sticker (0.05%) was included among the treatments.

After drying for 12 h, the seeds were sown in polystirene trays with 128 cells, containing plant substrate made of pine bark and coconut fiber. Trays were kept in glasshouse. When the seedlings developed their fourth to fifth leaves (26 days after sowing), they were transplanted to the planting beds.

Fertilization

Fertilizer doses for Brassica crops were based on soil analysis and recommendations of the Soil Fertility Commission of the State of Minas Gerais (1999). The recommended nitrogen (N), phosphorus (P), and potassium (K) doses for the crops were: 150 kg ha^{-1} N (urea), 100 kg ha^{-1} P_2O_5, and 100 kg ha^{-1} K_2O. Nitrogen and potassium doses were split-applied: at planting, 30 days after planting, and 45 days after planting. In addition, 1 g of boric acid (17.5% B) was applied per pit.

Bioassay

The bioassay was performed 45 days after transplanting the seedlings to the planting beds. Leaf disks (8 cm diameter) were placed over filter paper lightly moistened with distilled water in 9 cm Petri dishes.

In each Petri dish, 10 second instar *P. xylostella* larvae were placed on a leaf disk. The Petri dishes were closed with plastic film, to keep moisture and prevent the larvae escape (Thuler *et al.* 2007). The insects were obtained from the creation-stock of the Laboratory of Entomology (IFTM) and kept according to Barros and Vendramim (1999).

After three days, the following biological characteristics of the insect were daily evaluated: larval viability (LV), pupal viability (PV), larval stage duration (LD), pupal stage duration (PD), pupal weight (PW) and total mortality (TM) - sum of mortality in all stages.

When the insects reached the pupal stage, they were weighted and transferred to ELISA® plates for individualization and observation of the emergence. After emerging, the sex of the adult was observed, and the sex ratio was calculated.

Statistical Analysis

The data were subjected to Deviance test for the generalized models at 5% probability, where the modelling was done using the probability distribution of the response variable. The total mortality, larval and pupal viability follow binomial distribution, the larval and pupal duration follow Poisson distribution and the variable pupal weight follows a Gaussian distribution. For the cases where the effect of the treatments was significant, the generalized Tukey test was used to compare the averages. The R Core Team software was used to carry out the analyses.

RESULTS AND DISCUSSION

Larval, pupal viability and pupal duration

The biological parameters of diamondback moth showed significant differences between the treatments tested, indicating that the use of the rhizobacterias R14, EN4 and C210, altered the biological cycle of the insect. In Figure 1, from a linear function supported by the Deviance test, it is possible to observe a significant difference between the treatments.

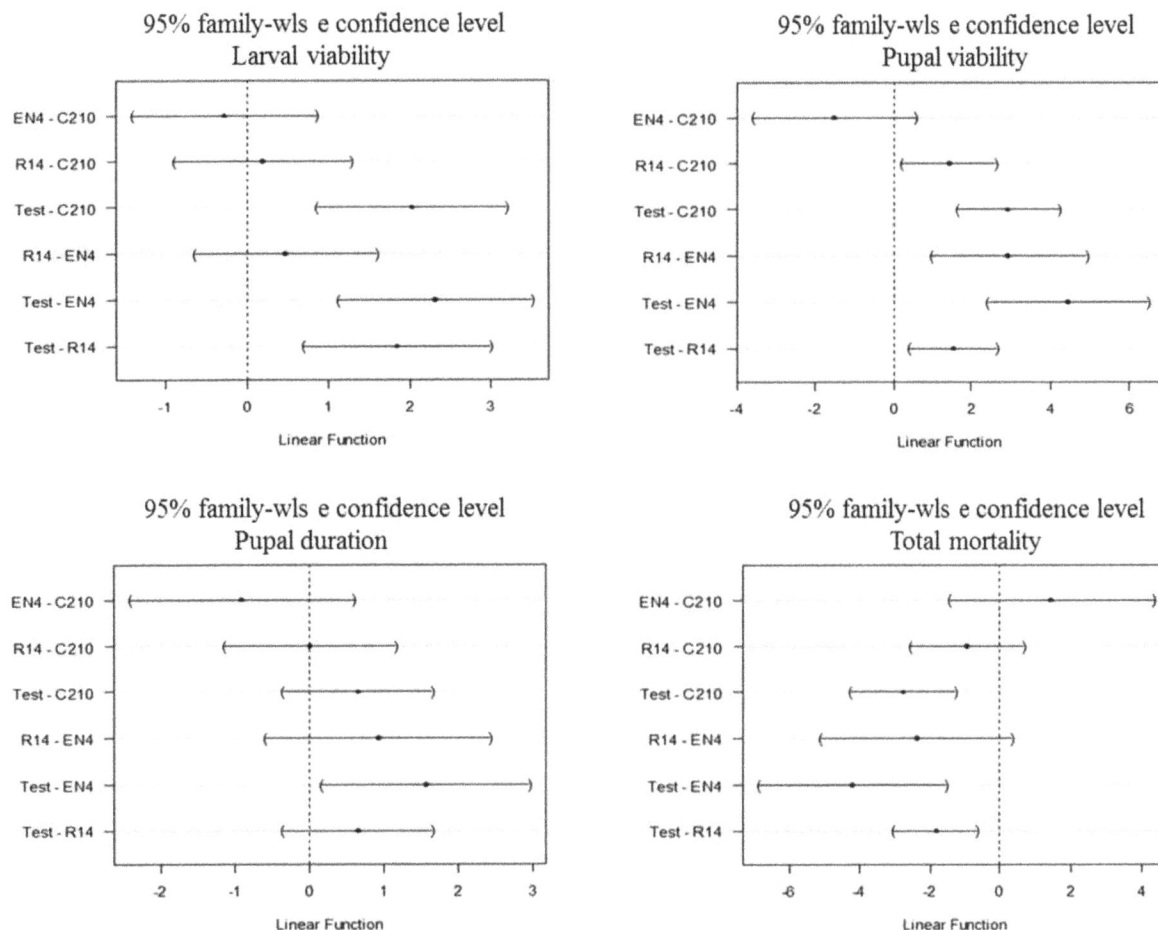

Figure 1. Generalized Tukey test for the biological characteristics of *Plutella xylostella* fed on leaves of broccoli plants inoculated and non-inoculated with isolates of rhizobacteria.

The values observed for pupal duration, and pupal and larval viability (Figure 2) were reduced and similar among the rhizobacteria evaluated, however they differed significantly from the control (p<0.05), proving that the treatments were efficient to the control the insect at these stages. This characteristic is important at the field level, because with the reduction of larval and pupal viability, the number of individuals reaching the maturity will be reduced, as well as the next generation of the insect in the field. Also, with a lower number of caterpillars, there are less damages and losses caused to crops, consequently, as highlighted by Maroneze and Gallegos (2009).

The lowest diamondback moth viability (larval and pupal stage) was caused by the isolate EN4 (26%), when compared to R14 (36%), C210 (38%) and control (65%), which suggests that the use of this EN4 rhizobacterium can be more effective in controlling diamondback moth. The influence of this rhizobacterium (EN4) on post-embryonic viability suggests that there is a cumulative effect of reductions in larval and pupal viability, resulting in a greater increase in total mortality. Similar results were observed by Crialesi *et al.* (2017), who showed that EN4 generally caused low larval and pupal viability, as well as reduction in pupal duration.

Evaluating the effect of plant growth promoting rhizobacteria on the development of *P. xylostella* on cabbage, Thuler *et al.* (2006) observed that post-embryonic stages viabilities were affected by isolates of *K. ascorbata* (EN4) and of *Alcaligenes piechaudii* (EN5), reducing by 80 and 50%, respectively, the viability of the such stages. In this study, the authors also highlighted that *Bacillus thuringiensis* pv. *kurstaki* (HPF14), *Bacillus amyloliquefaciens* (PEP81) and *Bacillus megaterium* pv. *cerealis* (RAB7) isolates also showed potential for use as an insect control strategy.

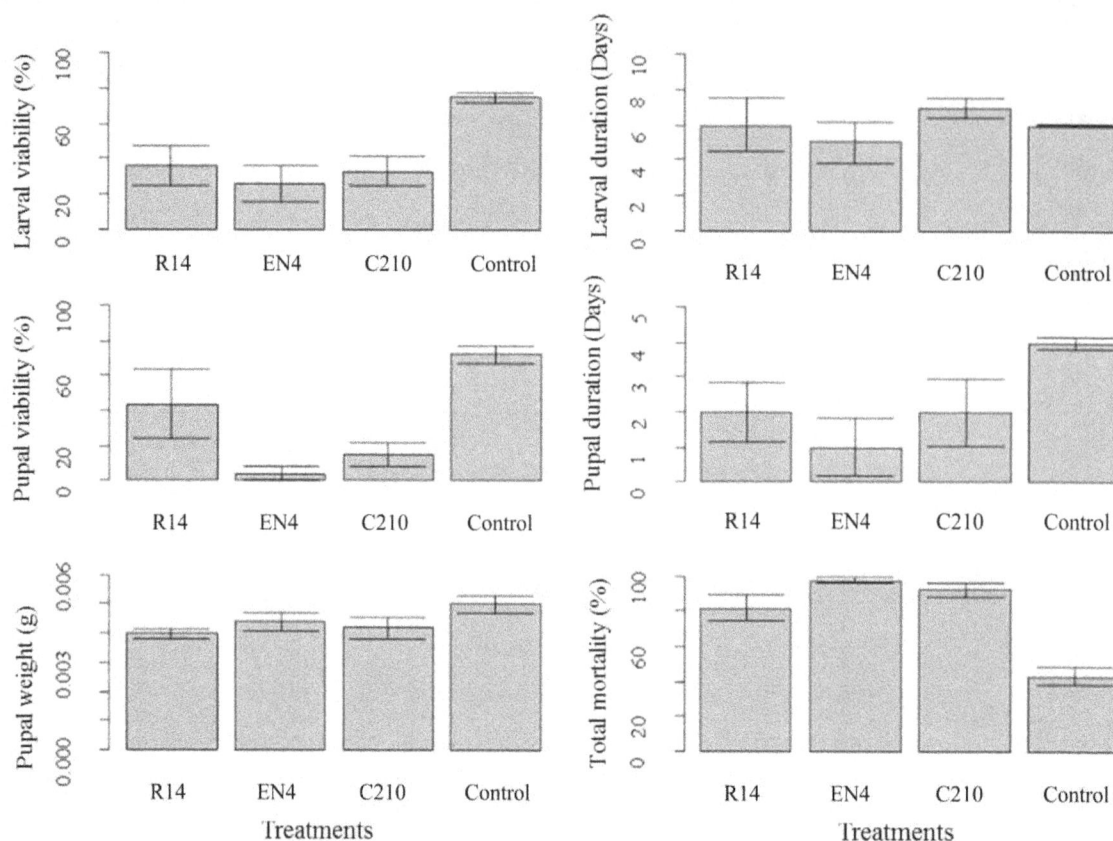

Figure 2. Means and standard error for biological characteristics of *Plutella xylostella* fed on leaves of broccoli plants inoculated and non-inoculated with isolates of rhizobacteria.

Larval duration and pupal weight

The larval stage duration was not affected by the rhizobacteria isolates. According to Hernandez and Vendramim (1997), a putative increase on this stage duration may be related to a small amount of food intake due to one or several factors, which decreased the feeding time on the leaf treated with rhizobacteria or yet as a result of any nutritional imbalance.

The pupal weight was not affected by the rhizobacteria either, suggesting that the caterpillars that were viable continued feeding normally, regardless of the treatments with rhizobacteria. This behaviour can be positive, because it shows that the insect doesn't avoid the treated plants, thus it will be affected by the rhizobacteria. Thuler *et al.* (2009) when evaluating diamondback moth resistance to glucosinolates, observed that the weight of the pupa is highly correlated with other parameters like the viability and duration of the larval and pupal stages.

Mortality rates

All the rhizobacteria evaluated caused high corrected mortality rates of diamondback moth, being 98% for

EN4, 92.5% for C210 and 82% for R14, with no differences among them (p<0.05), but significantly higher than the control (43.2%). These results point out a direct influence of the rhizobacteria treatments on the diamondback moth biology. The raise of the diamondback moth mortality due to seed treatment with rhizobacteria isolates will lead to reductions on its population over the course of several generations of this insect.

Other studies found similar results when evaluating the effects of rhizobacteria treatments on control of insects. Medeiros *et al.* (2005) observed high larval mortality (LM) and low pupal viability (PV) of diamondback moth on cabbage when *Enterobacter cloacae* (ENF14), *Alcaligenes piechaudii* (EN5) or *K. ascorbata* (EN4) isolates were applied. The isolates *Enterobacter cloacae* (ENF14), *Alcaligenes piechaudii* (EN5) or *K. ascorbata* (EN4) were also observed to have control action against diamondback moth (Silva *et al.* 1999). In field experiments, the cucumber seed inoculation with a suspension of the *Bacillus pumillus* (INR-7) and *Serratia marcescens* (90-166) isolates was more efficient than insecticides to reduce the population of *Diabrotica undecimpunctata* (Barber) and *Acalymma vittatum*

(F.) (Coleoptera: Chrysomelidae) (Zehnder *et al.* 1997).

Our results are supported by other previous studies and contribute to improve the knowledge about the use of rhizobacteria as a biological option to control insects in agriculture. This procedure can be easily incorporated to any strategy of broccoli cropping, and has the potential to reduce the use of chemical insecticides, environmental disturbances and human exposition to hazardous products, and increase cropping profits.

CONCLUSIONS

All rhizobacteria isolates evaluated caused low larval and pupal viability, decreased larval and pupal stage duration, and resulted in a high total mortality.

These isolates can be indicated as biological control agents in a management strategy to control diamondback moth in broccoli.

Acknowledgments
The authors would like to thank IFTM Campus Uberaba for the infrastructure available, and FAPEMIG and CNPq for granting scientific initiation scholarship to the students involved.

REFERENCES

Barros, R.; Vendramim, J.D. 1999. Efeito de cultivares de repolho, utilizadas para criação de *Plutella xylostella* (L.) (Lepidoptera: Plutellidae), no desenvolvimento de *Trichogramma pretiosum* Riley (Hymenoptera: Trichogrammatidae). Anais da Sociedade Entomológica do Brasil, 28: 469-476. http://dx.doi.org/10.1590/S0301-80591999000300012.

Bernardes, R.S.; Souza, S.A.; Mussury, R.M.; Santos, F.P.; Moura, L.O. 2016. Efeito de extrato aquoso de *Campomanesia xanthocarpa* sobre o ciclo de vida da Traça-das-Crucíferas (*Plutella xylostella L.*) (lepidoptera: plutellidae). Cadernos de Agroecologia, 11(2): 1-7.

Cardoso, M.O.; Pamplona, A.M.S.R.; Michereff Filho, M. 2010. Recomendações técnicas para o controle de Lepidópteros praga em couve e repolho no Amazonas. Circular n° 35: Embrapa Amazônia Ocidental, 16p.

Carvalho, C.; Kist, B.B.; Treichel, M. 2016. Anuário brasileiro das hortaliças. Santa Cruz do Sul: Editora Gazeta Santa Cruz, 64 p.

Charleston, D. S.; Kfir, R.; Vet, L. E. M.; Dicke, M. 2005. Behavioural responses of diamondback moth (*Plutella xylostella*) (Lepidoptera: Plutellidae) to extracts derived from *Melia azedarach* and *Azadirachta indica*. Bulletin of Entomological Research, 95(5): 457-465. https://www.ncbi.nlm.nih.gov/pumed/161975 66.

Comissão de Fertilidade do Solo do Estado de Minas Gerais. 1999. Recomendações para o uso de corretivos e fertilizantes em Minas Gerais: 5ª aproximação. Viçosa: CFSEMG, 359 p.

Crialesi, P. C. B.; Thuler, R. T.; Iost Filho, F. H.; Thuler, A. M. G.; Lemos, M. V. F.; De Bortoli, S. A. 2017. Plant Growth Promoting Rhizobacteria (PGPR) and *Plutella xylostella* (L.) (Lepidoptera: Plutellidae) interaction as a resistance inductor factor in *Brassica oleracea* var. capitata. Plant Science Today, 4 (3): 121-132. http://dx.doi.org/10.14719/pst.2017.4.3.305

De Bortoli, S. A.; Polanczyk, R. A.; Vacari, A. M.; De Bortoli, C. P.; Duarte, R. T. 2013. *Plutella xylostella* (Linnaeus, 1758) (Lepidoptera: Plutellidae): tactics for integrated pest management in Brassicaceae. In: Soloneski, S., Larramendy, M. (Eds.). Weed and pest control - conventional and new challenges. Rijeka: InTech, p. 31-51. http://dx.doi.org/10.5772/54110.

Delen, N.; Tosun, N. 2004. Fungicidas: Modos de ação e resistência. Parte 2: Fungicidas com modos de ação específicos. In: Luz, W.C. (Ed), Revisão Anual de Patologia de Plantas. Berthier, 12 (2): 27-90.

Empresa Brasileira De Pesquisa Agropecuária - EMBRAPA. 2013. Sistema brasileiro de classificação de solos. 3ª ed. Brasília, DF, 357 p.

Federici, B.A.; Park, H.W.; Bideshi, D.K. 2010. Overview of the basic biology of *Bacillus thuringiensis* with emphasis on genetic engineering of bacterial larvicides for mosquito control. Open Toxinol. J., 3: 83-100.

Filgueira, F. A. R. 2013. Novo Manual de Olericultura: Agrotecnologia moderna na produção e comercialização de hortaliças. Viçosa: UFV, 421 p.

Georghiou, G. P.; Lagunes-Tejada, A. 1991. Theocurrence of resistance to pesticides in arthropods. Food and Agriculture

Organization of the United Nations. Rome: AGPP/MISC/91-1. 318p.

Gomes, E.C.S.; Perez, J.O.; Barbosa, J. 2009. Resistência induzida como componente do manejo de doenças da videira.Engenharia Ambiental - Espírito Santo do Pinhal, 6(2): 114-120.

Hernandez, C. R.; Vendramim, J. D. 1997. Avaliação da bioatividade de extratos aquosos de Meliaceae sobre *Spodoptera frugiperda* (J.E. Smith). Revista Engenharia Agrícola, 72(1): 305-318. http://dx.doi.org/10.1590/S1519-566X2003000400018.

Kano, C.; Salata, A.C.; Higuti, A.R.O.; Godoy, A.R.; Cardoso, A.I.I.; Evangelista, R.M. 2010. Produção e qualidade de couve-flor cultivar Teresópolis Gigante em função de doses de nitrogênio. Horticultura Brasileira, 28(4): 453-457. http://dx.doi.org/10.1590/S0102-05362010000400013.

Lalla, J.G.; Laura, V.A.; Rodrigues, A.P.D.C.; Seabra Júnior, S.; Silveira, D.S.; Zago, V.H.; Dornas, M.F. 2010. Competição de cultivares de brócolos topo cabeça única em Campo Grande. Horticultura Brasileira, 28(3): 260-363. http://www.scielo.br/pdf/hb/v28n3/v28n3a20.pdf.

Maia, A, H, N. 2005. Definindo Estratégias de Manejo da Resistência de Pragas a Toxinas Bt Expressas em Culturas Trangênicas: o Papel dos Modelos de Simulação. Embrapa Meio Ambiente. Jaguariúna, p.1-5.

Maroneze, D. M.; Gallegos, D. M. N. 2009. Efeito de extrato aquoso de *Melia azedarach* no desenvolvimento das fases imatura e reprodutiva de *Spodoptera frugiperda* (J. E. Smith, 1797) (Lepidoptera: Noctuidae). Semina: Ciências Agrárias, 30(3): 537-550.

Medeiros, F.H.V.; Silva, G.; Mariano, R. L. R.; Barros, R. 2005. Effect of bacteria on the biology of Diamondback moth (*Plutella xylostella*) on cabbage (*Brassica oleraceae var. Capitata*) cv. Midori. Anais da Academia Pernambucana de Ciência Agronômica, 2: 204-212.

Mendes, R.T. Adubação borácica na cultura do brócolis. 2016. 67f. Dissertação (Mestrado em Produção Vegetal) – Universidade Estadual de Goiás – Campus Ipameri.

Renwick, J. A. A.; Haribal, M.; Gouinguené, S.; Städler, E. 2006. Isothiocyanates stimulating oviposition by the diamondback moth, *Plutella xylostella*. Journal of Chemical Ecology, 32(4): 755-766. http://dx.doi.org/10.1007/s10886-006-9036-9.

Silva, G., Araújo, D.V., Costa, V.S.O., Mariano, R.L.R., Barros, R.; Gomes, A.M.A. 1999. Potencial de bactérias endofíticas para controle biológico da traça das crucíferas (*Plutella xylostella* L., 1758) (Lepidoptera: Plutellidae) em repolho. In: Mota, R.A., Faustino, M.A.G. & Almeida, M.G.A. (Eds.). Proceedings, IX Congresso de Iniciação Científica, Recife, PE., p.70

Thuler, R.T.; Barros, R.; Mariano, R.L.R.; Vendramim, J.D. 2006. Efeito de bactérias promotoras do crescimento de plantas (BPCP) no desenvolvimento de *Plutella xylostella* (L.) (Lepidoptera: Plutellidae) em couve. Científica, 34(2): 217-222.

Thuler, R.T.; Bortoli, S.A.; Hoffmann-Campo, C.B. 2007. Classificação de cultivares de brássicas com relação à resistência à traça-das-crucíferas e à presença de glucosinolatos. Pesquisa Agropecuária Brasileira, 42(4): 467-474. https://dx.doi.org/10.1590/S0100-204X2007000400003.

Thuler, R. T. 2009. Criação de *Plutella xylostella (L.)*. In: Criação de Insetos: da base à biofábrica. ed. Jaboticabal: O autor, p. 58-68.

Uberaba em Dados, Prefeitura Municipal de Uberaba. Edição 2009, 23 p. Disponível: <http:www.uberaba.mg.gov.br/portal/acervo/desenvolvimento/arquivos/uberaba_em_dados/Edicao_2009/Capitulo01.pdf>. Acesso 16.04.2017.

Villas Boas, G.L.; Branco, M.C.; Medeiros, M.A.; Monnerat, R.G. França, F.H. 2004. Inseticidas para o controle da traça das crucíferas e impactos sobre a população de parasitoides. Horticultura Brasileira, 22(4): 696-699. http://dx.doi.org/10.1590/S0102-05362004000400006.

Zehnder, G.; Kloepper, J.; Yao, C.; Weial, G. 1997. Induction of systemic resistance in cucumber against cucumber beetles (Coleoptera: Chrysomelidae) by plant growth-promoting rhizobacteria. Journal of Economic Entomology, 90(2): 381-396. https://doi.org/10.1093/jee/90.2.391

3

Milk Production, Grazing Behaviour and Biomass Quality in Native Tropical Pastures Grazed to Different Stocking Rate during Two Years

Jesús Jarillo-Rodríguez[1], Epigmenio Castillo-Gallegos[1*],
Luis Ramírez y Avilés[2], Braulio Valles de la Mora[1]
and Eliazar Ocaña-Zavaleta[1].

[1]Centro de Enseñanza, Investigación y Extensión en Ganadería Tropical de la
FMVZ-UNAM. Facultad de Medicina Veterinaria y Zootecnia, Apartado
Postal136, Martínez de la Torre C.P. 93600, Veracruz, México E-mail:
jarillorj22@hotmail.com; pime11302002@yahoo.com.mx;
braulio_36@hotmail.com; eocanazvel58@gmail.com
[2]Facultad de Medicina Veterinaria y Zootecnia de la Universidad Autónoma de
Yucatán. Mérida, Yucatán, Km 15.5 Carretera Mérida-Xmatkuil. Mérida,
México. E-mail: raviles@correo.uady.mx
*Corresponding author

SUMMARY

The objective of this study was to determine the effect of stocking rate (SR: 2, 3 and 4 cows/ha) upon standing dry matter variables, its quality and ingestive behavior of F1 Holstein x Zebu cows that grazed rotationally and considering the climatic seasons (S): Rainy, winter and dry in the north-central region of the State of Veracruz. A completely randomized design was used. The analysis of variance considered the fixed effects of SR, S and their interaction, being the season the repeated measurement and the experimental unit the subject. The standing dry matter (kg/ha) was affected by S (7140 ± 637, 5341 ± 524 and 4512 ± 719, respectively), but not by SR. Grazing time (min/cow/d) was similar between 2 (428 ± 71) and 4 (400 ± 83) cows/ha, both different from 3 cows/ha (462 ± 55). Ruminating time (min/cow/day) was affected by SR, averaging 372 ± 104, 364 ± 96, and 315 ± 90 min/cow/d, respectively. In situ dry matter digestibility (%) went from 57.6 ± 1.3 to 79.2 ± 1.6 for leaves and from 62.6 ± 2.0 to 78.8 ± 0.5 for stems; either component increased their values as SR increased. Daily milk yield per cow was not affected by SR. It is concluded that que there is a significant effect of SR upon grazing and ruminating time, but these variables were not statistically related to pasture variables and individual milk yield, mostly due to the fact that forage availability never limited pasture intake in any of the three stocking rates utilized.
Palabras clave: Daily milk yield, biting rate, climatic period, standing dry matter, pasture management.

INTRODUCTION

The objective of pasture management in animal production systems is to optimize the proportion of available forage consumed by the grazing animal (Dillon et al., 2005). To reach this objective, it is necessary to understand climate, animal and pasture factors that affect dry matter intake during grazing. Management of stocking rate (SR) aims to manipulate pasture structural components (through numbers of grazing animals) in order to strike a balance between animal product per animal and per hectare. By manipulating pasture structure and dry matter yield, a high dry matter intake can be obtained, resulting in a high level of individual animal performance (Ungar and Noy-Meir, 1988).

Under tropical conditions, some pasture structural components, such as plant height, are related to animal ingestive behavior (Stobbs, 1973; Silva and Carvalho, 2005), and are altered when stocking rate is modified (Kennedy et al., 2007). In general, a high stocking rate reduces dry matter yield by reducing regrowth capacity, and also alters botanical composition, since survival of pasture plant species is affected by stocking rate. Eventually, pasture structural modifications occur that modify animal ingestive behavior (Gibb, 2006), resulting in decreased pasture intake and animal performance (Langer, 1979).

In the humid tropics of Mexico, native gramma pastures are the main source of feed (Amendola et al., 2005) for grazing cattle (Ocaña, 2003), but very little research has been conducted on pasture production and animal performance. Even less effort has been devoted to studying changes in pasture structure and its effect on ingestive behaviour of grazing cattle (Castillo, 2003), some studies indicate that ingestive behavior can be modified by plant structure (Orantes-Zebadúa et al., 2014) and time in erect pasture (Ortega et al., 2009); for example, more leafy plant structure decreases grazing time and increases the quantity of dry matter intake (Nochebuena et al., 1994). The present study was designed to measure the effects of stocking rate on pasture forage production and nutritional quality, and on animal grazing behaviour.

MATERIALS AND METHODS

The study was conducted from September 1, 2005 to July 30, 2007, and was divided into two periods: 2005-2006 and 2006-2007. The study area was located at the Centre for Teaching, Research and Extension in Tropical Animal Husbandry (CEIEGT), of the Faculty of Veterinary Medicine, at the National University of Mexico (20° 02'N, 97° 06'W; 112 masl), in the municipality of Tlapacoyan, Veracruz, Mexico.

Climate and soils

The study was conducted over two periods (2005-2006 and 2006-2007), each one beginning in the rainy season (September and October), continued in the winter season (January and February), and ending in the dry season (May and June). Total annual rainfall was 1211 mm and 1811 mm for the first and second periods, respectively (Table 1). The soil in the area is classified as an acid Ultisol (pH 4.5-5.2), with a clay-lime texture, low N (0.032%) and P (2.32 ppm, Bray II) content, low cation exchange capacity and no Al toxicity (23.18%).

Table 1. Temperature and rainfall for the three seasons during two experimental periods.

Period	Season	Temperature (°C)			Rainfall (mm)
		Mean	Max	Min	
Year 1	Rainy	25.9	38.4	16.8	557
	Windy	20.0	33.2	8.2	314
	Dry	25.6	40.2	12.5	340
Year 2	Rainy	26.4	31.8	21.3	1118
	Windy	21.1	25.3	16.8	377
	Dry	27.8	33.7	22.1	316

Year 1= 2005-2006, Year 2= 2006-2007

Animals, pastures and treatments

Thirty multiparous F1 (Holstein x Cebu) cows were divided into 3 groups of 10 animals (489.9±67.5 kg live weight), balanced according to parity, milk yield in the previous lactation and live weight, and assigned randomly to graze 3 stocking rate treatments.

In a 10.8 ha area of native pasture, three stocking rate treatments were established: 2 (low), 3 (medium) and 4 (high) cows ha^{-1}. A 3-day grazing/27-day recovering rotational grazing cycle was established, and during the study twenty four grazing cycles were performed. There were ten paddocks per stocking rate, on areas of 5.0, 3.3 and 2.5 ha, respectively. Prior to initiating the study, the main botanical component was native grasses (86.9%), primarily Paspalum notatum Fluegge) and Axonopus compressus (Swartz) Beauv) and introduced grasses (Cynodon nlemfuensis Vanderyst, Brachiaria arrecta (Hack. ex T. Durand & Schinz) Stent and Brachiaria mutica (Forsk.) Stapf

Ten different cows grazed on each stocking rate treatment during the experimental periods. Two paddocks in each stocking rate treatment were randomly chosen and used for performing measurements, in two continuous grazing cycles at the end of each season. These paddocks were grazed on successive 3-day periods. Cows were not supplemented during the rainy and dry seasons, but received 1.14 kg of concentrate (14% crude protein) per cow per day during the windy season.

Pasture measurements and ingestive behaviour

On the second day of the three day grazing period, the herbage mass present (HMP) before and after grazing was assessed using the comparative yield method (Hendricksen and Minson, 1980; Haydock and Shaw, 1975), with 5 double-sampling quadrats (50 x 50 cm) and 100 visual comparison quadrats. During the second period (2006 – 2007) duplicate measurements were made in each paddock, by dividing it into 2 sections of equal size. Ten subsamples from the total HMP before (BG) and after grazing (AG) were separated into leaf blades (LF), stems (ST: leaf sheath + true stem) and dead material (DM). From these data the percentage of use (USE) of LF and ST was calculated as:

$$USE = [(LFBG – LFAG)/(LFBG)] \times 100 \qquad (1)$$

Where: LFBG = Leaves before grazing; LFAG = Leaves after grazing

Botanical composition (BC) was estimated before and after grazing in the last two grazing cycles during each season, using the dry-weight-rank method (Mannetje and Haydock, 1963) on one hundred quadrats per paddock. Data for native and introduced grasses are shown, as these components represent more than 90% of the forage available in the pastures. Samples of HMP were analyzed for crude protein (CP; N x 6.25) using the Kjeldahl method (AOAC, 1980) neutral detergent fibre (NDF), acid detergent fibre (ADF) and lignin (LIG) (Van Soest, 1991), using an ANKOM digester and 48-h in situ ruminal dry matter (DM) degradability (ISDMD; Orskov and McDonald, 1979).

Observations of ingestive behavior were made on the second day of the 3-day grazing period. At 10-minute intervals over 21 h, "sweeping" visual observations were performed (Gary et al., 1970) and recorded as to whether each cow was grazing, ruminating or engaged in other activities. From these data, the total number of times of each activity per cow was computed and the result multiplied by 10, which provides an estimate of the number of minutes per day that each cow was engaged in each activity (Gary et al., 1970).

Biting rate was estimated using "focal" visual observations (Gary et al., 1970) during the last 10 minutes of each hour, when the time for each cow to take 20 bites was recorded. Data were converted to bites per minute (Penning and Rutter, 2004). The cows were not observed from 08:00 h to 11:00 h when they were in the milking shed.

Treatment design and statistical analyses

A completely randomized design was applied and the experimental unit for pasture variables was a paddock (two paddocks in each stocking rate treatment). For the case of ingestive behaviour variables, individual animals were the experimental units (Jamieson and Hodgson, 1979), with 10 cows per treatment. A grazing animal can be considered as an independent random effect if its individual behavior does not differ from that of the herd (Rook and Penning, 1991) thus the data were analyzed under this assumption.

There were twenty four 30-day grazing cycles performed throughout the experiment, and of those, twelve were chosen to sample the randomly selected paddocks with respect to HMP and its components. In the 2005-2006 period: Grazing cycles 1 and 2, 5 and 6, and 9 and 10, for the rainy, windy and dry seasons, respectively. In the 2006-2007 period: Grazing cycles 13 and 14, 17 and 18, and 21 and 22 for the same seasons. In the first period, there was a single sampling on each stocking rate x paddock x grazing cycle combination, while in the second period, each paddock was divided in two equal sections where two simultaneous samplings were done. The grazing cycle was considered the repeated measurement on the paddock.

The herbage mass present and its components were analyzed with a model that considered the fixed effects of stocking rate and grazing cycle and the random effect of paddock. Grazing cycle was used as a continuous variable, so the model included its linear, quadratic and cubic single effects as well as interactions of these effects with stocking rate in order to test if HMP behavior was similar between stocking rates, as grazing cycles advanced in time. The PROC MIXED of SAS for repeated measurements was used. Of the covariance structures tested (CS, AR(1), ARH(1), ARMA(1,1), HF, TOEP, TOEPH and UN) only CS, AR(1) and ARMA(1,1) converged, and of these three, AR(1) was the one with the lowest AIC_C value, indicating it was the most suitable to the data (Littell et al., 1998).

Animal behavior variables were analyzed with models that considered the fixed effects of stocking rate, season and their interactions; the individual animal, on which repeated measurements were taken was the replicate, considered a random effect, and the repeated measurement was each of the six combinations of growth cycle (2005-2006 and 2006-2007) by season (rainy, winter and dry). The procedure PROC MIXED in SAS for repeated measurements was used. The autoregressive [AR(1)] covariance structures was the most appropriate for ingestive behavior variables (Littell et al., 1998). Model effects and differences between least squares means were considered significant if P<0.05.

RESULTS AND DISCUSSION

Available dry matter and pasture nutritional value indicators

Stocking rate did not affect HMP (leaves, stems and total DM) for any season during 2005-2006. As well, there was no interaction between stocking rate and seasons (Table 2). Yet, in 2006-2007, during the rainy season, all variables were higher at 3 cows ha^{-1} than at the other stocking rates. During the windy season, all values appeared to decline as stocking rate increased, although stocking rates with 3 and 4 cows ha^{-1} were statistically similar; no differences were detected during the dry season. During 2005-2006, the variable "leaves" was different (P<0.0009) for the windy and dry seasons (P<0.02). For "stems", the windy season was different in comparison to the rainy (P<0.005) and dry seasons (P<0.008). During 2006-2007, the variables leaves, stems and total dry matter were statistically different (P<0.0001).

Pasture use showed no statistical differences between stocking rate and seasons, yet in 2005-2006 it was slightly more than 40% at 4 cows ha^{-1} (Table 3). During the rainy season, there was little difference in the use of leaves and stems, but as the year progressed, the preference for leaves over stems became more pronounced. This was more obvious at the lightest stocking rate. In 2006-2007, the response to stocking rate was not so obvious (Table 3), but the preference for leaves over stems remained.

During both experimental periods, the contribution of native (*Paspalum notatum* Fluegge and *Axonopus compressus* (Swartz) Beauv) and introduced grasses (*Cynodon nlemfuensis* Vanderyst, *Brachiaria arrecta* (Hack. ex T. Durand & Schinz) Stent and *Brachiaria mutica* (Forsk.) Stapf to herbage mass was affected by stocking rate. The proportion of native grasses increased, while introduced grasses decreased with increasing stocking rate (Table 4). The greatest change occurred with the increase from 2 to 3 cows ha^{-1} (around 17%), compared to 4% when going from 3 to 4 cows ha^{-1}.

Table 2. Dry matter (LSM and ± s.e.) of leaves, stems (true stem + leaf sheath) and total standing DM (kg ha^{-1}) of native grass-based pastures stocked at 2, 3 and 4 cows ha^{-1} during the rainy, windy and dry seasons during two consecutive grazing periods.

Component	2005-2006			2006-2007		
	2 cows ha^{-1}	3 cows ha^{-1}	4 cows ha^{-1}	2 cows ha^{-1}	3 cows ha^{-1}	4 cows ha^{-1}
Leaves, stems and total standing DM						
Rainy season						
Leaves	3190±260[a1]	2901±75[a]	3007±540[a]	2517±89[a]	3514±196[b]	2630±257[ab]
Stems	2076±161[a]	2234±102[a]	2491±454[a]	2919±221[a]	3821±327[b]	2291±310[ac]
Dead material	2328±491[a]	1159±194[a]	1605±542[a]	1495±365[a]	1520±201[a]	1147±270[a]
Total	7594±793[a]	6294±184[a]	7103±1501[a]	6931±498[a]	8855±384[b]	6068±461[ac]
Winter season						
Leaves	1935±263[a]	2184±465[a]	1232±33[a]	2422±214[a]	2384±287[ab]	1467±197[b]
Stems	1520±219[a]	1732±476[a]	1119±161[a]	1976±218[a]	1478±206[ab]	820±64[b]
Dead material	1975±322[a]	1555±430[a]	2343±399[a]	3244±538[a]	2078±389[b]	583±46[c]
Total	5430±727[a]	5471±792[a]	4694±354[a]	7642±958[a]	5940±83[b]	2870±232[c]
Dry season						
Leaves	2229±300[a]	2658±766[a]	1812±385[a]	727±17[a]	915±237[a]	1045±256[a]
Stems	2476±542[a]	2451±527[a]	1720±20[a]	1022±216[a]	791±179[a]	1107±277[a]
Dead material	2599±380[a]	2116±476[a]	895±425[a]	1309±136[a]	681±120[a]	519±40[a]
Total	7304±1184[a]	7225±1157[a]	4427±523[a]	3058±405[a]	2387±481[a]	2671±566[a]

[1]Means followed by the same letter within a row during a specific period and season are not significantly different (P≤0.05).

The contribution of native (NG; 86.9 %) and introduced (IG; 2.9 %) grasses in the pastures, as determined before and after grazing, is summarized in Table 4. The contribution of NG to botanical composition increased positively with stocking rate during all three seasons of 2005-2006. In 2006-2007, during the rainy and dry seasons, IG contributed more at the lower stocking rate than with 3 or 4 cows ha^{-1}. Native grasses showed a similar trend with stocking rate during 2006-2007, while introduced grasses did not show a defined trend. During the rainy and dry seasons their contribution decreased with increasing stocking rate, while during the windy season they remained the same.

Table 3. Percentage use of the same plant components before grazing, in native grass-based pastures stocked at 2, 3 and 4 cows ha^{-1} during the rainy, winter and dry seasons during two consecutive grazing periods.

Component	2005-2006			2006-2007		
	2 cows ha^{-1}	3 cows ha^{-1}	4 cows ha^{-1}	2 cows ha^{-1}	3 cows ha^{-1}	4 cows ha^{-1}
	Percentage use					
	Rainy season					
Leaves	14.8±10.7 [a1]	26.2±10.3[a]	40.6±10.6[a]	8.4±3.3[a]	33.8±8.6[b]	27.7±8.9[ab]
Stems	15.7±11.2[a]	30.5±6.0[a]	46.6±8.7[a]	11.7±4.7[a]	30.8±8.7[b]	11.5±6.3[b]
Total	18.6±11.2[a]	19.3±7.0[a]	32.5±11.6[a]	8.8±5.0[a]	35.6±5.3[b]	15.4±6.0[b]
	Winter season					
Leaves	30.0±8.4[a]	46.2±10.4[a]	43.6±14.9[a]	43.7±3.7[a]	36.9±9.9[ab]	20.6±8.4[b]
Stems	24.8±11.0[a]	43.3±9.8[a]	49.0±17.9[a]	30.9±8.3[a]	26.0±8.9[a]	6.1±5.3[b]
Total	19.5±9.7[a]	31.7±10.0[a]	49.1±8.9[a]	40.0±5.2[a]	30.4±10.5[a]	10.0±4.7[b]
	Dry season					
Leaves	42.6±7.0[a]	40.7±15.8[a]	36.5±14.3[a]	50.6±1.9[a]	39.1±10.2[a]	34.3±14.8[a]
Stems	17.2±8.4[a]	27.9±14.8[a]	39.3±16.0[a]	37.5±8.7[a]	13.1±8.1[a]	21.4±11.9[a]
Total	11.6±5.6[a]	27.0±8.8[a]	28.2±7.7[a]	38.2±5.3[a]	27.7±8.5[a]	23.3±9.1[a]

[1]Means followed by the same letter within a row during a specific period and season are not signifantly different (P≤0.05).

Table 4. Percent contribution (mean ± s.e.) of native (NG) and introduced (IG) grasses to botanical composition, during the rainy, windy and dry seasons, in pastures grazed at 2, 3 and 4 cows ha^{-1} during two grazing periods in the humid tropics of Mexico.

Component	2005-2006			2006-2007		
	2 cows ha^{-1}	3 cows ha^{-1}	4 cows ha^{-1}	2 cows ha^{-1}	3 cows ha^{-1}	4 cows ha^{-1}
	Rainy season					
NG	71.5±3.5[a1]	80.7±2.4[b]	81.6±2.9[b]	71.2±4.8[a]	89.8±1.2[b]	90.8±1.4[b]
IG	15.2±3.2[a]	7.1±0.9[a]	9.2±2.4[a]	8.0±1.7[a]	2.4±0.6[b]	1.9±0.3[b]
	Windy season					
NG	74.9±2.3[a]	78.6±2.6[a]	89.1±1.8[b]	78.3±2.9[a]	90.3±1.5[b]	92.9±1.8[b]
IG	9.2±1.7[a]	4.5±0.5[a]	1.6±0.7[a]	1.5±0.2[a]	4.2±1.2[a]	3.1±1.0[a]
	Dry Season					
NG	61.7±3.5[a]	88.4±1.4[bc]	91.0±2.4[c]	67.6±2.0[a]	86.5±1.5[b]	91.9±1.4[b]
IG	10.2±1.2[a]	1.6±0.3[a]	1.9±1.1[a]	11.8±2.0[a]	4.5±0.8[b]	3.4±1.4[b]

[1]Means followed by the same letter within a row during a specific period and season are not significantly different (P≤0.05).

Overall, stocking rate had no effect on pasture quality variables during the study (Tables 5, 6). Leaves had more crude protein and lower NDF concentrations than stems throughout the experiment. During 2005-2006, leaf crude protein was higher at all stocking rates during the windy (P<0.0007) and dry seasons (P<0.005) than during the rainy season (P<0.03). Stems showed differences between rainy and dry seasons (P<0.01), and ISDMD was greater during the rainy season compared to the windy (P<0.04) and dry seasons (P<0.02). However, crude protein and leaf ISDMD levels were higher overall during 2006-2007, and declined as the year progressed. At all stocking rates in all seasons, stem ISDMD was comparable to or slightly higher than that for leaves.

Under set-stocked conditions, as stocking rate increases the quantity of HMP and bite size normally

decline (Stobbs, 1973; Jones and Jones, 2003). To avoid a reduction in DM intake, animals compensate by increasing both biting rate and grazing time (Jones and Jones, 2003). Therefore, increasing the stocking rate increases the severity of grazing (Burns and Sollenberger, 2002), affecting both standing forage and its degree of utilization (O'Donovan et al., 2004). In the present study where rotational grazing was applied, stocking rate did not affect standing DM at the commencement of grazing, nor utilization rates of leaves and stems during the first year. Grazing intensity was therefore higher on pastures stocked at higher rates. During the regrowth period, the more heavily grazed pastures compensated by producing more growth than the pasture stocked at the lower rate so that pre-grazing DM levels were equivalent at all stocking rates.

Table 5. Nutritional value (mean ± s.e.) of leaves and stems (true stem + leaf sheath) of native grasses in pastures grazed at 2, 3 and 4 cows ha[-1] during the rainy, windy and dry seasons during 2005-2006 in the humid tropics of Mexico.

Component[1] (% of DM)	Leaves Cows ha[-1]			Stems Cows ha[-1]		
	2	3	4	2	3	4
Rainy season						
CP	7.7±0.4[a2]	8.2±0.2[a]	8.7±0.7[a]	5.2±0.3[a]	6.0±0.4[a]	5.6±0.1[a]
NDF	68.6±5.8[a]	71.9±6.2[a]	71.9±5.8[a]	73.6±1.2[a]	80.2±2.5[a]	79.7±1.4[a]
ADF	42.9±1.4[a]	45.5±1.1[a]	43.2±0.7[a]	44.1±1.1[a]	45.1±0.0[a]	44.6±0.4[a]
LIG	8.7±2.0[a]	10.6±2.4[a]	9.9±1.6[a]	9.5±0.4[a]	9.1±0.2[a]	9.7±0.4[a]
ISDMD	59.3±0.9[a]	57.6±1.3[a]	61.4±2.4[a]	62.9±0.8[a]	62.6±2.0[a]	63.7±3.3[a]
Windy season						
CP	11.9±1.5[a]	13.0±1.4[a]	11.6±0.3[a]	5.1±0.2[a]	7.4±0.8[a]	7.9±0.4[a]
NDF	63.8±7.5[a]	62.6±4.8[a]	62.3±6.5[a]	72.5±0.7[a]	69.9±0.5[a]	69.7±0.5[a]
ADF	35.5±0.1[a]	33.1±0.4[a]	35.2±1.0[a]	36.5±0.7[a]	34.1±0.5[b]	34.2±0.7[ab]
LIG	7.0±1.3[a]	7.8±1.7[a]	9.7±0.5[a]	6.0±0.1[a]	6.1±0.8[a]	7.7±0.5[a]
ISDMD	65.4±2.0[a]	67.4±1.2[a]	69.2±4.1[a]	65.3±3.1[a]	69.7±2.0[a]	71.0±0.2[a]
Dry season						
CP	9.0±0.6[a]	10.8±0.5[a]	11.9±0.3[a]	6.3±0.2[a]	7.0±1.3[a]	8.2±0.8[a]
NDF	62.8±6.5[a]	60.9±6.4[a]	61.7±4.8[a]	70.2±3.0[a]	66.5±2.6[a]	67.3±2.8[a]
ADF	34.7±1.4[a]	30.3±0.8[b]	31.3±0.9[ab]	33.2±1.8[a]	31.3±2.6[a]	32.6±0.9[a]
LIG	8.1±2.5[a]	6.7±1.0[a]	7.±1.0[a]	6.5±0.6[a]	5.9±0.01[a]	6.5±0.9[a]
ISDMD	62.8±1.0[a]	69.5±0.8[a]	70.1±3.1[a]	66.0±0.6[a]	68.4±0.2[a]	70.1±0.7[a]

[1]CP=crude protein; NDF=neutral detergent fibre; ADF=acid detergent fibre; LIG=lignin; ISDMD=in situ present herbage mass digestibility.
[2] Means followed by the same letter within a row during a specific period and season are not significantly different (P≤0.05).

Table 6. Nutritional value (%) (mean ± s.e.) of leaves and stems (true stem + leaf sheath) of native grasses in pastures grazed at 2, 3 and 4 cows ha[-1] during the rainy, windy and dry seasons during 2006-2007 in the humid tropics of Mexico.

Component[1] (% of DM)	Leaves			Stems		
	Cows ha[-1]			Cows ha[-1]		
	2	3	4	2	3	4
Rainy season						
CP	14.5±0.9[a2]	15.9±0.3[a]	15.4±0.9[a]	7.9±0.5[a]	10.1±0.1[b]	9.4±0.8[ab]
NDF	66.8±1.7[a]	66.6±1.1[a]	65.9±0.3[a]	71.2±1.1[a]	68.8±0.8[a]	68.3±1.4[a]
ADF	32.9±1.8[a]	31.9±0.3[a]	31.0±0.7[a]	35.9±1.2[a]	33.2±0.3[a]	33.2±1.4[a]
LIG	7.7±0.8[a]	7.2±0.4[a]	7.1±0.2[a]	6.5±0.4[a]	5.7±0.1[a]	5.4±0.2[a]
ISDMD	79.2±1.6[a]	79.0±2.1[a]	79.4±0.9[a]	71.9±1.7[a]	76.6±1.4[a]	76.2±2.2[a]
Windy season						
CP	13.2±1.1[a]	14.2±0.4[a]	14.8±0.2[a]	7.4±0.3[a]	7.3±0.5[a]	9.0±0.2[a]
NDF	69.9±0.5[a]	68.4±0.9[a]	67.4±0.6[a]	67.8±0.3[a]	65.2±0.7[a]	63.4±0.1[a]
ADF	36.7±0.9[a]	36.8±1.0[a]	33.0±0.1[b]	36.4±0.1[a]	33.0±0.8[a]	31.1±0.3[a]
LIG	8.9±0.9[a]	8.0±1.1[a]	7.9±0.3[a]	7.6±0.7[a]	5.2±1.1[b]	4.0±0.8[b]
ISDMD	70.7±1.0[a]	74.6±1.6[b]	76.7±1.5[b]	73.4±2.0[a]	76.7±0.9[a]	78.8±0.5[a]
Dry season						
CP	11.4±0.5[a]	12.6±0.5[a]	13.2±1.0[a]	6.5±0.2[a]	7.2±0.3[a]	7.9±0.5[a]
NDF	67.8±0.6[a]	68.8±0.3[a]	67.3±0.6[a]	66.8±0.6[a]	65.7±0.3[a]	64.4±0.3[a]
ADF	32.7±0.2[a]	35.0±0.7[a]	32.4±0.2[a]	32.9±0.6[a]	29.1±0.9[a]	28.7±0.6[a]
LIG	6.4±0.4[a]	7.6±0.8[a]	7.2±0.4[a]	5.8±0.1[a]	4.9±0.1[a]	5.2±0.2[a]
ISDMD	69.8±0.8[a]	75.0±2.1[a]	70.1±0.6[a]	66.9±1.2[a]	76.4±1.5[a]	77.5±0.3[a]

[1]CP=crude protein; NDF=neutral detergent fibre; ADF=acid detergent fibre; LIG=lignin; IPHMD=*in situ* present herbage mass digestibility.
[2]Means followed by the same letter within a row during a specific period and season are not significantly different (P≤0.05).

However, utilization rates of leaves and stems varied during the study. While levels were generally high in 2005-2006, with a peak of 49% for stems at 4 cows ha[-1] during the windy season, this contrasts with the utilization rate of 6% for the same treatment in the same season during 2006-2007 (Table 3). During 2005-2006, utilization rates of leaves and stems were generally comparable, in contrast to the pattern during 2006-2007, when utilization of leaves was generally higher than that for stems. Thus, animals preferentially select leaves rather than stems. Surprisingly, ISDMD values for stems were at least equal to those for leaves at all stocking rates during most seasons in both years, despite stems having less crude protein and higher NDF levels than leaves. This could explain why the cows did not preferentially graze leaves versus stems. Pasture availability, in relation to grazing time per day, decreased as stocking rate increased, even though during the rainy season total and leaf DM availability

remained similar. During the windy and dry seasons, grazing time increased with stocking rate, while availability decreased with stocking rate (i.e., animals compensated for the reduction in available DM by increasing their grazing time).

However, DM availability does not necessarily significantly affect grazing time or biting rate (Kennedy et al., 2007), as it was the general trend for the results of this study. Here, grazing time was long and ruminating time short, which can be related to pasture quality (Van Vuuren et al., 2005). Larger grazing time values have been reported, but it has the same relationship with ruminating time as those obtained in the present experiment (Stobbs, 1970). Due to the short period of paddock recovery (27 days), pasture quality variables such as less lignified cell walls (7.2±0.3 %) make the ingested forage less resistant to degradation by rumination. However, grazing time was shortened due to the reduction in pasture availability.

The interaction between stocking rate and season for pasture DM availability is related to the morphological response of the plants to increased stocking rate. In pastures with low stocking rates, plant height is greater than at high stocking rates (Kennedy *et al.*, 2007), and tiller density is lower at low stocking rates (Van Vuuren *et al.*, 2005) or with less grazing frequency (McCarthy *et al.*, 2007), as is the number and size of leaves. In *Paspalum notatum*, as cutting height decreases, leaf appearance rate increases and conversely, leaf extension rate decreases (i.e., more, but shorter, leaves appear) (Ayala *et al.*, 2000). This may be a compensatory mechanism leading to equivalent degrees of pasture availability among stocking rates.

Pasture utilization was similar for the three stocking rates, but different among seasons, indicating that utilization remains high as long as favourable climatic conditions (rainy=25.5%) for plant growth exist, even though during the windy and dry seasons leaves were defoliated more (36.8 and 40.6%, respectively). Pasture utilization was greater with higher stocking

rate, at least during the second year. Increased utilization with increased stocking rate suggests a leafier pasture structure (Nochebuena *et al.*, 1994; Hirata, 2000).

Trends in fiber and lignin in leaves and stems was similar to crude protein, which was likely related to the age of regrowth. In a 35-42 day study of regrowth in Bahiagrass (*P. notatum*), increases in lignin and NDF of 6.2% and 71%, respectively, have been observed (Mears and Humphreys, 1974).

Ingestive behaviour

During the rainy season, grazing time decreased as stocking rate increased from 2 to 4 cows ha^{-1}. However, during the windy and dry seasons in both years, the longest grazing time was recorded for 3 cows ha^{-1}. While time spent grazing was constant among seasons at 2 cows ha^{-1} (P>0.05), it increased at 3 cows ha^{-1} and decreased at 4 cows ha^{-1} during the windy and dry seasons. The average time spent grazing was 420 min d^{-1}, while cows ruminated an average of 367 min d^{-1}.

Table 7. Effects of stocking rate on ingestive behaviours of F1 Holstein x Cebu cows that grazed native grass-based pastures in the humid tropics of Mexico.

Parameter	2005-2006			2006-2007		
	2 cows ha^{-1}	3 cows ha^{-1}	4 cows ha^{-1}	2 cows ha^{-1}	3 cows ha^{-1}	4 cows ha^{-1}
	Rainy season					
Grazing time (min)	476±8[a1]	446±8[b]	353±12[c]	403±8[a]	368±14[b]	382±9[ab]
Ruminating time (min)	316±16[a]	321±13[a]	305±14[a]	451±9[a]	444±10[a]	429±14[a]
Biting rate (bites min^{-1})	49±1[a]	48±1[a]	50±1[a]	49±1[a]	48±1[a]	48±1[a]
	Windy season					
Grazing time (min)	406±9[a]	460±9[b]	420±11[a]	406±11[a]	456±12[b]	367±13[c]
Ruminating time (min)	365±12[a]	398±19[a]	310±12[b]	398±10[a]	381±9[a]	248±10[b]
Biting rate (bites min^{-1})	51±1[a]	45±1[b]	48±1[c]	44±1[a]	49±1[b]	51±1[bc]
	Dry season					
Grazing time (min)	401±11[a]	480±8[b]	427±13[a]	404±15[a]	465±11[b]	439±12[bc]
Ruminating time (min)	411±13[a]	396±14[a]	331±16[b]	379±17[a]	336±12[b]	392±18[a]
Biting rate (bites min^{-1})	49±1[a]	51±1[a]	51±1[a]	48±1[a]	47±1[ab]	45±1[b]

[1] Means followed by the same letter within a row during a specific period and season are not significantly different (P≤0.05).

Biting rate

Biting rate varied from 44 to 51 bites min^{-1} (Table 7). A reduction in biting rate is suggested by a reduction in forage allowance, as was observed during the windy season in 2005-2006 and the dry season in 2006-2007. Biting rate can vary from 70-80 bites min^{-1} at the beginning of grazing to 40-50 bites min^{-1} at the end of a grazing session (Juarez-Lagunes et al., 1999). In the research station used for the present study, a comparable mean biting rate with 2 cows ha^{-1} has been reported (46 ± 0.7 bites min^{-1}) (Castillo, 2003). However, other reports have stated that these biting rates are low (Juarez-Lagunes et al., 1999; Kennedy et al., 2007). Under artificial conditions, the effect of density and tension resistance of grass shoots to grazing was evaluated and found a time per bite of 1.5 to 2.0 seconds (40-30 bites min^{-1}) (Stobbs, 1974). As such, biting rate has been highly correlated with shoot height (Benvenutti et al., 2006). However, the present study did not show such an association; pasture height decreased as stocking rate increased, although biting rate means were identical among stocking rates.

In general, grazing patterns were equivalent for all stocking rates, seasons and during both years (Figure 1). There was a period of intense grazing activity when cows were returned to the paddocks after milking at 11:00 h. Peak activity occurred at 18:00 h, followed by a rapid cessation of grazing. There was a second, smaller grazing episode around midnight and an additional brief grazing episode before cows were taken to the milking barn at 8:00 h.

In the present experiment, mean grazing time (396±79 to 436±84 min d^{-1}) was comparable to the 420 min d^{-1} recorded on good quality pastures with adequate DM (Van Vuuren et al., 2005). Grazing time was much lower than the 605 and 603 min d^{-1} for the 4.5 and 5.5 cows ha^{-1}, respectively, recorded by Kennedy et al. (2007). Grazing times of 395 - 586 min d^{-1} for heifers in native grass pastures, with or without fertilization, have been reported (Benvenutti et al., 2006).

Using the same observational technique on pastures equivalent to those in the present study and in the same experimental station, grazing times of 457 ± 13 min d^{-1} and 395 ± 11 min d^{-1} have been reported on native pastures and those associated with Arachis pintoi Krapovickas and Gregory (Castillo, 2003).

Among stocking rates, the mean total number of bites per day in different seasons fluctuated between 17,600 and 24,500 with no set pattern. Means of 20,000, 21,400 and 19,400 bites cow^{-1} day^{-1} were recorded for 2, 3 and 4 cows ha^{-1}, respectively. Since available DM at the commencement of grazing was similar at all stocking rates, unless bite size declined as stocking rate increased, utilization rates would have to increase dramatically with increases in stocking rate. Daily patterns of grazing behaviour in this study differed from those normally recorded. Intense grazing peaks usually occur early in the morning and late afternoon, with a resting period and water intake in the middle of the day, especially in tropical areas (Van Vuuren et al., 2005). Milking herd management affected grazing patterns, with the longest period of grazing in the afternoon, from 12:00 h to 19:00 h, when the herd returned to pasture after milking. Animals removed from pasture for milking strongly affected grazing behaviour, with a predominance of grazing over other activities during the afternoon grazing session (Benvenutti et al., 2006). These authors found that milking cows under continuous grazing normally have 3 or 4 periods of intense grazing activity each day, compared with the 3 periods recorded here.

In the present study, the first grazing period ended at 19:00 h. When pasture quantity and quality is good, grazing patterns follow daylight variations, but cattle shift grazing to the darker hours of the day under adverse conditions (Van Vuuren et al., 2005). Cows in the present study spent a significant amount of time grazing at midnight, which could have compensated for grazing time lost when cows were being milked.

Pasture management could have a strong impact upon pasture composition, which can affect animal behaviour and productive performance (Boval et al., 2007). The pasture species that contribute most to botanical composition as well as DM production can be the basis for maintaining animal production at an optimal level (Gibb et al., 1997). Botanical composition in the present experiment differed with stocking rate, with low stocking rate always showing a higher contribution of taller introduced grasses than the medium and high stocking rates, in which short native grasses predominated. This suggests that the introduced grasses and native species had less tolerance for higher grazing pressures and were not able to recover rapidly during the 27-day regrowth periods. Long-term studies are needed to determine if introduced species disappear from comparable pastures over the time.

Rumination patterns alternated with grazing patterns, with limited activity during grazing periods, and peaks of activity when cows were resting after grazing (Figure 1). While grazing was the main activity during the day, time spent ruminating increased dramatically after 18:00 h. This continued until the midnight grazing period, followed by a resting and ruminating period from 03:00 h to 06:00 h.

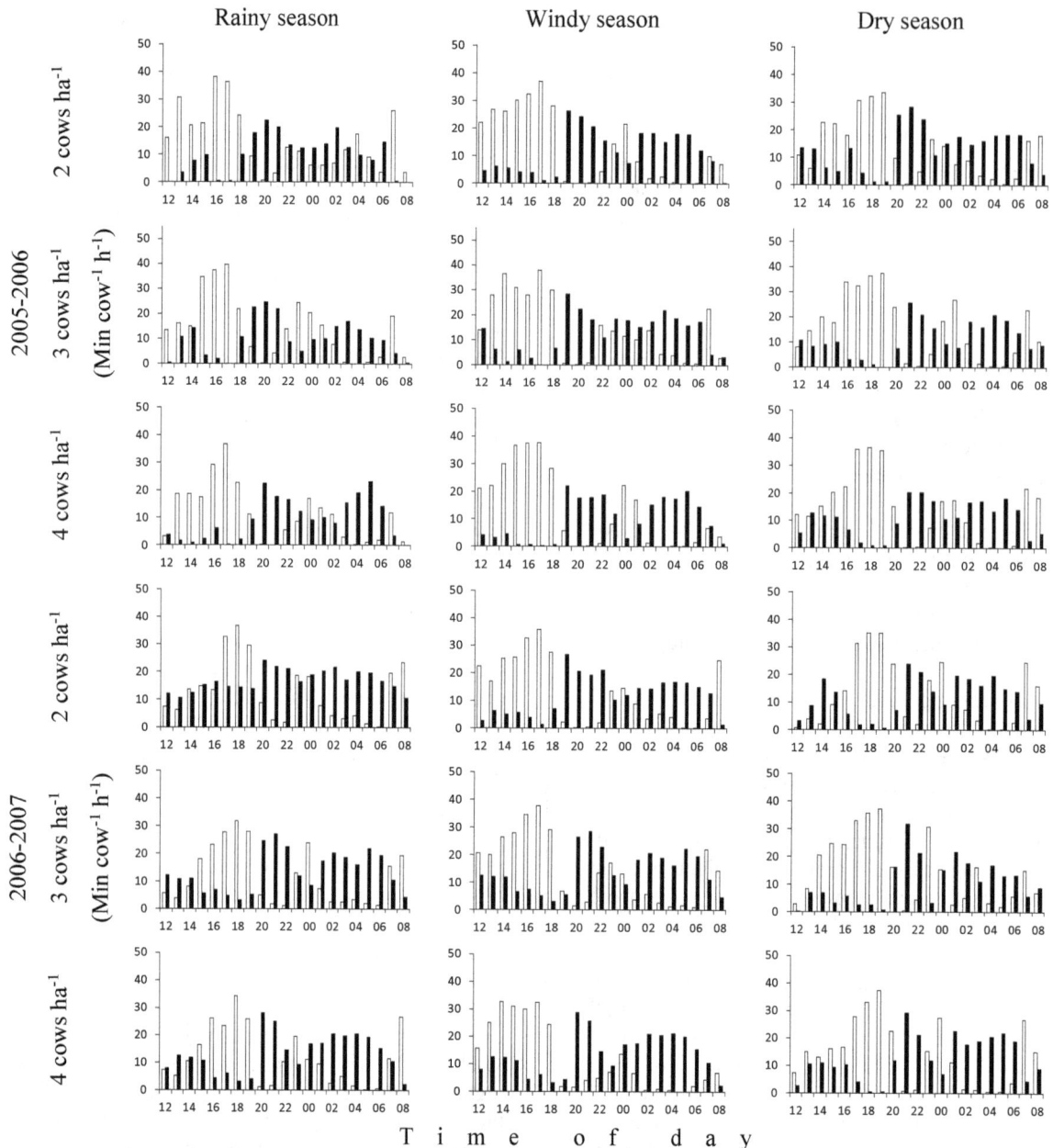

Figure 1. Effects of stocking rate and season on daily grazing (□) and ruminating (■) times in pastures during 2005-2006 and 2006-2007. Observations were not made when the cows were in the milking shed (08:00-11:00h).

Under temperate conditions, at two stocking rates (4.5 and 5.5 cows ha⁻¹) for two 49-day rotational grazing cycles (February and April), increasing the stocking rate in the first cycle significantly increased grazing time, while the reverse happened during the second period (Kennedy et al., 2007). An identical trend was observed during the present experiment. Grazing time showed an opposite trend during the first period, while there was no difference between stocking rates during the second period. Grazing time in the present study was comparable between 2 and 3 cows ha⁻¹ and significantly increased for 4 cows ha⁻¹.

The latter may be related to leaf availability which decreased as stocking rate increased. Leaf availability increased with stocking rate during the first rotation and decreased during the second one, but without statistical differences between stocking rate (Kennedy et al., 2007).

Milk production

Daily milk yield per cow (8.16±2.58 kg) was similar (P>0.05) among stocking rates. Therefore, milk yield per lactation (1894.9±904.3 kg) was also similar (P>0.05). Nevertheless these values stand above the typical milk yield per cow values for the tropics. Orantes-Zebadúa *et al.*, (2014) surveyed 246 dual-purpose farmers from Chiapas, México, and found that the mean milk yield per cow was 4.49 l at an approximate stocking rate of 1.5 animal units per ha.

When dry matter availability in the pasture is large enough, there are no limits to intake of pasture by the grazing cow, and there are no limits to animal selectivity and nutritive value of ingested dry matter (Benítez-Bahena *et al.*, 2010). As shown by Table 8, the increase in stocking rate lead to a decrease in standing dry matter. However, this decrease did not limit the cow's dry matter intake and consequently milk yield remained the same between stocking rates. On the other hand, the level of pasture utilization and in general the ingestive behavior variables did not show differences among stocking rates, and this coincides with the similar milk production among stocking rates.

The standing dry matter can in itself influence individual milk yield (Castillo *et al.*, 2014). A study reports that when the SDM was larger than 2,500 kg/ha the daily milk yield per cow remained unchanged, but below 2,000 kg/ha, intake and grazing time decreased (Cowan and O´Grady, 1976). In the present investigation the three levels of stocking rate had SDM amounts that did not limit intake and thus, individual milk yields.

Overall, the best response in the literature with respect to nutritional value, higher stocking rate and best weight gain per ha has been obtained with an availability of 5 kg de MS/100 kg de PV (Menezes *et al.*, 2016). If this latter value is applied to the cows of the present research that weighed on average 531 kg, results in an optimal availability of 26.55 kg of SDM per cow. In our study, the amount of SDM per cow per day was larger than 56.4 kg per cow per day. So pasture availability was enough in all stocking rates for the cows to reach maximum daily intakes, and therefore, similar individual milk yields.

CONCLUSIONS

There was no apparent significant effect from stocking rate on the ingestive behaviour of cows. As well, there was no clear relationship among any pasture variables and ingestive behaviour, except the proportion of native and introduced grasses. However, neither was there a difference in milk yield among stocking rates, which was nevertheless higher than typical grazing conditions of the tropics. Available forage significantly increased during the rainy season, yielding longer grazing periods and higher bite rates. But if pasture response variables were not affected by SR, ingestive behaviour depended on the number of cows per hectare or instantaneous stocking rate. Standing dry matter and its quality were not determinate factors for milk yield in this research.

Table 8. Area per paddock, live weight (LW) per cow, live weight per paddock, total standing dry matter (TSDM) available per paddock, and dry matter available per 100 kg live weight, per cow, lactation length day, milk per cow by day and lactation in a native pasture, grazed by 2, 3 and 4 cows ha⁻¹, in the Mexican humid tropics.

Concept	Stocking rate (cows/ha)		
	2	3	4
Number of cows/ha	2	3	4
Area/paddock (ha)	0.5	0.33	0.25
LW/cow (kg)	541	531	501
LW/paddock (kg)	5410	5310	5010
Kg/DM/paddock	2600.9 ± 994.6	2038.5 ± 1167.2	1693.4 ± 855.4)
DM/day (kg)	866.9 ± 331.5	679.5 ± 389.1	564.5 ± 285.1)
DM/100 kg LW (kg)	16.3 ± 5.7	12.8 ± 6.9	10.6 ± 5.0)
DM/cow/day (kg)	86.6 ± 33.1	67.9 ± 38.9	56.4 ± 28.5
Lactation length (days)	249.8 ± 108.7	205.2 ± 66.4	236.0 ± 85.5)
Milk/cow/day (kg)	7.8 ± 2.1	8.6 ± 2.7)	8.0 ± 2.9
Milk/lactation (kg)	2030.5 ± 1177.0	1737.3 ± 671.6)	1917.1 ± 864.3

Mean and standard desviation

REFERENCES

Amendola, R., Castillo, E., Martínez, P.A. 2005. Forage Resource Profiles. Mexico. In: Food and Agriculture Organization (ed.), Country Pasture Profiles. Rome, Italy. On CD.

Association of Official Analytical Chemists. 1980. Official Methods of Analysis. 13th Edn. Washington, D.C., USA.

Ayala, T.A.T., Acosta, G.L., Deregibus, V.A., Moauro, P.M. 2000. Effects of grazing frequency on the production, nutritive value, herbage utilization, and structure of a *Paspalum dilatatum* sward. New Zealand Journal Agricultural Research. 43:467-472.

Benítez-Bahena, Y., Bernal-Hernández, A., Cortés-Díaz, E., Vera Castillo, G., Carrillo Anzures, F. 2010. Producción de forraje de guaje (*Leucaena* spp.) asociado con zacate (*Brachiaria brizantha*) para ovejas en pastoreo Revista Mexicana de Ciencias Agrícolas. 1(3): 397-411.

Benvenutti, M.A., Gordon, I.J., Poppi, D.P. 2006. The effect of the density and physical properties of grass stems on the foraging behaviour and instantaneous intake rate by cattle grazing an artificial reproductive tropical sward. Grass Forage Science. 61:272-281. https://doi.org/10.1111/j.1365-2494.2006.00531.x

Boval, M., Fanchone, A., Archimede, H., Gibb, M.J. 2007. Effect of structure of a tropical pasture on ingestive behaviour, digestibility of diet and daily intake by grazing cattle. Grass Forage Science. 62:44-54. https://doi.org/10.1111/j.1365-2494.2007.00560.x

Burns, J.C., Sollenberger, L.E. 2002. Grazing behavior of ruminants and daily performance from warm-season grasses. Crop Science. 42:873-881. doi:10.2135/cropsci2002.8730

Castillo, G.E. 2003. Improving a native pasture with the legume *Arachis pintoi* in the humid tropics of Mexico. [Doctoral thesis]. Wageningen, The Netherlands, Wageningen University,

Castillo, G.E., Rascón, C.R., García, G.D., Jarillo, R.J., Aluja, S.A., Mannetje, L. 't. 2014. Calidad de la pastura ingerida y comportamiento ingestivo de vacas en una asociación grama nativa/*Arachis pintoi* en el trópico húmedo veracruzano. Revista Mexicana Ciencias Pecuarias. 5(4): 487-504.

Cowan, R.T., O′Grady, P. 1976. Effect of presentation yield of a tropical grass-legume pasture on grazing time and milk yield of Friesian cows. Tropical Grasslands.10:213-218.

Dillon, P., Roche, J.R., Shalloo, L., Horan, B. 2005. Optimising financial return from grazing in temperate pastures. Proceeding of a satellite workshop of the XX International Grassland Congress, Cork, Ireland. pp. 131-147.

Gary, L.A., Sherritt, G.W., Hale, E.B. 1970. Behaviour of Charolais cattle on pasture. Journal Animal Science. 30:203-206. https://doi.org/10.2527/jas1970.302203x

Gibb, M. 2006. Grassland management with emphasis on grazing behaviour. In: Fresh Herbage for Dairy Cattle. Elgersma, A., Dijkstra, J., Tamminga, S. (eds.). IGER, Aberystwyth, UK. pp. 141-157.

Gibb, M.J., Huckle, C.A., Nuthall, R., Rook, A.J. 1997. Effect of sward surface height on intake and grazing behaviour by lactating Holstein-Friesian cows. Grass Forage Science. 52:309-321. https://doi.org/10.1111/j.1365-2494.1997.tb02361.x

Haydock, K.P., Shaw, N.H. 1975. The comparative yield method for estimating the dry matter yield of pasture. Australian Journal Experimental Agricultural Animal Husbandry. 15:663-670. http://dx.doi.org/10.1071/EA9750663

Hendricksen, R., Minson, D.J. 1980. The feed intake and grazing behaviour of cattle grazing a crop of *Lablab purpureus* cv. Rongai. Journal Agricultural Science. 95:547-554. https://doi.org/10.1017/S0021859600087955

Hirata, M. 2000. Effects of nitrogen fertiliser rate and cutting height on leaf appearance and extension in bahiagrass (*Paspalum notatum)* swards. Tropical Grasslands. 34:7-13.

Jamieson, W.A., Hodgson, J. 1979. The effect of daily herbage allowance and sward characteristics upon the ingestive behaviour and herbage intake of calves under strip-grazing management. Grass Forage Science. 34:262-264. https://doi.org/10.1111/j.1365-2494.1979.tb01478.x

Jones, R.M., Jones, R.J. 2003. Effect of stocking rates on animal gain, pasture yield and composition, and soil properties from setaria-nitrogen and setaria-legume pastures in coastal south-east Queensland. Tropical Grasslands. 37:65-83.

Juarez-Lagunes, F.I., Fox, D.G., Blake, R.W., Pell, A.N. 1999. Evaluation of tropical grasses for milk production by dual-purpose cows in tropical Mexico. Journal Dairy Science. 82:2136-2145. https://doi.org/10.3168/jds.S0022-0302(99)75457-3

Kennedy, E., O'Donovan, M.O., Murphy, J.P., Delaby, L., O'Mara, F.P.O. 2007; Effect of spring grazing date and stocking rate on sward characteristics and dairy cow production during midlactation. Journal Dairy Science. 90:2035-2046. https://doi.org/10.3168/jds.2006-368

Langer, R.H.M. 1979. How grasses grow. Edward Arnold Publishers Ltd., London, UK.

Littell, R.C., Henry, P.R., Ammerman, C.B. 1998. Statistical analysis of repeated measures data using SAS procedures. Journal Animal Science. 76:1216-1231. DOI 10.2527/1998.7641216x

Mannetje, L.'t., Haydock, K.P. 1963. The dry-weight-rank method for the botanical analysis of pasture. Journal British Grassland Society. 18:268-275. https://doi.org/10.1111/j.1365-2494.1963.tb00362.x

McCarthy, S., Horan, B., Rath, M., Linnane, M., O'Connor, P., Dillon, P. 2007. The influence of strain of Holstein-Friesian dairy cow and pasture-based feeding system on grazing behaviour, intake and milk production. Grass Forage Science. 62:13-26. https://doi.org/10.1111/j.1365-2494.2007.00557.x

Mears, H., Humphreys, L.R. 1974. Nitrogen response and stocking rate of Pennisetum clandestinum pastures: I. Pasture nitrogen requirement and concentration, distribution of dry matter and botanical composition. Journal Agricultural Science. 83:451-467. https://doi.org/10.1017/S0021859600026940

Menezes, D.A., Oliveira, G.E.N., Vinhas, L.C., Moura, A.L.A., Brandão-Ferreira, Í.C.C., Nogueira, E., Ferreira da Silva, F., Junges, L. 2016. Herbage allowance effects on the characteristics of Brachiaria brizantha cv. Marandu pastures and the production and economic viability of Nellore heifers. Semina: Ciencias Agrárias. Londrina. 37, (4), 1: 2301-2312. http://dx.doi.org/10.5433/1679-0359.2016v37n4Supl1p2301

Nochebuena, N.G., Valles, M.B., De Lucía, S.G.R. 1994; Comportamiento del pastoreo y producción de leche en vacas F1 (Holstein-Cebú) en una pradera de zacate elefante en Veracruz, México. Avances en Investigación Agropecuaria. 3(1):16-27.

O'Donovan, M., Delaby, L., Peyraud, J.L. 2004. Effect of time of initial grazing date and subsequent stocking rate on pasture production and dairy cow performance. Animal Research. 53:489-502. DOI: 10.1051/animres:2004036

Ocaña, Z.E. 2003. Efecto de la carga animal sobre gramas nativas, características del suelo y producción de leche y becerros de vacas Holstein x Cebú en pastoreo intensivo en el trópico mexicano. [Tesis de Maestría].

Montecillo, México. Colegio de Postgraduados,

Orantes-Zebadúa, M.A., Platas-Rosado, D., Córdova-Avalos, V., De los Santos-Lara, M.C., Córdova-Avalos, A. 2014. Caracterización de la ganadería de doble propósito en una región de Chiapas, México. Ecosistemas recursos agropecuarios. 1(1):49-58.

Orskov, E.R., McDonald, I. 1979. The estimation of protein degradability in the rumen from incubation measurements weighted according to rate of passage. Journal Agricultural Science. Cambridge. 92:499-503. https://doi.org/10.1017/S0021859600063048

Ortega, R.L., Castillo, H.J.E., Rivas, P.F.A. 2009. Conducta ingestiva de bovinos Cebú adultos en Leucaena manejada a dos alturas diferentes (Ingestive behavior of adult Cebu cattle grazing Leucaena at two different heights. Técnica Pecuaria México. 47(2):125-134.

Penning, P.D., Rutter, S.M. 2004. Ingestive behaviour. In: Herbage Intake Handbook. Penning, P.D. (ed.). British Grasslands Society. Reading, UK. 151-175.

Rook, A.J., Penning, P.D. 1991. Synchronization of eating, ruminating and idling activity by grazing sheep. Applied Animal Behaviour Science. 32:157-166. https://doi.org/10.1016/S0168-1591(05)80039-5

Silva, S.C., Carvalho, P.C. 2005. Foraging behaviour and herbage intake in the favourable tropics/subtropics. In: Grassland: A global resource. McGilloway, D.A. (ed). Wageningen Academic Publishers, Netherlands. pp. 81-96.

Soder, K.J., Rook, A.J., Sanderson, M.A., Goslee, S.C. 2007. Interaction of plant species diversity on grazing behavior and performance of livestock grazing temperate region pastures. Crop Science. 47:416-425. doi:10.2135/cropsci2006.01.0061

Stobbs, T.H. 1970. Automatic measurement of grazing time by dairy cows on tropical grass and legume pastures. Tropical Grasslands. 4:237-244.

Stobbs, T.H. 1974. Rate of biting by Jersey cows as influenced by yield and maturity of tropical grasses. Tropical Grasslands. 8:81-86.

Stobbs, TH. 1973. The effect of plant structure on the intake of tropical pastures I. Variation in the bite size of grazing cattle. Australian Journal

Agricultural Research 24:809-819. https://doi.org/10.1071/AR9730809

Ungar, E.D., Noy-Meir, I. 1988. Herbage intake in relation to availability and sward structure: Grazing processes and optimal foraging. Journal Applied Ecology 25:1045-1062. DOI: 10.2307/2403765

Van Soest, P.J., Robertson, J.B., Lewis, B.A. 1991. Methods of dietary fibers, and nonstarch polysaccharides in relation to animal nutrition. Journal Dairy Science. 74:3583-3597. https://doi.org/10.3168/jds.S0022-0302(91)78551-2

Van Vuuren, A.M., Van den Pol-Van Dasselar, A. 2006. Grazing systems and feed supplementation. 2005 Proceeding Workshop on Fresh Herbage for Dairy Cattle, The Key to a Sustainable Food Chain. Wageningen University. pp. 85-101.

4

Effectiveness of Fungicides for Rice Blast Control in Lowland Rice Cropped in Brazil

C.Ogoshi[1*], F.S. Carlos[2], A.R. Ulguim[3], A.J. Zanon[3], C.R.C. Bittencourt[2] and R. D. Almeida[2]

[1]*Agricultural Research and Rural Extension Enterprise of Santa Catarina (EPAGRI), Experimental Station of Caçador, Caçador, SC, Brazil*
E-mail: claudioogoshi@epagri.sc.gov.br
[2]*Rio Grandense Rice Institute (IRGA), Rice Experiment Station, Cachoeirinha, RS, Brazil*
[3]*Federal University of Santa Maria (UFSM), Santa Maria, RS, Brazil*
Corresponding author

SUMMARY

Rice cultivation is fundamental to Brazil's economy, which is the largest grain-producing country outside the Asian continent. However, several factors harm sustainable rice production, including diseases incidence. The blast is the major rice disease, and annually, losses caused by this disease would be sufficient to feed millions of people. Due to the complexity in rice blast management, fungicide application has been most used by rice growers, however, despite the high number of registered fungicides, is questioned the real efficiency in field conditions. This work aimed to analyze the effectiveness of fungicides in rice blast control and grain yield maintenance in irrigated rice. Experiments were carried out in a randomized block design, with four replications, in 2015/2016 and 2016/2017 crop seasons, at the Rio Grande do Sul Rice Experiment Station located in Cachoeirinha city, RS, Brazil. We tested the efficiency of nine fungicides registered in Brazil with different target sites for rice blast control. The analyzed variables were the Area Under the Disease Progress Curve (AUDPC) in leaves; neck blast incidence; panicle blast severity and grain yield. In general, only the tricyclazole and trifloxystrobin + tebuconazole fungicides were effective in rice blast control with direct reflection on grain yield.
Key words: *Pyricularia oryzae*; *Oryza sativa* L.; sustainability; disease management.

INTRODUCTION

Rice (*Oryza sativa* L.) is the staple food of more than three billion people around the world (Espe *et al.*, 2016). Brazil is the most significant producer country outside the Asian continent, with an annual production of approximately 12 Mt of grains (USDA, 2017). The State of Rio Grande do Sul (RS) in Brazil is the most important national producer, with more than 70% of grain production, where they have grown annually around 1.1 million hectares (CONAB, 2017). Rice yields quadrupled since the beginning of cultivation in

RS in 1906 (1.8 t.ha^{-1}) to the last five years (7.5 t.ha^{-1}) (CONAB, 2017).

Due to the projection of increase in the world population, it is necessary to invest in research and technology transfer to reduce the gap between the irrigated rice yield observed in research stations and average yield observed in the field (Espe *et al.*, 2016; Ribas *et al.*, 2016). Among the main factors that cause loss of productivity of rice in Brazil, we highlight the rice blast (*Magnaporthe oryzae* Couch; anamorph *Pyricularia oryzae* Cavara).

Rice blast is the most critical disease affecting rice cultivation around the world and can cause 100% losses in productivity if adequate management measures are not adopted (Prabhu *et al.*, 2009). Currently, it is estimated that the rice blast causes declines of 30% in world rice production, and only those losses would be enough to feed more than 60 million people (Nalley *et al.*, 2016).

The integrated rice blast disease management recommended in Brazil involve cultural methods and chemical control. The most important of which is: sowing at the suggested date, balanced nitrogen fertilization, irrigation uniformity, incorporation of cultural remains of rice into the soil, weed management, use resistant varieties and fungicides application (Pak *et al.*, 2016; Pooja and Katoch, 2014; Soares *et al.*, 2014; Prabhu *et al.,* 2009).

Use of resistant varieties to plant diseases management is an adequate way for the sustainable production of any crop (Mundt, 2014). However, rice blast resistant-varieties have not been presenting stable resistance over the years, being this resistance overcome in three to five years after its release (Prabhu *et al.,* 2009). This loss of resistance is mainly because of the high pathogen genetic variability (Prabhu *et al.*, 2009; Zhou *et al.*, 2007; Prabhu and Filippi, 2006), occurring selection of new virulent races that overcome the resistance. Therefore, fungicides application in rice blast management is essential, and this practice is preferred by growers of irrigated rice in Brazil and the world (Chen *et al.*, 2015; Pooja and Katoch, 2014).

Currently, there are a large number of fungicides registered for blast control in Brazil (AGROFIT, 2017). However, many of them are not efficient for this disease in the field. This situation has been caused the applications of fungicide in an irrational way, increasing the rice production cost, harming the environment and putting food safety at risk through the chemical residues in grains.

Also, studies published in scientific journals evaluating fungicides efficiency in blast control under irrigated rice conditions in Brazil are scarce or conducted and published years ago, not reflecting in the pathogen populations currently present in the environment. Therefore, the purpose of this paper is to fill this gap in information about which fungicides control rice blast in Brazil, assisting technicians and growers in the management of this disease.

This work aimed to analyze the fungicides effectiveness in rice blast control and grain yield maintenance in irrigated rice.

MATERIALS AND METHODS

Experiments were carried out in a randomized block design, with four replications, in the 2015/2016 and 2016/2017 crop seasons, at the Rio Grande do Sul Rice Experiment Station (IRGA) located in the city of Cachoeirinha, RS, Brazil. The experimental area is located at 29°57'02 "S 51°05'02 "W and with seven meters of elevation above sea level. The edaphoclimatic characteristics of the area are favorable for the natural occurrence of rice blast.

The experimental units were composed of plots with dimensions of 5.0 m x 1.53 m, totaling an equivalent area to 7.65 m^2. The treatments consisted of fungicides used for the management of the disease, as described in Table 1. In the 2015/2016 crop season, the fungicide picoxystrobin + cyproconazole was replaced by thiophanate-methyl + mancozeb. Spraying of all other fungicides was similar in all crop season (Table 1).

Guri Inta CL variety was planted at 100 kg.ha^{-1} seeds density, with a population of 400 plants.m^2. Sowing occurred on November 17, 2015, and November 19, 2016. These sowing dates were performed at the end of the recommended period to favor rice blast occurrence.

The soil fertilization was carried out during sowing in the amount of 400 kg.ha^{-1} with formulation 04-18-08, according to soil analysis. Covering fertilization was 120 kg.ha^{-1} of nitrogen in urea form, divided twice, once during the V3 development stage and once in the R0 development stage (Counce *et al.*, 2000). All other crop management followed the bulletin with the technical recommendations by researchers to cultivation irrigated rice in the south of Brazil (SOSBAI, 2016).

Table 1. Fungicides evaluated for the control of rice blast in irrigated rice in the 2015/2016 and 2016/2017 crop season.

Common Name	Chemical Group	Dosage (kg or L/ha)	Target Site and Code (FRAC)	FRAC Code	Season	
					2015/2016	2016/2017
Control	-	-	-		X	X
Tetraconazole	Triazole	0.5	G1	3	X	X
Azoxystrobin	Strobilurin	0.4	C3	11	X	X
Tricyclazole	Benzothiazole	0.3	I1	16.1	X	X
Trifloxystrobin + Tebuconazole	Strobilurin + Triazole	0.75	G1 + C3	3 + 11	X	X
Epoxiconazole + Kresoxim-Methyl	Triazole + Estrobilurin	0.75	G1 + C3	3 + 11	X	X
Tebuconazole	Triazole	0.75	G1	3	X	X
Kasugamycin	Antibiotic	1.5	D3	24	X	X
Picoxystrobin + Cyproconazole	Strobilurin + Triazole	0.4	G1 + C3	3 + 11	X	
Thiophanate-methyl + Mancozeb	Benzimidazole + Dithiocarbamate	2.5	M + B1	1+ M03		X

Two fungicides sprays in 2015/2016 season were performed. The first occurred at the end of the R2 stage and, the second 15 days after the first. For the 2016/2017 season, we performed three fungicides sprays. The first spray was in V8 vegetative stage when began the first symptoms of rice blast in leaves, and the other spray in same stages applied in previous season, that is, the final stage of R2 and 15 days after. The fungicide applications was carried out with a precision costal pulverizer, pressurized to CO_2 with a constant pressure of 40 PSI, with a spray volume of 200 liters per hectare.

During the 2015/2016 crop season, we evaluated the neck blast incidence, panicle blast severity, and grain yield. We collected randomly 15 panicles within the plot at 20 days before harvest for incidence and severity evaluation. The neck blast incidence was quantified counting number of plants showing disease symptoms, and for panicle blast severity we used the scale proposed by Silva-Lobo et al. (2012). Grain yield was determined by harvesting 5 m^2 of the plot and then expressed in kg.ha^{-1}, with grain moisture adjusted to 13%.

During the 2016/2017 crop season, in addition to all variables evaluated in the 2015/2016 season, we assessed the leaf blast, beginning with the appearance of disease first symptoms in leaves, and in a seven-day interval using a proposed scale by IRRI (2013). With leaf blast data, we calculated the Area Under the Disease Progress Curve (AUDPC) using the formula proposed by Campbell and Madden (1990).

All data was subjected to ANOVA and mean of variables that demonstrated significance by F-test ($p<0.05$) were compared by using Duncan test at 5% probability of error. SAS program version 9.0 (SAS, 2000) was used.

RESULTS

Crop Season 2015/2016

There was statistical significance for all tested variables (Table 2). The neck blast incidence showed that only tricyclazole, trifloxystrobin + tebuconazole and picoxystrobin + cyproconazole fungicides control the disease around 58%, differing from the control (Table 2). As for the panicle blast incidence, the fungicide tetraconazole did not affect the disease severity, being equal to the control, whose severity was superior to 65% (Table 2). On the other hand, the fungicides kasugamycin, tebuconazole, and epoxiconazole + kresoxim-methyl had an intermediate effect in disease severity control, being in an average 17% more effectivecompared to control. Finally, the fungicides tricyclazole, trifloxystrobin + tebuconazole, picoxystrobin + cyproconazole and azoxystrobin showed the best control of panicle blast severity with severity values around 30% (Table 2).

Rice blast reduced the grain yield since the average yield in control was only 1,899 kg/ha (Table 2), not differing from treatments with tetraconazole, epoxiconazole + kresoxim-methyl, and tebuconazole. The fungicides azoxystrobin, kasugamycin, and picoxystrobin + cyproconazole showed no difference among them and were superior to control. On the other hand, the treatments that provided the highest grain yield were by applying the fungicides trifloxystrobin + tebuconazole and tricyclazole, showing on average 310% higher grain yield than control without fungicide application (Table 2).

The ANOVA evidenced the absence of significance for AUDPC variable, whereas for other variables there was significance by F-test (Table 3). The panicle blast incidence was reduced only by tricyclazole and trifloxystrobin + tebuconazole, being tricyclazole superior to all the other fungicides, showing half of the disease incidence in relation to control (Table 3). In comparison to 2015/2016 crop season, the panicle blast incidence was higher, and the disease was observed in all analyzed panicles in control without fungicide application (Table 3). On the other hand, the absence of significance for AUDPC indicates that even though the rice blast had occurred on the leaf, it did not show significant progress (Table 3).

For rice blast severity in the panicle, we observed a similar response with disease incidence, whose treatments tricyclazole and trifloxystrobin + tebuconazole showed on average 20% lower disease severity compared to control (Table 3). Also, the fungicides kasugamycin and azoxystrobin showed lower panicle blast severity compared to control, but superior about the best treatments tricyclazole and trifloxystrobin + tebuconazole (Table 3).

Crop Season 2016/2017

As in the 2015/2016 crop season, grain yield was profoundly affected by rice blast in 2016/2017 season. We observed that the fungicides trifloxystrobin + tebuconazole and tricyclazole were on average 228% higher than control in grain yield, not different from each other (Table 3). Also, azoxystrobin and kasugamycin were superior to control but lower than the trifloxystrobin + tebuconazole and tricyclazole treatments (Table 3).

Table 2. Neck blast incidence, panicle blast severity and grain yield with fungicides application in 2015/2016.

Fungicides	Incidence (%)	Severity (%)	Grain yield (kg/ha)
Tetraconazole	96.67 A[1]	67.17 A	1514.10 C
Control	98.33 A	66.58 A	1899.17 C
Epoxiconazole + Kresoxim-Methyl	96.67 A	49.33 B	2529.88 C
Tebuconazole	91.67 A	50.42 B	2541.75 C
Azoxystrobin	78.33 A	29.92 C	3068.52 B
Kasugamycin	86.67 A	49.09 B	3484.99 B
Picoxystrobin + Cyproconazole	66.67 B	31.65 C	3770.66 B
Trifloxystrobin + Tebuconazole	58.34 B	33.58 C	5516.22 A
Tricyclazole	50.00 B	27.83 C	6251.58 A

[1]Means in same columns followed by the same letter do not differ significantly (p≤0.05) according to Duncan's test.

Table 3. The Area Under the Disease Progress Curve (AUDPC) in leaves, neck blast incidence, panicle blast severity and grain yield with fungicides application in 2016/2017.

Fungicides	AUDPC	Incidence (%)	Severity (%)	Grain yield (kg/ha)
Tetraconazole	137.62NS[1]	100.00 A[2]	59.08 A	3928.80 C
Tebuconazole	128.88	100.00 A	51.75 A	4231.80 C
Control	131.12	100.00 A	50.66 A	4183.24 C
Epoxiconazole + Kresoxim-Methyl	135.25	100.00 A	48.75 A	3916.55 C
Thiophanate-methyl + Mancozeb	136.50	95.00 A	40.33 A	4903.86 C
Kasugamycin	134.75	90.00 A	31.67 B	6060.72 B
Azoxystrobin	128.62	94.99 A	30.33 B	6094.23 B
Trifloxystrobin + Tebuconazole	132.63	73.33 B	16.17 C	9135.95 A
Tricyclazole	129.13	50.00 C	14.42 C	9903.79 A

[1] Not significant (p>0.05) according to F-test. [2] Means in the same columns followed by the same letter do not differ significantly (p≤0.05) according to Duncan's test.

DISCUSSION

Rice blast is the major irrigated rice disease (Nalley *et al*., 2016; Chen *et al*., 2015; Kunova *et al*., 2012; Prabhu *et al*., 2009). This disease can cause productivity losses of up to 100% (Filippi and Barata, 2014; Kunova *et al*., 2014; Prabhu *et al*., 2009). Mainly when it attacks the neck or panicles of rice, which directly affects the filling of grains. Chemical control with fungicides has been the most used by rice growers in the state of Rio Grande do Sul in Brazil. Since most of the available vaieties are susceptible to rice blast (Ogoshi, 2015), the use of fungicides, in this case, is vital for blast management in order to achieve high grain yield (Chen *et al*., 2015; Kunova *et al*., 2014; Prabhu and Filippi, 2006).

Currently, there are 49 fungicides registered in the Ministry of Agriculture, Livestock and Food Supply (MAPA) for rice blast control in Brazil (AGROFIT, 2017), is recommended for seed treatment and aerial spraying. These fungicides are composed of a single active ingredient or by mixtures of two or more active ingredients, which have eight different target sites following classification by FRAC (2017).

Despite the high number of registered fungicides available for rice blast control in Brazil (49 fungicides for seed treatment and aerial spraying), we observed that many of them are not effectivet, especially for the neck and panicle blast control. This situation directly affects grain yield and can cause severe economic losses to rice growers. Given this, the knowledge of the actual effectiveness of fungicides currently available in the market is essential for success in the management of rice blast in lowlandconditions.

In the present work, we tested several fungicides for blast control in irrigated rice during two crop seasons; we chose these fungicides according to different modes of action, representing fungicides registered in MAPA for aerial application. Among the nine fungicides tested, only the tricyclazole fungicides and the trifloxystrobin + tebuconazole mixture were efficient in panicle blast control, reflecting directly on high grain yield (Tables 2 and 3).

The fungicides azoxystrobin and picoxystrobin + cyproconazole showed a reduction in panicle blast severity. However, we did not observe the same result in grain yield. Chen *et al*. (2015) analyzed the effect of azoxystrobin and kresoxim-methyl fungicides for blast control in irrigated rice in China and found that both fungicides had disease control superior to 73% of efficiency and resulted in high grain yield, these results are in contrast of our results. Others reports showed that fungicide azoxystrobin is efficient in the rice blast control with direct reflection on grain yield in countries

such as Italy, Taiwan, and the United States (Chen *et al*., 2015; Kunova *et al*., 2014; Kunova *et al*., 2012; Jin *et al*., 2009; Fang *et al*., 2009).

These positive results of the azoxystrobin fungicide in blast control in lowland rice are divergent to that found in our study, possibly due to the difference in the *M. oryzae* populations in Brazil compared to other countries. This inference may be related to the low sensitivity of several isolates of the pathogen to fungicide azoxystrobin. Since for *M. oryzae* population, which infects wheat in Brazil, strains of this fungus resistant to azoxystrobin have already been found (Castroagudín *et al*., 2015; Oliveira *et al*., 2015). However, this hypothesis still needs to be confirmed for rice in Brazil.

In general, the fungicide that promoted the best response for rice blast control in our work was tricyclazole. This fungicide is the standard for control this disease in Brazil and several other countries and, your application alone or in combination with other fungicides, has shown satisfactory results in rice blast control (Chen *et al*., 2015; Kunova *et al*., 2014; Ganesh *et al*., 2012; Kunova *et al*., 2012; Prabhu *et al*., 2002). Tricyclazole was superior to the fungicide tebuconazole in the panicle blast control and avoided a yield loss in irrigated rice in Brazil (Prabhu *et al*., 2002).

Magar *et al*. (2015) analyzed the effectiveness of several fungicides in rice blast control in irrigated rice. They showed that the fungicide tricyclazol + hexaconazole were the most effective for control of leaf and neck blast exceeding 79%, being reflected directly in high yields. Tricyclazole was the most efficient in the rice blast control in irrigated rice in India, in addition to reducing the disease severity, the fungicide application also increased the tillering number per plant, panicle length, number of grains per panicle and, grain yield (Pandey, 2016).

The fungicide trifloxystrobin + tebuconazole showed, in our present work, the same disease control effectiveness in comparison to tricyclazole. Our results revealed that trifloxystrobin + tebuconazole is an alternative option for rice blast management in Brazil, mainly for active ingredient rotation and, consequently, preventing a possible fungus resistance to fungicides.

Pramesh *et al*. (2016) analyzed the effectiveness of trifloxystrobin + tebuconazole in rice blast and sheath blight control in irrigated rice and verified that this fungicide was most effective in control of both diseases in India. Silva-Lobo (2004) evaluated the effectiveness of fungicides in blast control in upland rice in Brazil and found that fungicide trifloxystrobin

+ tebuconazole showed a lower severity of panicle blast, similar to tricyclazole.

In our work, tricyclazole and trifloxystrobin + tebuconazole were the most effective in rice blast control with direct reflection on high grain yield. Therefore, we highlight the importance thet of technicians and rice growers correctly choose the fungicide for sustainable management of blast in lowland rice cultivation in Brazil.

CONCLUSIONS

Tricyclazole and trifloxystrobin + tebuconazole were the most effective in rice blast control with direct reflection on high grain yield.

Acknowledgements
Authors are thankful to Rio Grandense Rice Institute (IRGA) for financial support and the field technicians who assisted us in experiments execution.

REFERENCES

Agrofit. 2017. Sistema de Agrotóxicos Fitossanitários. Available on: http://agrofit.agricultura.gov.br/agrofit_cons/principal_agrofit_cons. Accessed 26 may 2017.

Campbell, C.L and Madden L.V. 1990. Introduction to plant disease epidemiology. John Willey & Sons, New York NY, 532 pp.

Castroagudín, V.L., Ceresini P.C., de Oliveira S.C., Reges J.T.A., Maciel J.L.N., V Bonato A.L., Dorigan A.F., McDonald, B.A. 2015. Resistance to QoI Fungicides Is Widespread in Brazilian Populations of the Wheat Blast Pathogen *Magnaporthe oryzae*. Phytopathology 105: 284–94. DOI: 10.1094/PHYTO-06-14-0184-R.

Chen, Y., Yang, X.,Yuan, S.K., Li, Y.F., Zhang, A.F.,Yao, J., Gao, T.C. 2015. Effect of azoxystrobin and kresoxim-methyl on rice blast and rice grain yield in China. Annals of Applied Biology 166: 434–443. DOI: 10.1111/aab.12202.

Counce, P.A., Keisling, T.C., Mitchell, A.J. 2000. A uniform, objectives, and adaptive system for expressing rice development. Crop Science 40: 436–443.

Fang, M., Yan, L., Wang, Z., Zhang, D., Ma, Z. 2009. Sensitivity of *Magnaporthe grisea* to the sterol demethylation inhibitor fungicide propiconazole. Journal of Phytopathology. 157: 568–572. DOI: 10.1111/j.1439-0434.2009.01576.x.

Filippi, M.C and Barata, G. 2014. Induction of resistance to rice leaf blast by avirulent isolates of *Magnaporthe oryzae*. Revista de Ciências Agrárias 57: 388–395. http://dx.doi.org/10.1590/S0100-41582007000500003

FRAC, 2017. FRAC Code List 2017: Fungicides sorted by mode of action (including FRAC Code numbering). Available on: http://www.frac.info/docs/default-source/publications/frac-code-list/frac-code-list-2015-finalC2AD7AA36764.pdf?sfvrsn=4.

Ganesh, N.R., Gangadhara, N.B., Basavaraja, N.T., Krishna, N.R. 2012. Fungicidal management of leaf blast disease in rice. Global Jornal of Bio-Science & Biotechnology 1: 18–21.

IRRI, 2013. Standard Evaluation System for Rice. 5th ed. IRRI, Manila. 55 pp.

Jin, L.H., Chen, Y., Chen, C.J., Wang, J.X., Zhou, M.G. 2009. Activity of Azoxystrobin and SHAM to Four Phytopathogens. Agricultural Sciences in China 8: 835–842. https://doi.org/10.1016/S1671-2927(08)60285-0.

Kunova, A., Pizzatti, C., Bonaldi, M., Cortesi, P. 2014. Sensitivity of nonexposed and exposed populations of *Magnaporthe oryzae* from rice to tricyclazole and azoxystrobin. Plant Disease 98: 512–518. https://doi.org/10.1094/PDIS-04-13-0432-RE.

Kunova, A., Pizzatti, C., Cortesi, P. 2012. Impact of tricyclazole and azoxystrobin on growth, sporulation and secondary infection of the rice blast fungus, *Magnaporthe oryzae*. Pest Management Science 69: 278–284. DOI: 10.1002/ps.3386.

Soares, L.C.S., Raphael, J.P.A., Bortolotto, R.P., Nora, D.D., Gruhn, E.M. 2014. Blast disease in rice culture. Brazilian Journal of Applied Technology ofr Agricultural Science 7: 109–119. DOI: 10.5935/PAeT.V7.N2.13

Magar, P.B., Acharya, B., Pandey, B. 2015. Use of Chemical Fungicides for the Management of Rice Blast (*Pyricularia Grisea*) Disease at Jyotinagar, Chitwan, Nepal. International Journal of Applied Sciences and Biotechnology 3: 474-478. DOI: 10.3126/ijasbt.v3i3.13287 .

Mundt, C.C. 2014. Durable resistance: A key to sustainable management of pathogens and pests. Infection Genetics Evolution 27: 446–455. DOI: 10.1016/j.meegid.2014.01.011.

Nalley, L., Tsiboe, F., Durand-Morat, A., Shew, A., Thoma, G. 2016. Economic and environmental impact of rice blast pathogen (*Magnaporthe oryzae*) alleviation in the United States. PLoS One 11: 1–15. DOI: 10.1371/journal.pone.0167295.

Ogoshi, C. 2015. Epidemia de Brusone do Arroz no Estado do Rio Grande do Sul. Lavoura Arrozeira 465: 13–15. Available on: http://www.apec.unesc.net/IV_EEC/sessoes_te maticas/Economia rural e agricultura familiar/Panorama da produ??o de arroz no Rio Grande do Sul.pdf.

Oliveira, S.C. de, Castroagudín, V.L., Leodato, J., Maciel, N., Augusto, D., Pereira, S., Ceresini, P.C. 2015. Resistência cruzada aos fungicidas IQo azoxistrobina e piraclostrobina no patógeno da brusone do trigo *Pyricularia oryzae* no Brasil. Summa Phytopathologica 41: 298–304. DOI: 10.1590/0100-5405/2072.

Pak, D., You, M.P., Lanoiselet, V., Barbetti, M.J. 2016. Azoxystrobin and propiconazole offer significant potential for rice blast (*Pyricularia oryzae*) management in Australia. European Journal of Plant Pathology 148: 247–259. DOI: 10.1007/s10658-016-1084-6.

Pandey, S. 2016. Effect of fungicides on leaf blast and grain yield of rice in Kymore region of Madhya Pradesh in India. Bangladesh Journal of Botany 45: 355–361.

Pooja, K and Katoch, A. 2014. Past, present and future of rice blast management. Plant Science Today 1: 165–173. DOI: 10.14719/pst.2014.1.3.24.

Prabhu, A.S and Filippi, M.C. 2006. Resistência da cultivar no manejo integrado da brusone.. In: Prabhu, A.S and Filipi, M.C. (eds). Brusone em arroz: controle genético, progresso e perspectivas. EMPBRAPA Arroz e Feijão, Santo

Antônio de Goiás. pp. 323–379.

Prabhu, A.S., Filippi, M.C., Silva, G.B., Silva Lobo, V.L., Morais, O.P. 2009. An unprecedented outbreak of rice llast on a newly released cultivar BRS Colosso in Brazil. In: Xiaofan,W and Valent, B. (eds.). Advances in Genetics, Genomics and Control of Rice Blast Disease. Springer Netherlands, Dordrecht. pp. 257–266

Prabhu, A.S., Guimarães, C.M., Silva, G.B. 2002. Manejo da Brusone no Arroz de Terras Altas. Circ. Técnica EMBRAPA, 52: 1–6.

Pramesh, D., Muniraju, K.M., Mallikarjun, K., Guruprasad, G.S. 2016. Bio-efficacy of a Combination Fungicide against blast and sheath blight diseases of paddy. 14: 1–8. DOI: 10.9734/JEAI/2016/28893.

Silva-lobo, V.L., Filippi, M.C.C., Silva, G.B., Venancio, W.L., Prabhu, A.S. 2012. Relação entre o teor de clorofila nas folhas e a severidade de brusone nas panículas em arroz de terras altas. Tropical Plant Pathology 37: 83–87. DOI: 10.1590/S1982-56762012000100011.

SOSBAI. 2016. Arroz irrigado: recomendações técnicas da pesquisa para o Sul do Brasil. Sociedade Sul-Brasileira de Arroz Irrigado, Pelotas, RS, 200 pp.

Webster, R.K and Gunnell, P.S. 1992. Compendium of rice diseases. St. Paul: American. Phytopatological Society.

Zhou, E., Jia, Y., Singh, P., Correll, J.C., Lee, F.N. 2007. Instability of the *Magnaporthe oryzae* avirulence gene AVR-Pita alters virulence. Fungal Genetics and Biology 44: 1024–1034. DOI: 10.1016/j.fgb.2007.02.003.

5

Soil Moisture on Emergence and Initial Development of *Bidens Pilosa*

Márcia Maria Mauli[1]*, Lúcia Helena Pereira Nóbrega[1],
Antonio Pedro Souza Filho[2], Adriana Maria Meneghetti[3],
Danielle Medina Rosa[1] and Michelle Tonini[1]

*[1] Universidade Estadual do Oeste do Paraná- UNIOESTE – Cascavel – PR,
Brazil. Email: marcia.m.mauli@gmail.com
[2]EMBRAPA Amazônia Oriental – Belém – PA, Brazil
[3]Universidade Tecnologica Federal do Paraná – UTFPR – Santa Helena – PR,
Brazil*
Corresponding author

SUMMARY

Some species are used as cover crops, so they can produce and release products from their secondary metabolism. When these products are released in the environment, they are exposed to direct or indirect influences. Thus, this trial aimed to evaluate the influence of soil moisture content in plant decomposition, possibly caused by allelochemicals release. The influence of soil field capacity was evaluated by the decomposition of black oats (*Avena strigosa* Schreb), turnip (*Vicia villosa* Roth) and hairy vetch (*Raphanus sativus* L.) residues, with 70 and 50% of water availability and the control. Pots (1.0 kg) were filled with sterilized and unsterilized soil, plus 30 g of plant residue, which remained in decomposition for four weeks before the experiment start. After that, five seeds of beggartick (*Bidens pilosa* L.) were sowed and seedling emergence was daily evaluated for 10 d. The emergence speed index and emergence speed were calculated. Five seedlings were transplanted at the 8th d and evaluated during 30 d, to obtain the initial development of plants, and to determine fresh and dry mass. Greatest growth inhibition of the weed tested occurred with 70% available water for plant emergence and with 50% for initial plant development.
Keywords: allelochemicals; available water on soil; cover crops; weeds.

INTRODUCTION

The correct species choice of cover crops in different systems of regional production has been an important practice. Such heterogeneity contributes to greater biodiversity and therefore, creates more favorable conditions for the development of many natural enemies and environment balance for both, amount of organisms and soil nutrients. These are ideal conditions for crop development and yield (Colozzi and Andrade, 2006; Rains *et al.*, 2011; Embrapa, 2004; Silva, 2012). In addition, many plants used as cover crops have chemical compounds that, when released into the environment, may affect the growth and development of other organisms around. This interference is known as allelopathy (Rice, 1974). The use of plants with proven allelopathic properties is a promising alternative in agriculture, specially to control weeds in trading crops. Cover crops management has significantly reduced the intensity of weed infestation, as well as changed the composition of local weeds (Silva, 2012; Nunes *et al.*, 2014; Mauli *et al.*, 2015; Martins *et al.*, 2016). When cover crops are used correctly, they can improve physical, chemical and biological properties of the soil, and the use of herbicides can be reduced or even eliminated. Although this is a promising alternative for weed control, it has not been adequately studied (Silva, 2012).

Each species has a specific ability to produce allelopathic substances, as well as its chemical nature, concentration, and susceptibility to allelochemicals that are released by other plants (Ferreira *et al.*, 2000; Bonanomi *et al.*, 2006). These substances may act indirectly or directly. When acting indirectly, changes occur in soil properties, nutritional conditions, variation of microorganism populations and their activity. When the action is direct, the substances act on membranes of the target plant. Therefore, they allow both bonding and penetration of compounds into cells, and thus interfere on their metabolism (Seigler, 2006).

In the soil, chemical compounds can strongly interact with colloids and suffer microbial transformations responsible for deactivation or degradation (Colozzi and Andrade, 2006). However, the decomposition of these substances is slower in a sterile soil, which highlights the involvement of microorganisms in the process. The phytotoxic potential is controlled by microbial activity, responsible for transformations and degradation of active forms, or even their activation in soil. Although some steps are abiotic, the participation of microorganisms is evident and essential for the dynamics and activity of several substances in soil. Some substances must be transformed to act as allelochemicals (Pedrol *et al.*, 2006; Colozzi and Andrade, 2006). Several microorganisms are specialized to convert, synthesize, eliminate, or use secondary metabolites as a source of carbon and energy. At the same time, microorganisms release bioactive compounds from wastes in decomposition and contribute to eliminate these natural toxins from the soil through decomposition. Allelochemicals can be immobilized during solid phase; they establish equilibrium with the soil solution and may suffer degradation or even leach, especially in sandy soils (Moreira and Siqueira, 2006). Understanding the occurrence, activity and/or behavior of microorganisms and natural compounds in processes influenced by soil moisture can support the background to design more efficient crop rotation systems with these residues (Pedrol *et al.*, 2006; Souza Filho, 2006; Mauli *et al.*, 2011). Some species, such as black oats (*Avena strigosa* Schreb), turnip (*Vicia villosa* Roth) and hairy vetch (*Raphanus sativus* L.) are already used as cover crops, with reports of potentially allelopathic activity on weeds when used in crop rotation. These species have allelopathic potential that can act on germination and/or development of other plants, affecting the affected organism in several ways, such as reducing photosynthesis, protein synthesis, membrane permeability and inhibition of enzymatic activity, among several others (Santos and Reis, 2003; Embrapa, 2004; Weih *et al.*, 2008, Rains *et al.*, 2011). Thus, the objective of this study was to evaluate the influence of soil water availability on the decomposition of black oats (*Avena strigosa* Schreb), turnip (*Vicia villosa* Roth) and hairy vetch (*Raphanus sativus* L.), cover crops already used in some regions, and the initial germination and development of *Bidens pilosa* L.

MATERIAL AND METHODS

To verify the effect of plant residue decomposition on *B. pilosa*, an assay was established in a greenhouse to test two water availability levels (70 and 50%) in soil. Initially, the capacity of water retention in pots was determined according to Ruiz *et al.* (2003) methodology. Pots of 1.0 kg capacity were filled up with soil from cropping land at 0 to 20 cm-depth in Catanduvas city, Paraná (chemical characteristics shown in Table 1); soil was sieved and air-dried. Pots were watered and kept at 70 and 50% of available water; control treatment received 100% of available water. Soil was covered with 30 g of dry mass of the following cover crops: Black oats (*A. strigosa*), turnip (*V. villosa*) and hairy vetch (*R. sativus*). The soil with leaves on top mulch were decomposed for four weeks in a greenhouse with an average temperature of 26 °C. After this period, five seeds of beggartick (*B. pilosa*) were sowed and seedling emergence was evaluated daily for 10 d. Emerged seedlings were selected and the emergence rate speed (ERS), that is, number of plants emerge d^{-1} (Maguire, 1962) and emergence speed (ES), or number of days that took seedlings to emerge, were also calculated (Edmond and Drapalla, 1958). Five beggartick seedlings 8 d old were transplanted to evaluate their initial development. They were submitted to the same experimental conditions of emergence test described above, and were evaluated for 30 d. Then, each plant was separated in roots and shoots, weighed on a 0.001-g precision scale and taken to the oven at 65 °C until constant weight, to quantify dry mass. To check for microorganisms' activity, during emergence and initial development, a sterilized and non-sterilized soil were used. Pots were daily weighted, adding the amount of water required for each test.

Data was submitted to analysis of variance. Means were compared by Tukey tests at 5% probability. Initially, data normality and homogeneity of variances were verified. Transformations were carried out according to Pimentel (2000), and Banzatto and Kronka (2006) with Minitab® 14 and SISVAR software (Ferreira, 2008). Data in percentage were transformed into $arcsen\sqrt{\frac{x}{100}}$, and those that showed a non-normal distribution were submitted to $\sqrt{x}$, as necessary.

Tabla 1. Chemical characteristics of the soil used in the experiment, taken from a depth of 0-20 cm. Catanduvas – PR, Brazil (2011).

Characteristics of soil*	Values	Classification**
pH (CaCl$_2$)	5.5	Medium
Organic matter (g dm^3)	54.95	Very high
Exchangeable aluminum (cmol$_c$ dm^3)	0.00	Very low
Potential acidity (cmol$_c$ dm^3)	4.28	-
Base sum (cmol$_c$ dm^3)	16.34	High
Cation exchange capacity (cmol$_c$ dm^3)	20.62	High
Base saturation (%)	79.24	High
P (cmol$_c$ dm^3)	21.40	Very high
K (cmol$_c$ dm^3)	1.14	Very high
Ca (cmol$_c$ dm^3)	12.25	High
Mg (cmol$_c$ dm^3)	2.95	High
Cu (mg/dm^3)	13.77	High
Zn (mg/dm^3)	18.32	High
Fe (mg/dm^3)	36.00	Medium
Mn (mg/dm^3)	351.00	High

* Analysis performed in the Soils Laboratory of the Cooperativa Central de Pesquisa Agrícola (COODETEC).
** Classification according to Alvarez *et al.* (1999).

RESULTS AND DISCUSSION

Values for beggartick seedlings emergence, submitted to 70 and 50% of water availability in sterilized soil are shown on Table 2.

The emergence inhibition percentage of beggartick was significant on black oats and hairy vetch. In both residues, inhibition at 70% water availability was higher than at 50%. However, on black oats, this result was superior to 85%. Concerning turnip, there was no significant difference.

The rate of emergence was significant for the three species studied, and presented a lower rate when submitted to 70 and 50% in relation to the control. However, when we compared 70 with 50%, the data did not differ statistically from each other. This means that the number of emergent seedlings per day decreased when there was more water availability in the soil, as observed in the control (100%).

The emergence speed was significant for black oats and hairy vetch plant residues. The ES was lower under 70% water availability when submitted to black oats, but, when analyzed at 50% and compared to the control, there were no significant differences. For hairy vetch residues, 70 and 50% of water availability had lower ES when compared to control, but did not present statistical difference between them. Thus, seedlings submitted to black oats at 70% water availability and hairy vetch at 70 and 50% took more days to emerge than those ones at 100% water availability (Table 1). The lowest water availability in soil may have resulted in higher allelochemical concentrations, and consequently, it might promote greater influence on the beggartick emergence. One advantage of this delay in weeds emergence is the increased period of time to control them properly. This would make the operation more efficient under field conditions (Moraes *et al.*, 2011).

The values of seedling emergence in beggartick at 70 and 50% water availability, in a not sterilized soil are shown in Table 3. The beggartick inhibition percentage was significant for black oats and presented data with regular dispersion. However, at 70% water availability there was 85.69% inhibition that differed from 50% water availability. For turnip and hairy vetch plant residues, there was no significant difference.

The emergence speed index for seedlings of beggartick was significant on unsterilized soil for the three plants residues studied. The number of seedlings emerged per day showed similar behavior in black oats and turnip, however, the response to 50, 70 and 100% (control) water availability showed different behaviors among themselves. At 70% water availability, the lowest emergence speed index was recorded. For the hairy vetch, there was a significant difference among the control and 70 and 50% water availability, being the last two with the lowest emergence speed index values, but they did not differ among themselves.

The emergence speed of berggartick seedlings on unsterilized soil was significantly different for black oats, turnip and hairy vetch plant residues. The beggartick seedling emergence speed showed the same behavior for the three plant residues, however, significant differences were shown with each treatment and the control, and between both treatments, where 70% the emergence speed showed the lowest value.

Based on the results, the greatest emergence inhibition of beggartick seedlings occurs at 70% water availability in soil. As a hypothesis, this may be associated to the presence and activity of organisms, which are conditioned by physical (mineralogy, moisture, aeration and temperature) and chemical (pH and fertility) components of soil. For example, when the soil is dry, the bacteria reduce their activity or stay under dormancy for some time (Colozzi and Andrade, 2006; Moreira and Siqueira, 2006), consequently, allelochemicals could have been retained in soil particles, since the phytotoxic potential is controlled by microbial activity, responsible for transformation and degradation of active forms or even their activation (Pedrol et al., 2006; Colozzi and Andrade, 2006).

Table 2. Percentage of emergence inhibition, emergence speed index (ESI) and emergence speed (ES) of beggartick seedlings submitted to plant residue of black oat, turnip and hairy vetch, at 70 and 50% of water availability in sterilized soil.

Water availability	Black oats	Turnip	Hairy vetch
	Emergence inhibition (%)		
70%	85.69 a	67.00a	57.08 a
50%	49.92 b	64.23a	55.41 b
Mean	67.91	65.61	56.24
Coefficient of Variation	7.11	11.57	1.73
F value	25.00 *	0.33 [n.s.]	26.00 *
	Emergence speed index (seedlings per day)		
Control	0.81 a	0.80 a	0.82 a
70%	0.11 b	0.24 b	0.29 b
50%	0.36 b	0.22 b	0.37 b
Mean	0.42	0.43	0.48
Coefficient of Variation	5.63	19.86	18.31
F value	41.87 *	48.82 *	28.96 *
	Emergence speed (days)		
Control	0.80 a	0.81a	0.81 a
70%	0.16 b	0.44a	0.50 b
50%	0.55 a	0.45a	0.41 b
Mean	0.50	0.54	0.58
Coefficient of Variation	27.48	9.46	10.98
F value	16.69 *	3.61 [n.s]	32.95 *

Means followed by different letters in the same column, differ by Tukey test ($p \leq 0.05$).
n.s.= no significant, * = Significant.

Table 3. Emergence inhibition percentage, emergence speed index (ESI) and emergence speed (ES) of beggartick seedlings under plant residues of black oats, turnip and hairy vetch, at 70 and 50% water availability in unsterilized soil

Water availability	Black oats	Turnip	Hairy vetch
Emergence Inhibition Percentage			
70%	85.69 a	78.54 a	71.37 a
50%	57.08 b	58.00 b	49.92 b
Mean	71.39	67.81	60.66
Coefficient Variation	12.27	22.38	25.02
F value	16.00 *	13.00 *	13.99 *
Emergence Speed Index (seedlings per day)			
Control	0.89 a	0.88 a	0.89 a
70%	0.11 c	0.16 c	0.23 b
50%	0.27 b	0.37 b	0.25 b
Mean	0.43	0.47	0.42
Coefficient Variation	15.04	24.02	28.37
F value	124.08 *	32.72 *	24.82 *
Emergence Speed (Days)			
Control	0.89 a	0.89 a	0.89 a
70%	0.17 c	0.24 c	0.29 c
50%	0.51 b	0.53 b	0.50 b
Mean	0.52	0.55	0.57
Coefficient Variation	9.84	15.46	12.30
F value	10.82 *	42.42 *	4.35 *

Means followed by different letters in the same column, differ by Tukey test ($p \leq 0.05$). * = Significant.

Soil sterility (sterilized soil) did not show major differences in this case, except for turnip, with greater effects in unsterilized soil, which could be explained by studies related to the sensitivity to the presence of microorganisms in the soil.

The obtained values for the initial development of beggartick in plant residues of black oats, turnip and hairy vetch at 70 and 50% water availability, in sterilized soil, are shown in Table 4. The three studied plant residues influence significantly on beggartick shoot fresh mass (SFM). In all residues, 70 and 50% of water availability differed from the control and presented the lowest fresh mass. However, 70 and 50% water availability showed similar results. Root fresh mass was significantly different in treatments with turnip and hairy vetch residues. The turnip, submitted to 70 and 50% water availability, was significantly different from the control, but not at each level of water availability. The hairy vetch, under 50% water availability, differed from the control, however, under 70% did not differ from the control nor under 50% water availability.

Shoot dry mass showed significant changes under black oats and turnip residues, with significant differences under 70 and 50% water availability compared to the control. However, both water availability levels did not differ between them. With hairy vetch residues, both treatments and the control differed among themselves, however, the 50% water availability treatment showed the lowest shoot dry mass.

Root dry mass showed significant changes under the studied plant residues. On black oats residues, root dry mass at 70% water availability differed from the control and showed the lowest value. However, it did not differ from the 50% treatment, which also did not differ from the control. Turnip residues, under 70 and 50% water availability treatments, differed from the control and showed the lowest root dry mass. But both water availability levels did not differ significantly from each other. Root dry mass under hairy vetch, at 50% water availability differed from the control and showed the lowest root dry mass, but it did not differ from the 70% water availability level, and was not significantly different from the control. The use of this possible allelopathic activity stands out as an alternative to complement the chemical control in suppressing weeds in the agro-ecosystem (Weih et al., 2008).

On Table 5, the values for the initial plant growth of beggartick are presented at 70 and 50% water availability in an unsterilized soil.

Table 4. Fresh and dry mass of shoot and root beggartick plants (g plant^{-1}) grown in soil mixed with black oats, turnip and hairy vetch plant residues, at 70 and 50% water availability in sterilized soil.

Water availability	Black Oats	Turnip	Hairy Vetch
	Shoot fresh mass (g plant^{-1})		
Control	0.77 a	0.77 a	0.77 a
70%	0.47 b	0.33 b	0.45 b
50%	0.60 b	0.47 b	0.20 b
Mean	0.61	0.52	0.47
Coefficient Variation	14.20	13.25	22.27
F value	8.85 *	30.54 *	21.92 *
	Root fresh mass (g plant^{-1})		
Control	0.24 a	0.24 a	0.24 a
70%	0.11 a	0.07 b	0.13 ab
50%	0.13 a	0.06 b	0.04 b
Mean	0.16	0.12	0.13
Coefficient Variation	33.58	25.64	38.85
F value	5.29 not significant	32.40 *	10.67 *
	Shoot dry mass (g plant^{-1})		
Control	0.20 a	0.20 a	0.20 a
70%	0.10 b	0.06 b	0.10 b
50%	0.12 b	0.10 b	0.06 c
Mean	0.14	0.12	0.12
Coefficient Variation	17.76	9.83	3.31
F value	13.08 *	100.00 *	78.00 *
	Root dry mass (g plant^{-1})		
Control	0.16 a	0.16 a	0.16 a
70%	0.05 b	0.04 b	0.08 ab
50%	0.08 ab	0.06 b	0.03 b
Mean	0.10	0.08	0.09
Coefficient Variation	35.35	38.27	35.27
F value	8.09 *	14.59 *	13.13 *

Means followed by different letters in the same column, differ by Tukey test (p $\leq$ 0.05). * = Significant.

Shoot fresh mass of beggartick plants in an unsterilized soil, under black oats residues, presented significantly difference from the control and treatments of 50 and 70% water availability, with lower mass, but with no significant differences between residue treatments. The water availability treatments and control differed among themselves under turnip residues, but at 50%, it was observed the lowest shoot fresh mass. With hairy vetch residues, 50% water availability differed from the control and showed the lowest shoot fresh mass, however, it did not differ from 70%, and also it did not differ from the control.

Root fresh mass differed among the cover crops residues. The lowest mass values were observed under black oats and turnip residues, and 70 and 50% water availability differed from the control, but they did not differ from each plant residue. The 50% water availability treatment differed from the control and showed the lowest mass under hairy vetch residues,

however, it did not differ from the 70% water availability treatment, and it also did not differ from the control.

Shoot dry mass differed in the three analyzed residues. Results showed some similarity in black oats, turnip and hairy vetch. The 70 and 50% water availability treatments presented lower mass values when compared to the control, however, both treatments were not significantly different. Root dry mass showed significant changes only in black oats, with low data homogeneity; however, 70 and 50% water availability treatments were not significantly different. The probable effects caused by allelochemicals release might be useful in crop rotation or intercropping, in the context of integrated management of weeds (Erasmo et al., 2004). However, further studies are needed to indicate a reduction of specific allelochemicals produced by a precise group of microorganisms, in order to elucidate these results.

Table 5. Fresh and dry mass of shoot and root beggartick plants (g per plant^{-1}) grown in soil mixed with black oats, turnip and hairy vetch plant residues, at 70 and 50% water availability in an unsterilized soil.

Water availability	Black oats	Turnip	Hairy vetch
Shoot fresh mass (g plant^{-1})			
Control	0.77 a	0.77 a	0.77 a
70%	0.26 b	0.40 b	0.32 ab
50%	0.34 b	0.07 c	0.13 b
Mean	0.45	0.41	0.41
Coefficient Variation	24.37	16.86	37.90
F value	17.97 *	73.83 *	13.05 *
Root fresh mass (g plant^{-1})			
Control	0.24 a	0.24 a	0.24 a
70%	0.09 b	0.10 b	0.06 ab
50%	0.08 b	0.07 b	0.13 b
Mean	0.13	0.14	0.14
Coefficient Variation	20.88	17.50	33.86
F value	30.29 *	41.16 *	10.31 *
Shoot dry mass (g plant^{-1})			
Control	0.20 a	0.20 a	0.20 a
70%	0.05 b	0.09 b	0.06 b
50%	0.07 b	0.05 b	0.05 b
Mean	0.10	0.11	0.10
Coefficient Variation	9.37	13.09	19.14
F value	99.00 *	71.28 *	47.36 *
Root dry mass (g plant^{-1})			
Control	0.16	0.16 a	0.16 a
70%	0.04 b	0.07 a	0.08 a
50%	0.06 b	0.08 a	0.09 a
Mean	0.09	0.10	0.10
Coefficient Variation	35.35	37.69	38.45
F value	8.94 *	5.95 not significant	6.39 not significant

Means followed by different letters in the same column, differ by Tukey test ($p \leq 0.05$). * = Significant.

Soil sterilization did not have a significant influence in this trial. The cause of variation was the available water in soil. This may have occurred because the pot with plant material remained for four weeks to be decomposed; this period, determined in previous evaluations, may have provided a favorable environment for microorganisms' development (Unpublished data). Probably, when beggartick seeds or seedlings were added, microorganisms might have been already present in both cases, sterilized and non-sterile soil. This could provide similar results in relation to the decomposition of plant residues and, consequently, the release of allelochemicals.

Thus, cover crops could be managed during rainy season, since soil water content might influence directly the activity of microorganisms on allelochemicals activation and thus it could increase the inhibitory effects on beggartick. However, more studies need to be done about the microorganisms present in the soil and how these cover plants can act on them.

CONCLUSIONS

Seventy percent water availability promoted the decomposition of plant residues of black oats, turnips and hairy vetch plants left on top mulch soil, and that caused greater inhibitions of emergence and the early development of beggartick weed (*Bidens pilosa*).

REFERENCES

Banzatto, D.A., Kronka, S. Do N. 2006. Experimentação Agrícola. FUNEP. 4 ed., Jaboticabal, Brazil. 237 p.

Bonanomi, G., Sicurezza, M.G., Caporaso, S., Esposito, S., and Mozzoleni, S. 2006. Phytotoxicity dynamics of decaying plant materials. New Phytologist 169: 571-578.

Colozzi F., A., Andrade, D.S. 2006. Organismos do solo e atividade microbiana no plantio direto. Cap. 4. p. 39-53. In: Casão J., R. Siqueira, R., Mehta, Y. R., Passini, J.J. Sistema Plantio Direto com Qualidade. IAPAR, ITAIPU, Binacional, Londrina, Brazil. 212 p.

Edmond, J.B., and Drapalla, W.J. 1958. The effects of temperature, sand and soil, and acetone on germination of okra seed. Proceedings of the American Society for Horticultural Science 71: 428-443.

EMBRAPA (Empresa Brasileira De Pesquisa Agropecuária). 2004. Tecnologia da Produção de Soja – Paraná 2005. 1ª ed. Londrina: Embrapa Soja, 239 p.

Erasmo, E.A.L., Azevedo, W.R., Sarmento, R.A., Cunha, A.M., Garcia, S.L.R. 2004. Potencial de espécies utilizadas como adubo verde no manejo integrado de plantas daninhas. Planta Daninha 22: 337-342.

Ferreira, D.F. 2008. Sisvar: um programa para análise e ensino de estatística. Revista Científica Symposium 6: 6-41.

Ferreira, T.N., Schwartz, R.A., Streck, E.V. 2000. Solos: manejo integrado e ecológico – elementos básicos. EMATER/RS, Porto Alegre: Brazil. 95 p.

Maguire, J.D. 1962. Seeds of germination-aid selection and evaluation seedling emergence and vigor. Crop Science 2: 176-177.

Martins, D., Gonçalves, C.G., Silva Junior, A.C. 2016. Coberturas mortas de inverno e controle químico sobre plantas daninhas na cultura do milho. Revista Ciência Agronômica 47: 649-657.

Mauli, M.M., Nóbrega, L.H.P., Rosa, D.M., Lima, G.P. and Ralish, R. 2011. Variation on the Amount of Winter Cover Crops Residues on Weeds Incidence and Soil Seed Bank during an Agricultural Year. Brazilian Archives of Biology and Technology, Curitiba 54: 683-690.

Mauli, M. , Nóbrega, L.H.P., Souza Filho, A.P.S., Stein, L.D.N., Cruz-Silva, C.T.A, Pacheco, F.P. 2015. Seeds germination and early development of beggartick on extracted soil solution from an area with cover crops. African Journal of Agricultural Research 10: 3123-3133.

Moraes, P. V. D., Agostinetto, D., Panozzo, L. E., Tironi, S. P., Galon, L., e Santos, L. S. 2011. Alelopatia de cover crops na superfície ou incorporadas ao solo no controle de Digitaria spp. Planta Daninha 29: 963-973.

Moreira, F.M.S., Siqueira, J.O. 2006. Microbiologia e Bioquímica do Solo. Editora UFLA, Lavras, Brazil. 729 p.

Nunes, J.V.D., Melo, D., Nóbrega, L.H.P., Loures, N.T.P., Fariña Sosa, D.E.F. 2014. Atividade alelopática de extratos de plantas de cobertura sobre soja, pepino e alface. Revista Caatinga 27: 122-130.

Pedrol, N., González, L., and Reigosa, M. 2006. Allelopathy and abiotic stress. Cap. 9. In: Reigosa, M.J., Pedrol, N., and Gonzalez, L. (Eds). Allelopathy: A Physiological Process with Ecological Implications. Springer, Netherlands. 637 p.

Pimentel G., F. 2000. Curso de Estatística Experimental. 14. ed. Degaspari, Piracicaba, Brazil. 477 p.

Rains, G.C., Olson, D.M. and Lewis, W.J. 2011. Redirecting technology to support sustainable farm management practices. Agricultural Systems 104: 1:365-370.

Rice, E.L. 1974. Allelopathy. New York: Academic Press, United States. 333 p.

Ruiz, H.A., Ferreira, G.B., Pereira, J.B.M. 2003. Estimativa da capacidade de campo de latossolos e neossolos quartzarênicos pela determinação do equivalente de umidade. Revista Brasileira de Ciência do Solo 27: 389-393.

Santos, H.P., Reis, E.M. 2003. Rotação de culturas. Cap 1. p. 13-132. In: Santos, H.P.; Reis, E. M. Rotação de Culturas em Plantio Direto. 2 ed. Passo Fundo: Embrapa Trigo, 212 p.

Seigler, D.S. 2006. Basic pathway for the origin of allelopathy compounds. Cap. 2. In: Reigosa, M.J., Pedrol, N., and Gonzalez, L. (Eds). Allelopathy: A Physiological Process with Ecological Implications. Springer, Netherlands. 637 p.

Silva, P.S.S. 2012. Atuação dos aleloquímicos no organismo vegetal e formas de utilização da alelopatia na agronomia. Revista Biotemas, 25: 65-74.

Souza Filho, A.P. 2006. Alelopatia e as plantas. Belém: EMBRAPA Amazônia Oriental. 159 p.

Weih A., U.M.E., Didon A., A.C., Rönnberg-
      Wästljung B., C. and Björkman, M. 2008.
      Integrated agricultural research and crop
      breeding: Allelopathic weed control in cereals
      and long-term productivity in perennial
      biomass crops. Agricultural Systems 97: 99-
      107.

6

Risk Factors Associated with Abortion and Calf Preweaning Mortality in a Beef Cattle System in Southeastern Mexico

José C. Segura-Correa[1*], Victor M. Segura-Correa[2],
Juan G. Magaña-Monforte[1] and Jesús R. Aké-López[1]

[1]Facultad de Medicina Veterinaria y Zootecnia, Universidad Autónoma de Yucatán,
Km. 15.5 carretera Mérida-Xmatkuil, A.P.4-116, Itzimná, Mérida, Yucatán, México.
Email: jose.segura52@hotmail.com
[2]Centro de Investigación Regional del Sureste, INIFAP km 25 carretera Mérida-
Motul, C.P. 97454, Mocochá, Yucatán, México
*Corresponding author

SUMMARY

Data from 2438 calvings born to 682 cows recorded from 2004 to 2015 in an extensive production system were used, to investigate factors associated with abortion and calf mortality until weaning. Cows belonged to Brahman, Nellore, Guzerat and Brown Swiss x Zebu breed groups. Data were analyzed using binary logistic regression, and the statistical model included the effects of year and season of calving (or abortion), parity number, breed group of the cow and sex (only for preweaning mortality). Abortion rate was 0.99% varying from 0.61% to 1.94% among year groups. First parity and Nellore cows had the greatest abortion rates (1.66 and 1.46%, respectively). The calf mortality rate was 9.65%, varying from 3.18% to 14.65% across all years. The major factors associated ($P < 0.05$) with mortality of calves included year and season of calving, parity number and breed group of the cow. Nellore cows had the highest odds of preweaning mortality (OR=4.41). Cow parity number and season of calving were also associated with calf mortality. First parity cows had the major calf losses overall. In conclusion, closer attention to the management of first parity cows could reduce calf mortality.
Key words: Breed; parity; season; sex; tropics.

INTRODUCTION

Abortion in cows and mortality of calves are important causes of production losses and low profitability in livestock farms. Both traits reduce the number of calves for sale, the number of replacement heifers and cows productivity. In addition, they cause indirect losses through underutilization of equipment and infrastructure. According to Bagley (1999) cows suffer abortion rates of 1 to 2%; mainly caused by etiological agents such as bacteria, fungi and viruses (Khodakaram-Tafti and Ikede, 2005). Plasse *et al.*, (1998), in Venezuela, reported abortion rates of 0.6 to 4.5% in 14 beef cattle herds. In the tropics of Mexico, there are few reports on abortion and pre-weaning mortality rates of Zebu calves (Rodríguez and Escrivá, 1971; Gonzalez-Gonzalez and Segura Correa, 1989; Segura-Correa *et al.*, 2009), even though abortive diseases such as leptospirosis, bovine viral diarrhea virus, infectious bovine rinotracheitis are highly prevalent in the region (Segura *et al.*, 2003, Solis *et al.*, 2003, 2005). Mortality rate of calves can be influenced by sex of the calf, climate and management conditions, parity number of cow, vigor and birth weight, and genetics (Gonzalez-Gonzalez and Segura-Correa, 1989; Correa *et al.*, 2000; Riley *et al.*, 2001, 2004). Identification of the risk factors associated with abortion and calf mortality could aid in the optimization of herds productive efficiency. Differences in herd management, feeding systems, breed used and microclimatic conditions may causes variations in herd abortion and mortality rates. Therefore, the objective of this study was to estimate the abortion and preweaning mortality rates, and to determine the importance of some risk factors in an extensive beef cattle system under the tropical conditions of southern Mexico.

MATERIALS AND METHODS

Location

Data were obtained from a productive system located at the northeastern region of Yucatan, Mexico. Yucatan is located at the southeast of Mexico and it has a tropical sub-humid climate, with rain mainly in summer, and averages of temperature and annual rainfall of 25.8 C and 1105 mm, respectively ((INEGI, 2004).

Animals and management

Cows belonged to Brahman, Guzerat and Nellore breeds, and crosses of cows of those breeds with Brown Swiss sires. The proportion of Brown Swiss in crossbred cows was unknown. At the farm, calves were identified at birth, and they stayed with the dam until weaning (approximately at 8 months of age). Cows were managed under extensive conditions in *Panicum máximum* and *Brachiaria brizantha* grass paddocks at night (17 to 6 hours). When pasture was scare in the dry season (February to May) animals were provided with bales of forage. Pure breeds were kept in different lots. Reproduction of the herds was basically by natural mating and some were by artificial insemination. The calves were vaccinated against clostridia, bovine paralytic rabies and leptospira. External parasites were controlled with immersion baths when animals were seen with ticks and internal

parasites by deworming every six months. The herd was free of brucellosis.

Abortion and mortality traits

The overall traits of interest for the present study were abortion and calf mortality from birth until weaning (preweaning mortality). Date and reason of calf mortality were not recorded regularly; therefore, reasons of death were not included in the study. In the production system, veterinary intervention and artificial rearing was not practiced

Risk factors

Information on calf, cow identification, as well as occurrence of abortion, calving date, sex of the calf, occurrence of mortality until weaning, parity number and breed of cows were obtained from the records (n=2438) of 682 cows kept for the years 2004 to 2015. Because of low incidence of abortions year of abortion data were grouped in five categories (2004-2006, 2007-2008. 2009-2010, 2011-2012 and 2013-2015). In addition and for same reason, data from cows with 2 and 3, 4 and 5 and 6 or more calvings were grouped. Based on month of the year three season were established: dry (February to May), rainy (June to September) and windy (October to January). To determine the effect of year of calving on preweaning mortality, data from 2004 to 2006 were grouped in a single category, as were those data from 2014 and 2015, due to small number of observations for those years. Seasons of calving were established based on months of the year as previously described, based on temperature and amount of rain. In addition, the data of the cows with 8 or more parities were grouped in a single category. Information on weight at birth, weaning and reproductive traits of this herd has been reported in previous studies (Segura *et al.*, 2017a; Segura *et al.*, 2017b).

Statistical analysis

The logit of the probability of abortion (yes, no) or preweaning dead (yes, no) was modelled using the Bayes option of the GENMOD procedure of SAS (SAS, 2010) assuming binomial distributions for both traits. The default prior used by SAS program for the fixed-effect parameters was the normal distribution. We further used 5000 burn-in iterations, 100,000 iterations after burn- in and a thinning of 5. The statistical for abortion and preweaning mortality included the fixed effects of year of calving, season of calving, parity number, cow breed group and sex (only for preweaning mortality). Referent levels of the risk factors were year 2009, dry season, sixth parity cows, crossbred cows and females. Significance of risk factors was declared at P<0.05. Markov chain convergence was assessed by visually checking trace, correlation, and kernel density plots (available on request).

RESULTS AND DISCUSSION

Abortion

From the information available (n=2438) there were 24 abortions given an overall abortion rate of 0.98%. The overall abortion rate here reported is within the range of abortion losses (0.6 to 4.5%) reported by Plasse *et al.* (1998) in 14 beef herds in Venezuela. In addition, it is similar to the 1–2% acceptable values suggested by Bagley (1999) and slightly lower than that found by Segura *et al.,* 2009 (1.17%) in the same region of this study. Bagley (1999) stated that if abortion rate is greater than 3% it should be some concern about it.

Abortion could be caused mainly by infectious disease agents but also by trauma, toxins and plants. In southern Mexico, diseases associated to reproductive problems, such as leptospirosis, infectious bovine rhinotracheitis and bovine viral diarrhea are endemic in the region (Segura-Correa *et al.*, 2003; Solís-Calderón *et al.*, 2003, 2005). Abortion rates obtained from field data could be underestimated if abortions at early stage of gestation occurs, because the expulsion of the fetus and placental tissues may not be detected and recorded.

Abortion rates by risk factors (except year) are shown in Table 1 and their odds ratios in Table 2. Abortion rates varied from 0.61% to 1.94% among year groups and there were not season differences. Year of calving is a factor difficult to interpret; because its effect include many climatic and management changes among years, which make it difficult to interpret; however, it is recommended to be included in the statistical model to remove its confounding effect on other risk factors to be evaluated.

The lack of significant influence of season on abortion, here found, disagree with the results of Segura-Correa and Segura-Correa (2009) who found that the risk of abortion was lower in the dry seasons compared to the rainy and windy seasons (P=0.009). Those authors suggest that environmental conditions may play an important role in the presence of pathogens and in consequence on abortion.

Fist parity cows had higher frequencies of abortion (1.66%) than multiparous cows (Table 1); the odds being 3.55 times those of cows with 6 or more parities (Table 2). Higher abort in first parity cows may be explained because of hormonal and uterus recognition may not be well stablished. However, Segura-Correa and Segura-Correa (2009) reported that the risk of abortion was higher in second parity cows followed by the third and first parity cows, as compared to older cows.

In this study, Nellore cows had the highest abortion rate (1.46%), and 6.15 times higher risk than crossbred cows (Table 2). However, in a previous study, Segura-Correa and Segura-Correa (2009) in the same region did not find breed differences in abortion, corresponding the highest prevalence to the crossbred cows. Wijeratne and Stewart (1971) reported breed differences in abortions in dairy and beef cattle.

Table 1. Abortion rates by season of abortion, parity number and breed group of the cow in a beef cattle system in southeastern Mexico.

Risk factor	N	aborted	%	Exact P value
Season				
Dry	1087	12	1.11	0.1681
Rainy	623	3	0.48	-----
Windy	728	9	1.22	0.1043
Parity				
1	601	10	1.66	0.0591
2-3	834	7	0.84	0.4580
4-5	566	3	0.53	0.4992
>=6	434	4	0.92	-----
Breed group				
Brahman	497	4	0.80	0.3057
Nellore	1026	15	1.46	0.0423
Guzerat	626	2	0.32	----
Brown Swiss x Zebu	289	3	1.04	0.2917

Table 2. Posterior descriptive and statistics of regression coefficients and odds ratios by season, parity number and breed group of cow for abortion, in a beef cattle system in southeastern Mexico.

Risk factor	Beta	EE	OR	95% HPD interval
Season				
Dry	1.0296	0.7186	2.800	0.828, 12.401
Rainy	0	----	1	----
Windy	1.226	0.7510	3.000	0.865, 16.525
Parity				
1	1.2668	0.7176	3.459	0.937, 14.622
2-3	0.4998	0.7372	1.648	0.434, 7.083
4-5	0	----	1	----
>=6	0.3803	0.8504	1.463	0.308, 8.262
Breed group				
Brahman	1.0409	0.9528	2.832	0.435, 20.962
Nellore	1.8161	0.8483	6.148	1.377, 32.243
Guzerat	0	----	1	----
Brown Swiss x Zebu	1.0802	1.0842	2.945	0.418, 26.779

HPD = Highest posterior distribution (credible interval).

Preweaning mortality

Of the information available (n=2414), there were 233 deaths till weaning given a preweaning mortality rate of 9.65%. The preweaning mortality rate reported here for the cows managed under an extensive pasture system is within the range (1–12.6%) of calf losses (not including abortions) reported in 14 beef cows by Plasse *et al.* (1998).

All risk factors evaluated were significant on preweaning mortality (except sex of the calf).

Incidence and odds ratios by risk factors (except year of calving) for preweaning mortality are shown in Tables 3 and 4. There was no trend of mortality with years; incidences varying from 3.18% to 14.65% across all years. The largest odds of preweaning mortality relative to the year 2009 corresponded to years 2014 and 2007 with odds values of 20.99 and 4.78, respectively. As mentioned before, year of calving is a factor difficult to interpret; but it must be included in the statistical analysis of the data because is an important confounding factor.

Table 3. Preweaning mortality rates by season of calving, parity number, breed group and sex of calf of cows in a beef cattle system in southeastern Mexico.

Risk factor	N	Dead	%	Exact P value
Season				0.0001
Dry	1068	77	7.21	
Rainy	620	91	14.68	
Windy	728	67	9.20	
Parity				0.0001
1	594	86	14.48	
2	461	44	9.54	
3	366	22	6.01	
4	306	20	6.54	
5	257	18	7.00	
6	190	12	6.32	
7	122	17	13.93	
>=8	118	14	11.86	
Breed group				0.0181
Brahman	493	42	8.52	
Nellore	1013	84	8.29	
Guzerat	624	80	12.82	
Brown Swiss x Zebu	286	29	10.14	
Sex				0.1422
Female	1251	111	8.87	
Male	1165	124	10.64	

Table 4. Posterior descriptive and statistics of regression coefficients and odds ratios by season calving, parity number and breed group of cow for preweaning mortality, in a beef cattle system in southeastern Mexico.

Risk factor	Beta	EE	OR	95% HPD interval
Season				
Dry	0	----	1	----
Rainy	0.9523	0.1788	2.592	1.856, 3.735
Windy	0.2889	0.1905	1.335	0.941, 1.982
Parity				
1	1.3935	0.3563	4.029	2.038, 8.002
2	0.8385	0.3696	2.313	1.134, 4.723
3	0.3197	0.4031	1.377	0.629, 3.002
4	0.3461	0.4064	1.414	0.656, 3.207
5	0.3337	0.4146	1.396	0.625, 3.15
6	0	----	1	----
7	0.7235	0.4248	2.062	0.914, 4.721
>=8	0.3230	0.4455	1.381	0.575, 3.206
Breed group				
Brahman	0.9803	0.3005	2.665	1.486, 4.790
Nellore	1.4837	0.2827	4.409	2.626, 7.823
Guzerat	0.8965	0.2725	2.451	1.433, 4.147
Brown Swiss x Zebu	0	----	1	----
Sex				
Male	0.1476	0.1482	1.159	0.865, 1,549
Female	0	----	1	----

HPD = Highest posterior distribution (credible interval).

Mortality was greatest in the rainy season as compared to the dry season. The odds of mortality of a calf born in the rainy season was 2.59 times that of a cow calving in the dry season. The results of this study disagree with those of Rodriguez and Escrivá (1971) and Gonzalez-González and Segura-Correa (1989) in Brahman cattle in Mexico who did not observed differences among seasons. However, both reported a higher mortality rate in the rainy season as compared to the dry season as observed here. As mention before environmental conditions may play an important role in the incidence of pathogen agents, because warm and humid conditions favor their proliferation and distribution.

First parity cows had greatest calf losses (14.48%) with odds approximately 3-times those for multiparous cows. These results are similar to those reported by Gonzalez-González and Segura-Correa (1989) who found an influence of parity in mortality. However, the mortality rate for first parity cows observed by those authors was slightly lower 12.2%. The reasons of a greater mortality in first parity cows was not determined, in this study; however, the results suggest the need of greater attention and better management for those type of females. In addition, old cows tend to have higher mortality than middle age (parity) cows (Tables 3 and 4). Schmidek et al. (2013) reported that preweaning mortality was higher among calves born from cows aged ≤3 and ≥11 years at calving compared with cows aged 7 to 10 years. Therefore, herd owners

should give greater attention to first parity cows to reduce calf mortality ensuring that heifers achieve target body conditions score and body weight at first calving (Mee et al., 2008).

Sex of the calf was not an important risk factor on mortality till weaning as have been reported by others authors in Zebu cattle in Mexico (Rodríguez and Escrivá, 1971; Gonzalez-González and Segura-Correa, 1989). Schmidek et al. (2013) observed that male calves presented less vigor and higher preweaning mortality than Nellore female calves. Nevertheless, in a previous study with same data here used, birth weight of calves were similar among pure and crossbreed cows (Segura et al., 2017a).

Guzerat had the greatest mortality rate (12.82%) follow by crossbred cows (10.14%), and Nellore cows had the lowest mortality. However, adjusted OR (from binary logistic regression) showed lowest risk of mortality for Nellore cattle (Tables 3 and 4). This is explained because mortality rates in Table 3 are crude percentage, whereas breed OR values in Table 4 are adjusted by season, parity and sex confounding effects. The mortality for Brahman cows, in this study (8.52%, at 7 months), is higher than that reported (7.0% at 9 months) by Gonzalez-González and Segura-Correa (1989) in Tamaulipas, Mexico; and that of 5.1% at 7 months observed by Rodriguez and Escrivá (1971) in "La Huasteca Potosina", Mexico. In addition, the mortality rate, here found, is higher than that reported for Nellore cattle (4.6%) by Correa *et. al.* (2000) in Brazil. According to Benjaminsson (2007) among the factors that might affect mortality rate until weaning in cattle are parity number, sex of the calf, age at first calving, length of gestation, sire of calf and inbreeding. Here the importance of recording information on mortality and risk factors associated to it, in order to prevent calves losses. However, this should be accompanied with the recording of the reason of calf death. In addition, this study was carried out in a representative type farm; therefore, the results here obtained could be extrapolated to other farms in the region. Reducing preweaning mortality in a given farm will increase its profitability by weaning more calf for sale and replacement.

Some of the significant risk factors here studied are largely not under management control of the farm owner (v. gr. year of calving and season of calving). However, attention should be given to key determinants under the owner control, such as first parity cows, infectious diseases, age at first calving and body condition, to reduce abortion and preweaning mortality of zebu cattle, under the tropical conditions of this study.

CONCLUSION

The overall abortion rate here found is in the lower limit of values reported in other studies in the literature; however, preweaning mortality rate is relatively high, which means that it could be reduced with better management practices. Season, parity number and breed group were significant risk factors on preweaning mortality. However, only breed group effect was significant on abortion rate. Closer attention to the management of first parity cows could reduce calf mortality.

REFERENCES

Azevedo Jr, J., Petrini, J., Mourao, G.B., Ferraz, J.B.S. 2017. Preweaning calf survival of a Nellore cattle population. Journal of Agricultural Science. 9(8): 51-62.

Bagley, C.V. 1999. Abortion in cattle. Utah State University Extension. Animal health Fact sheet. http://extension.usu.edu/htm/publications/by=author/char=B/author=39.

Benjaminsson, B.H. 2007. Prenatal death in Icelandic cattle. Acta Veterinaria Scandinavica. 49(suppl 1), S16 doi:10.1186/1751-0147-49-S1-S16.

Bunter, K.L., Johnston, D.J., Wolcott, M.L., Fordyce, G. 2014. Factors associated with calf mortality in tropically adapted beef breeds managed in extensive Australian production systems. Animal Production Science. 54: 25-36.

Correa, E.S., Andrade, P., Euclides Filho, K., Alves, R.G.D. 2000. Evaluation of a beef cattle production system. 1. Reproductive performance. Brazilian Journal of Animal Science. 29(6): 2209-2215.

González-Gonzalez, G., Segura-Correa, J.C. 1989. Factores que afectan la mortalidad al nacimiento, destete y año de edad en ganado Brahman. Veterinaria México. 20(3): 259-263.

Khodakaram-Tafti, A., Ikede, B.O. 2005. A retrospective study of sporadic bovine abortions, stillbirths and neonatal abnormalities in Atlantic Canada, from 1990 to 2001. Canadian Veterinary Journal. 46:635-637.

Mee, J.F., Berry, D.P., Cromie, A.R. 2008. Prevalence of, and risk factors associated with, perinatal calf mortality in pasture-based Holstein-Friesian cows. Animal. 2(4): 613–620.

Plasse, D., Fossi, H., Hoogesteijn, R. 1998. Mortality in Venezuelan beef cattle. World Animal Review 90. http://www.fao.org/documents/pub_dett.asp?pub_id=21073&lang=en

Riley, D.G., Sanders, J.O., Knutson, R.E., Lunt, D.K., 2001. Comparison of F1 Bos indicus x Hereford cows in central Texas: 1 Reproductive, maternal and size traits. Journal of Animal Science. 79(6): 1431-1438.

Riley, D.G., Chase Jr., C.C., Olson, T.A., Coleman, S.W., Hammond, A.C. 2014. Genetic and non-genetic influences on vigor at birth and preweaning mortality of purebred and high percentage Brahman calves. Journal of Animal Science. 82(6): 1581-1588.

Rodríguez, O.L. and J. L. Escrivá. 1971. Factores que afectan el porcentaje de destetes en ganado Brahman: mortandad entre nacimiento y destete. Técnica Pecuaria en México. 22: 22-27.

Schmidek, A., Rodrigues Paranhos da Costa, M.J., Zerlotti Mercadante, M.E., Macedo de Toledo, L., dos Santos Gonçalves Cyrillo, J.N., Branco, R.H. 2013. Genetic and non-genetic effects on calf vigor at birth and preweaning mortality in Nellore calves. Revista Brasileira de Zootecnia. 42(6): 421-427.

Segura-Correa, V.M.; Sólis-Calderón, J.J., Segura-Correa, J.C. 2003. Seroprevalence of and risk factors for leptospiral antibodies among cattle in the state of Yucatan, Mexico. Tropical Animal Health and Production 35(4):293-299.

Segura-Correa, J. C. and V. M. Segura-Correa. 2009. Prevalence of abortion and stillbirth in a beef cattle system in southeastern Mexico. Tropical Animal Health and Production. 41(8):1773-1778. doi:10.1007/s11250-009-9376-x.

Segura Correa, J.C., Magaña-Monforte, J.G., Aké-López, J.R., Segura-Correa, V.M., Hinojosa-Cuellar, J.A., Osorio-Arce, M.M. 2017a. Breed and environmental effects on birth weight, weaning weight and calving interval of Zebu cattle in southeastern Mexico. Tropical and Subtropical Agroecosystems. 20(2): 297-305.

Segura-Correa, J.C., Magaña-Monforte, J.G., Ake-Lopez, J.R., Segura-Correa, V.M. 2017b. Season and parity number influence the conception rate of zebu breed cows in South-eastern Mexico. Livestock Research for Rural Development. 29, Article #215. http://www.lrrd.org/lrrd29/11/jose29215.html

Solis-Calderon, J.J., Segura-Correa, V.M., Segura-Correa, J.C., Alvarado-Islas, A. 2003. Seroprevalence of and risk factors for infectious bovine rhinotracheitis in beef cattle herds of Yucatan, Mexico. Preventive Veterinary Medicine. 57(4): 199-208

Solís-Calderon, J.J., Segura-Correa, V.M., Segura-Correa, J.C. 2005. Bovine viral diarrhoea virus in beef cattle herds of Yucatan, Mexico: Seroprevalence and risk factors. Preventive Veterinary Medicine. 72(3-4):253-262.

Wijeratne, W., Stewart, D. 1971. Population study of abortion in cattle with special reference to genetic factors. Animal Science. 13(2): 229-235. doi:10.1017/S0003356100029664.

7

Prevalence of Antibodies and Risk Factors to Bovine Viral Diarrhea in Non-Vaccinated Dairy Cattle from Southern Ecuador

V. Herrera-Yunga[1], J. Labanda[2], F. Castillo[3], A. Torres[3], G. Escudero-Sanchez[2], M. Capa-Morocho[4,5] and R. Abad-Guamán[2,5,*]

[1] Carrera de Zootecnia. Facultad de Ciencias Pecuarias, Escuela Superior Politécnica de Chimborazo (ESPOCH). Avenida Panamericana km ½ Riobamba, Ecuador
[2] Carrera de Medicina Veterinaria y Zootecnia, Universidad Nacional de Loja (UNL). La Argelia, 110150 Loja – Ecuador. E-mail: rodrigo.abad@unl.edu.ec
[3] Centro de Biotecnología, Universidad Nacional de Loja (UNL). La Argelia, 110150 Loja – Ecuador
[4] Carrera de Ingeniría Agronómica, Universidad Nacional de Loja (UNL). La Argelia, 110150 Loja – Ecuador
[5] Facultad de Ciencias Agropecuarias, Universidad Técnica de Ambato, Carretera Cevallos-Quero, 180350 Cevallos, Tungurahua, Ecuador
*Corresponding author

SUMMARY

The aim of this work was to determine the prevalence of antibodies and risk factors of bovine viral diarrhoea virus (BVDV) in non-vaccinated dairy cattle at the South of Ecuador. A cross-sectional study was carried out to identify risk factors for BVDV infection in 394 randomly selected dairy cows from 75 farms, which were tested for antibodies in milk samples using a commercial Kit ELISA (IDEXX). Epidemiological survey was conducted to determine the risk factors and signs associated with BVDV. Results of this test revealed that the BVDV herd prevalence was 63.5% and the BVDV individual prevalence was 27%. The utilization of artificial insemination (AI) was significantly associated with BVDV status (P > 0.001) where the use of AI increased 2.35 the odds of BVDV positivity (95% CI: 1.46 – 3.38). The cows with clinical signs (diarrhoea, abortions, and ocular and nasal discharge) were not predominantly positive to BVDV antibodies.

Keywords: BVDV, epidemiology, risk factor, prevalence, artificial insemination, antibodies.

INTRODUCTION

Bovine Viral Diarrhea (BVD) is caused by a small single-stranded RNA virus of positive polarity belonging to the genus Pestivirus of the family Flaviviridae. This virus affects cattle compromising their health and milk production that leads to important economic impairment (Pellerin *et al.,* 1994; Fourichon *et al.,* 2005; Weldegebriel *et al.,* 2009). The bovine viral diarrhea virus (BVDV) is a worldwide spread cattle pathogen. The genus contains a number of species including the two genotypes of bovine viral diarrhea virus (BVDV) (types 1 and 2) and the closely related classical swine fever and ovine border disease viruses (Ridpath *et al.,* 1994). The first way of transmission of BVDV-1 is through nasal, ocular, and genital secretion, and by semen from infected cattle (Guarino *et al.,* 2008).

The BVDV infection mainly affects pregnant cows causing abortions, stillborns, foetus mummification and calves birth with immune-tolerance to BVDV (Terpstra, 1985; Houe, 1995; Paton, 1995; Nettleton *et al.,* 1998). The cows infected with noncytopathic BVDV during the early gestational period are very likely to produce infected calves, which are mainly responsible for spreading BVDV via continuous viral shedding from all mucosal surfaces in herds (Bauermann *et al.,* 2014). Furthermore, Pestivirus infection occurs with leukopenia and immunosuppression because BVDV mostly attack the immune system cells, making these animals susceptible to other pathogens (Potgieter, 1995; Thabti *et al.,* 2002). This infection can be indirectly detected by antibody analysis of serum or milk from animals surrounding the infected groups (Houe, 1992; Niskanen, 1993; Beaudeau *et al.,* 2001).

In many countries the information about prevalence, incidence and associated risk factors have been the baseline for designing and implementing effective regional control actions that minimizes the adverse effects of BVDV infection on herd health and productivity (Rush *et al.,* 2001). Moreover, factor such as type of reproduction, elevation of daily herd, age of cows, and livestock production system have been associated with BVDV infection (Mainar-Jaime *et al.,* 2001; Talafha *et al.,* 2009; Saa *et al.,* 2012).

In the south region of Ecuador there are some reproductive problems that can be associated with BVD. A previous study on seroprevalence of BVDV infection was performed in the Central and North region of Ecuador (Saa *et al.,* 2012), but the risk factors related to this infection have not been clearly defined. Therefore, the aims of the present study were to know about the distribution of BVDV prevalence and to determine their risk factors in a population of non-vaccinated dairy herds from the South of Ecuador. In this region cattle have many reproductive problems that can be associated with BVD.

MATERIALS AND METHODS

This cross-sectional study was performed in the district of Loja to investigate prevalence and risks associated with the presence of antibodies against BVDV in the milk in dairy cows. The information of dairy herds was collected in the three peri-urban (El Valle, San Sebastian y Sucre) and 10 rural sub-districts (Chuquiribamba, El Cisne, Gualel, Jimbilla, Malacatos, San Lucas, Santiago, Taquil, Vilcabamba and Yangana) from February to April 2015, where the vaccinated cows were excluded from the study.

Sample size

Sample size was calculated as Aguilar-Barojas (2005) described. Due to the lack of updated information, the number of adult bovine units was taken from a projection for 2013 made by the National Institute of Statistics and Census (ESPAC, 2013). This projection estimates that the district of Loja will have approximately 49.829 adult bovine units. Sixty percent of them are categorized as dry cows and dairy production cows, 50% of which are related to our study due to the milk sample analysis. Therefore, the sample size was 394 cows from 75 dairy farms.

Data collection

The epidemiological survey was conducted before collecting the milk sample. The cattle farmers were interviewed using "close-ended" questions. For identification purposes, each cattle farmer and milk cow was assigned a unique identification code. The variables included in this study were:

1) Serological status: the presence or absence of BVDV antibodies were determined by sample and positive control (S/P), where the values ≥ 0.30 were considered positives and values ≤ 0.20 were considered negatives.
2) Livestock production systems: semi-intensive and extensive system
3) Quarantine: application and not application of quarantine
4) Breeding methods: artificial insemination and natural mating.
5) Clinical signs: diarrhea, ocular discharge, abortions, infertility, and runny nose.
6) Elevation of dairy herd: it was classified into Groups: 1) from 1400 to 2100 meters above sea level and Group 2) more than 2100 meter above the sea level.
7) Number of calving: it was considered from the first birth onwards.

8) Biosecurity: footbath, wheel-dip or neither was considered.

9) Breeds: all breeds that are exploited in the district of Loja were chosen. It comprises Holstein, Brown Swiss, Jersey and creole cattle.

Sample collection and serological examination

Aseptic teat end preparation (cleaned with ethanol 70%) and discard of the first 4 jets of milk were performed before taking the sample. A single milk sample (~10 mL) per cow was collected. Then, it was held on ice until it arrived to the investigation facilities, this took at maximum 8 hours. Next, the samples were stored at −20°C. To start the analyses in the laboratory, the samples were gradually defrosted to reach room temperature. The presence of BVDV antibodies was tested using the commercial Kit ELISA (IDEXX HerdChek® ELISA kit for BVDV-Ab, IDEXX laboratories, Westbrook, Maine, USA). Following the manufacturer's instructions for undiluted milk, the samples were tested and expressed as sample-to-positive (S/P) ratio where the cut point was set to ≥18U/ul. Antibody concentration was measured at 450 nm using a photometer Biotek ELx800. For measuring the optical density (D.O.), the GEN 5 software and a photometer were employed. The data obtained were calculated by applying the following formulas:

Negative Control Mean.

$$NC\bar{x} = \frac{NC1\ A450 + NC2\ A450}{2}$$

where:

NC $\bar{x}$: Negative control mean

NC1 A450: Negative control_1 has been read in optical densities at 450 nm

NC2 A450: Negative control_2 has been read in optical densities at 450 nm

Positive Control Mean

$$C\bar{x} = \frac{PC1\ A450 + PC2\ A450}{2}$$

where:

PC $\bar{x}$: Positive control mean

PC1 A450: Positive control_1 has been read in optical densities at 450 nm

PC2 A450: Positive control_2 has been read in optical densities at 450 nm

Test Sample.

$$S/P = \frac{Sample\ A450 - NC\bar{x}}{PC\bar{x} - NC\bar{x}}$$

where:

S/P: Sample/Positive ratio

NC $\bar{x}$: Negative control mean

PC $\bar{x}$: Positive control mean

Statistical Analysis

An ELISA sensitivity (Se) of 96.3% and a specificity (Sp) of 99.5% (IDEXX HerdChek® ELISA kit for BVDV-Ab, IDEXX laboratories, Westbrook, Maine, USA) were used to adjust the apparent prevalence (AP) by using the equation for true prevalence (TP) = (AP + Sp)/(Se + Sp) (Thrusfield, 2007). The 95% confidence interval (CI) for the prevalence was based on the normal approximation to the binomial distribution. These tests were performed utilizing R software [R Development Core Team (2014)]. The prevalence and CI were calculated using the Clopper-Pearson exact method (Clopper and Pearson, 1934) of the 'epi.prev' function from the Package epiR version 0.9-62 (Stevenson et al., 2013). Then, these data were presented using ArcGIS 10, version 10.4 software.

The observation (cow) is independent from each other. Then, logistic regression was used to analyze the association between the BVDV antibody status and the predictors. Univariable analysis was performed using all preselected variables. Logistic regression was made using SAS PROC LOGISTIC (SAS Studio version 3.4, Institute Inc, Cary, NC, USA). Multivariable model was not built because the significant variables (p < 0.1) selected for inclusion in the multivariable analysis were correlated (r > 0.8) between them. Next, the risk was calculated as odds ratio (OR). The OR's standard error and 95% confidence interval were calculated according to Altman (1991).

RESULTS

Three hundred and ninety-four cows were sampled across thirteen sub-districts from 75 dairy farms. These cows had diverse number of calving: 71 cows (18.0%) had one calving, 190 (48.2%) had two to three calving, and 133 (33.8%) had more than three calving. The percentage of farms with semi-intensive and extensive livestock production systems were 11% (8/75 farms) and 89% (67/75 farms), respectively. Most of the subjects were naturally bull serviced (74%) and only 26% of the cows were artificially inseminated. The true herd prevalence of BVDV in Loja and their sub-districts are presented in Figure 1. From the 394 examined milk samples, one hundred and four (26.4%) were BVDV positive by antibody ELISA. The true individual prevalence of BVDV was 27.0% (95% CI: 22.5 - 31.9%).

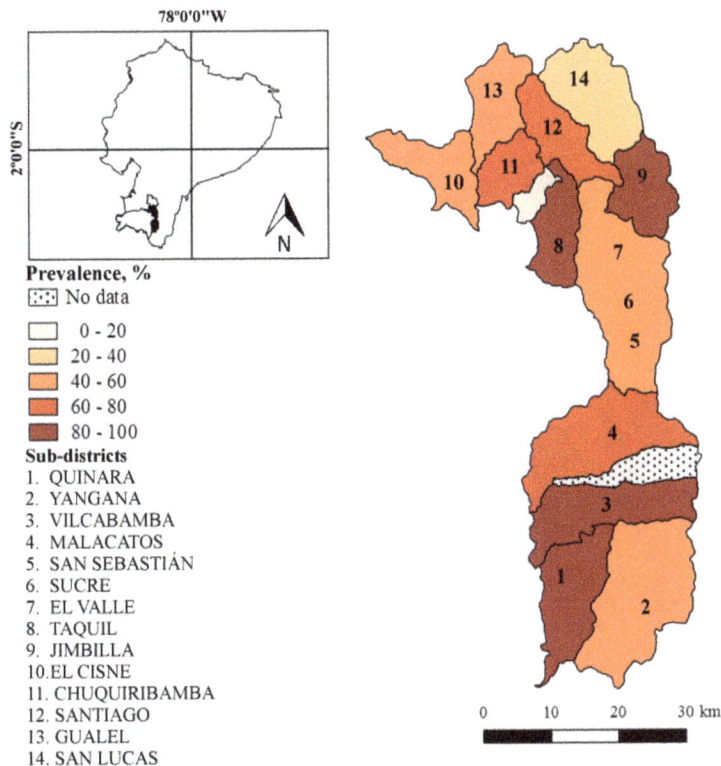

Figure 1. Map of Southern Ecuador indicating the true prevalence of each sub-district of Loja. True prevalence (%) of each sub-district of Loja, Southern Ecuador.

The overall prevalence of antibodies against BVDV in cattle ranged from 6.93% to 86.4% among the thirteen sub-districts studied. The BVDV antibodies occurrence in the peri-urban sub-districts was lower than that of the rural sub-district (12.1% vs. 29.8% respectively). The true herd prevalence was 63.5% (95% CI: 51.0 – 75.0%).

Intrinsic and extrinsic factors associated with the prevalence of antibodies against BVDV are showed in Tables 1, 2 and 3. Herd antibodies against BVDV tended to be associated with the livestock production systems and the use of some biosecurity measures (P = 0.10; OR: 5.02). The use of AI is an intrinsic risk factor associated with BVDV in the South Ecuador (P < 0.001; OR: 2.35). The AI is strongly related with Semi-intensive livestock production system and application of quarantine (P < 0.001).

A linear relationship with the parameter (beta) for estimating the elevation of daily herd (as continuous variable) was observed (P = 0.05; Table 3). The number of calving and the breeds of cow (as categorical variables) were not significant related to

BVDV (P > 0.40). Animals that didn't present the clinical signs showed the highest BVDV prevalence (Table 1). No clinical evidence was found in positive animals to BVDV antibodies. The vesicular stomatitis, nasal discharge and abortions were not related with BVDV prevalence (P > 0.35). The cows without diarrhea or ocular discharge showed a higher BVDV antibodies prevalence (P < 0.01) than the animals with these clinical signs.

DISCUSSION

As the cows with vaccination against BVDV were not sampled in this study, the presence of antibodies indicates a natural exposure to BVDV at some point of its life. Herd prevalence in this study was higher (63.5%) than those reported in other regions of Ecuador (36.2%; Saa *et al.*, 2012) and lower than that of Peru (96%; Stahl *et al.*, 2002). However, there is an important variation of the BVDV prevalence among different sub-district of Loja (from 6.93% to 86.4%). This prevalence variation could be attributable to factors such as population density and different management practices (Houe, 1995).

Table 1. Descriptive results for explanatory variables and clinical sings with BVDV status among 394 daily cattle from 75 farmers surveyed in Loja from February to April 2015

Variable	Category	N[1,2]	DVB true prevalence, % Positive	% (CI 95%)
Extrinsic factors[1]				
Livestock production systems	Semi-intensive	8	7	90.8 (48.9 – 100)
	Extensive	67	39	60.2 (47.0 – 72.7)
Biosecurity measures	yes	8	7	90.8 (48.9 – 100)
	no	67	39	60.2 (47.0 – 72.7)
Quarantine	yes	6	5	86.5 (36.9 – 100)
	no	69	41	61.5 (48.4 – 73.7)
Intrinsic factors[2]				
Parity	1	71	21	30.3 (19.6 – 42.9)
	2, 3	190	52	28.0 (21.6 – 35.3)
	≥4	133	41	31.6 (23.6 – 40.6)
Artificial insemination	yes	103	44	44.1 (33.9 – 54.6)
	no	291	70	24.6 (19.6 – 30.1)
Breed	Brown Swiss	28	6	21.8 (8.14 – 42.2)
	Brown Swiss mestizo	22	6	27.9 (10.7 – 51.9)
	Holstein Friesian	104	47	46.6 (36.4 – 57.2)
	Holstein Friesian mestizo	230	63	28.1 (22.2 – 34.6)
	Jersey	1	0	0 (0 – 100)
	Creole	9	2	22.7 (2.42 – 62.1)
Clinical Sings				
Sings	yes	161	63	40.3 (32.4 – 48.7)
	no	233	51	22.3 (17.0 – 28.4)
Stomatitis	yes	3	0	0 (0 – 73.3)
	no	391	114	29.9 (25.2 – 34.9)
Diarrhea	yes	31	4	12.9 (3.26 – 30.6)
	no	363	110	31.1 (26.2 – 36.3)
Aborts	yes	3	0	0 (0 – 73.3)
	no	391	114	29.9 (25.2 – 34.9)
Nasal discharge	yes	44	12	27.9 (15.1 – 44.1)
	no	350	102	29.9 (25.0 – 35.2)
Eye discharge	yes	187	40	21.8 (15.9 – 28.7)
	no	207	74	37.0 (30.0 – 44.0)

[1]Number=75 dairy farming. [2]Number = 394 cows

The results show that the semi-intensive livestock production system and application of quarantine tended to be a risk factor in the sampled herds. The semi-intensive livestock production system can be associated with a great herd size and herd density or the use of AI. This also consistent with previous studies (Houe et al., 1995; Valle et al., 1999), who have shown that the important risk factors for BVDV infection are herd size and herd density. Additionally, the infection prevalence tends to increase with the increment in the cattle density in the area.

The AI was the biggest risk factor related with the prevalence of antibodies against BVDV. In the same way, Saa et al. (2012), found the same relation on the north Ecuador. Moreover, several studies have found association between BVDV and bovine reproductive management such as contaminated semen and use of infected bulls (Houe, 1999; Lindberg and Alenius, 1999; Gard et al., 2007; Saa et al., 2012). Additionally, common risk factors in epidemiological studies about BVDV are the acquisition of new animals or being in contact with animals from other farms (Solis-Calderon et al., 2005; Luzzago et al., 2008; Talafha et al., 2009; Saa et al., 2012).

On the other hand, our identification of AI utilization as a BVDV risk factor could be indirectly associated with the spread of infection. This could be explained by the indirect transmission through the materials used during AI process or transmission by fomites from farm to farm transported by the AI technicians, who usually inseminate a number of cows in numerous farms per day. Several ways of indirect transmission of BVDV have been demonstrated such as plastic gloves used in rectal palpation (Lang-Ree et al., 1994), needles and nose tongs or contaminated vaccines (Houe, 1999)

Table 2. Odds ratio analysis for risk factors associated with BVDV status of dairy herds in Loja

Variable	Category	Odds ratio	95% CI	P-value
Extrinsic factors[1]				
Livestock production systems	Semi-intensive	5.02	0.585 – 43.2	0.10
	Extensive	1	Reference	
Biosecurity measures	yes	5.02	0.585 – 43.2	0.10
	no	1	Reference	
Quarantine	yes	3.42	0.378 – 30.8	0.25
	no	1	Reference	
Intrinsic factors[2]				
Parity	1	0.942	0.503 – 1.77	0.85
	2, 3	0.845	0.519 – 1.38	0.49
	≥ 4	1	Reference	
Artificial insemination	yes	2.35	1.46 – 3.78	<0.001
	no	1	Reference	
Breed	Brown swiss	0.954	0.156 – 5.85	0.96
	Brown swiss mestizo	1.31	0.211 – 8.18	0.77
	Holstein Friesian	2.89	0.572 – 14.6	0.20
	Holstein Friesian mestizo	1.08	0.219 – 5.39	0.92
	Jersey	1.00	0.030 – 33.3	0.99
	Creole	1	Reference	

[1]Number=75 dairy farming. [2]Number = 394 cows

Nevertheless, the altitude of daily farm was detected as a risk factor when this variable was considered as a continuous variable. This can be explained by the relation between the altitude of farms and the livestock system production. For instance, farms in Loja with semi-intensive livestock production system are located below 2000 m above sea level. Meanwhile, in central and north region from Ecuador the most intensive dairy production farms are located over 2000 m above sea level.

Table 3. Results of univariable logistic regression analysis for risk factors associated with BVDV status of dairy herds in Loja

Variables	Level	Estimate (ß)	P-value
Artificial insemination			
	Yes	0.366	0.003
	No		
Elevation	Continues	-0.0131	0.050

The cows with clinical signs (diarrhea, abortions, and ocular and nasal discharge) were no predominantly positive to BVDV antibodies. This further support the idea of these clinical signs as cofactors of BVDV infection (Pfeiffer *et al.,* 2002; Park *et al.,* 2006) and a high occurrence of subclinical BVDV infections (between 70% to 90% of BVDV infections) occur without manifestation of clinical signs (Ames, 1986).

Although, the exact mechanism of immunotolerance is unidentified, it is due that circulation of virus during of gestation is when immunocompetence is developing and is a prerequisite for persistence. Viral proteins are recognized as self-antigens resulting negative to antibody-antigen test (Grooms, et al., 2004).

CONCLUSIONS

The results suggest that the natural exposure to BVDV in the Southern Ecuadorian dairy cattle is common and the main risk factors associated with the BVDV infection are the artificial insemination and the livestock production system.

Acknowledgements

We are grateful to Universidad Nacional de Loja for the financial support.

REFERENCES

Aguilar-Barojas, Saraí. 2005. Fórmulas para el cálculo de la muestra en investigaciones de salud. Salud en Tabasco. 11: 333-338.

Altman D.G. 1991. Practical statistics for medical research. London: Chapman and Hall.

Ames, T. R. 1986. The causative agent of BVD: its epidemiology and pathogenesis. Veterinary Medicine. 81: 848-869.

Bauermann F.V., Falkenberg S.M., Vander Ley B., Decaro N., Brodersen B.W., Harmon A., Hessman B., Flores E.F., Ridpath J.F. 2014. Generation of Calves Persistenly infected with HoBi-Like Pestivirus and Comparison of Methods for Detection of These Persistent Infections. Journal of Clinical Microbiology. 52: 3845-3852

Beaudeau, F., Assie, S., Seegers, H., Belloc, C., Sellal, E., Joly, A. 2001. Assessing the within-herd prevalence of cows antibody-positive to bovine viral diarrhoea virus with a blocking ELISA on bulk tank milk. Veterinary Record. 149: 236–240.

Clopper C.J., Pearson E.S. 1934. The use of confidence of fiducial limits illustrated in the case of the binomial. Biometrika. 26: 404 - 413.

Fourichon, C., Beaudeau, F., Bareille, N., Seegers, H., 2005. Quantification of economic losses consecutive to infection of a dairy herd with bovine viral diarrhoea virus. Preventive Veterinary Medicine. 72: 177–181.

ESPAC, 2013. Encuenta de Superficie y Producción Agropecuaria Continua. Obtained from Ecuador en cifras: http://www.ecuadorencifras.gob.ec/documentos/web-inec/Estadisticas_agropecuarias/espac/espac%202013/PRESENTACIONESPAC2013.pdf

Gard, J.A., Givens, M.D., Stringfellow, D.A. 2007. Bovine viral diarrhea virus (BVDV): Epidemiologic concerns relative to semen and embryos. Theriogenology. 68: 434–442.

Grooms, D.L. 2004. Reproductive consequences of infection with bovine viral diarrhea virus, The Veterinary Clinics of North America, Food Animal Practice. 20: 5–19.

Guarino, H., Nunez, A., Repiso, M.V., Gil, A., Dargatz, D.A. 2008. Prevalence of serum antibodies to bovine herpesvirus-1 and bovine viral diarrhea virus in beef cattle in Uruguay. Preventive veterinary medicine, 85: 34-40.

Houe, H. 1992. Serological analysis of a small herd sample to predict presence or absence of animals persistently infected with bovine viral diarrhoea virus (BVDV) in dairy herds. Research in Veterinary Science. 53: 320–323.

Houe H. 1995. Epidemiology of bovine viral diarrhea virus. The Veterinary Clinics of North America, Food Animal Practice, 11: 521-547.

Houe H. 1999. Epidemiological features and economical importance of bovine virus diarrhoea virus (BVDB) infections. Veterinary Microbiology 64: 89-107.

Mainar-Jaime, R.C., Berzal-Herranz, B., Arias, P., Rojo-Vazquez, F.A. 2001. Epidemiological pattern and risk factors associated with bovine viral-diarrhoea virus (BVDV) infection in a non- vaccinated dairy-cattle population from the Asturias region, Preventive Veterinary Medicine. 52: 63–73

Nettleton PF, Gilray JA, Dlissi E. 1998. Border disease of sheep and goats. Veterinary Research. 29: 327-340

Niskanen, R. 1993. Relationship between the levels of antibodies to bovine viral diarrhoea virus in bulk tank milk and the prevalence of cows exposed to the virus. Veterinary Record. 133: 341–344.

Lang-Ree, J.R., Vatn, T., Kommisrud, E. Løken, T. 1994. Transmission of bovine viral diarrhoea virus by rectal examination. Veterinary Record. 135: 412–3.

Lindberg, A.L., Alenius, S. 1999. Principles for eradication of bovine viral diarrhoea virus (BVDV) infections in cattle populations. Veterinary Microbiology. 64: 197– 222.

Luzzago, C., Frigerio, M., Piccinini, R., Daprà, V., Zecconi, A. 2008. A scoring system for risk assessment of the introduction and spread of bovine viral diarrhoea virus in dairy herds in Northern Italy. The Veterinary Journal. 177: 236–241.

Park, S.J., Jeong, C., Yoon, S.S., Choy, H.E., Saif, L. J., Park, S.H., Lee, B.J. 2006. Detection and characterization of bovine coronaviruses in fecal specimens of adult cattle with diarrhea during the warmer seasons. Journal of Clinical Microbiology. 44: 3178-3188.

Paton DJ. 1995. Pestivirus diversity. Journal of Comparative Pathology. 112: 215-236.

Pfeiffer, D. U., Williamson, N. B., Reichel, M. P., Wichtel, J. J., & Teague, W. R. 2002. A longitudinal study of Neospora caninum infection on a dairy farm in New Zealand. Preventive Veterinary Medicine. 54: 11-24.

Pellerin Ch., Van Den Hurk J., Lecotme J., Tijseen P., 1994. Identification a New Group Of Bovine Viral Diarrhea Virus Strains Associated with Severe Outbreaks and High Mortalities. Virology. 203: 260-268.

Potgieter L. 1995. Immunology of bovine viral diarrhea virus. The Veterinary Clinics of North America, Food Animal Practice. 11: 501-520.

R Development Core Team, 2014. R: A Language and Environment for Statistical Computing. R Foundation for Statistical Computing, Vienna, Austria.

Ridpath, J. F., Bolin, S. R., & Dubovi, E. J. 1994. Segregation of bovine viral diarrhea virus into genotypes. Virology, 205, 66-74.

Rush, D.M., Thurmond, M.C., Munoz-Zanzi, C.A., Hietala, S.K. 2001. Descriptive epidemiology of postnatal bovine viral diarrhea virus infection in intensively managed dairy heifers. Journal of the American Veterinary Medical Association. 219: 1426–1431.

Saa, L. R., Perea, A., García-Bocanegra, I., Arenas, A. J., Jara, D. V., Ramos, R., Carbonero, A. 2012. Seroprevalence and risk factors associated with bovine viral diarrhea virus (BVDV) infection in non-vaccinated dairy and dual purpose cattle herds in Ecuador. Tropical animal health and production. 44: 645-649.

Stahl, H., Rivera, I., Vagsholm, J., Moreno-López, J. 2002. Bulk milk testing for antibody seroprevalence to BVDV and BHV-1 in a rural region of Perú, Preventive Veterinary Medicine. 56: 193–202

Stevenson, M., Nunes, T., Sanchez, J., Thornton, R.,

Reiczigel, J., Robison-Cox, J., Sebastiani, P. 2013. epiR: An R package for the analysis of epidemiological data. R package version 0.9-43.

Solis-Calderon, J.J., Segura-Correa, V.M., Segura-Correa, J.C. 2005. Bovine viral diarrhoea virus in beef cattle herds of Yucatan, Mexico: seroprevalence and risk factors. Preventive Veterinary Medicine. 72: 253–262.

Talafha, A.Q., Hirche, S.M., Ababneh, M.M., Al-Majali, A.M., Ababneh M.M. 2009. Prevalence and risk factors associated with bovine viral diarrhea virus infection in dairy herds in Jordan, Tropical Animal Health and Production: 41: 499–506.

Terpstra C. 1985. Border disease: a congenital infection of small ruminants. Progress in veterinary microbiology and immunology. 1: 175-198.

Thabti F, Fronzaroli L, Dlissi E, Guibert JM, Hammami S, Pepin M, Russo P. 2002. Experimental model of border disease virus infection in lambs: comparative pathogenicity of pestiviruses isolated in France and Tunisia. Veterinary Research. 33: 35-45.

Thrusfield, M. (Ed.) 2007. Veterinary Epidemiology, third ed. Blackwell Publication, Oxford, 610 pp.

Valle, P. S., Martin, S. W., Tremblay, R., Bateman, K. 1999. Factors associated with being a bovine virus diarrhoea (BVD) seropositive dairy herd in the Møre and Romsdal County of Norway. Preventive Veterinary Medicine. 40: 165-177.

Weldegebriel, H.T., Gunn, G.J., Stott, A.W. 2009. Evaluation of producer and consumer benefits resulting from eradication of bovine viral diarrhoea (BVD) in Scotland, United Kingdom. Preventive Veterinary Medicine. 88: 49–56.

8

Productivity of Cucumber (*Cucumis Sativus* L) and Post-Harvest Soil Chemical Properties in Response to Organic Fertilizer Types and Rates in an Ultisols

Kolawole E. Law-Ogbomo*and Agbonsalo U. Osaigbovo

Department of Crop Science, Faculty of Agriculture, University of Benin, PMB 1154, Benin City 300001, Nigeria. Email: kolalawogbomo@yahoo.com
Corresponding author

SUMMARY

Cucumber productivity in Nigeria is low due to low native fertility status of the soil among other factors. This field experiment was conducted in February to June of 2015 and 2016 at the Experimental Farm, Faculty of Agriculture, University of Benin, Benin City, Nigeria to evaluate the effects of cattle and poultry manures on the productivity of cucumber (*Cucumis sativus* L.) and their post-harvest effect on the chemical properties of the soil. The experiment was in a 2 x 4 split plot arrangement fitted into a randomized complete block design with three replications with the organic fertilizers (cattle and poultry manures) as main treatments the application rates (0, 5, 10 and 15 t ha^{-1}) as sub plots. Data were collected on growth and yield variables of cucumber. The results revealed that organic fertilizer types had no effect (P>0.05) on growth, yield and post-harvest variables except vine girth. The rate of application had effect on growth, yield and post-harvest soil chemical properties (P<0.05). The highest fruit yields were 3.99 and 3.66 t ha^{-1} observed on plants treated with 15 and 10 t ha^{-1} of organic fertilizer, respectively. Based on convenience and cost, 10 t ha^{-1} of poultry manure is thereby recommended for farmers.

Keywords: Cattle and poultry manures; growth and yield parameters; soil analysis.

INTRODUCTION

Cucumber (*Cucmis sativus* L.) is the family of Cucurbitaceae and an important fruity vegetable with great economic potentials, as medicinal plant and source of industrial raw material. Cucumber is a dependable laxative for those who suffer constipation. The juice of cucumbers is a valuable food in the treatment of hyperacidity, gastric and duodenal ulcers (Ernestina, 2001).It is a popular fresh market vegetable for making salads. Fruits are sliced into pieces and served with groundnut, vinegar or a salad dressing, on their own or served other vegetables. Young or unripe cucumber fruits are usually used as cooked vegetables or made into Chutney (Grubben and Denton, 2004). It is a rich source of minerals and vitamins (Eifediyi and Remison, 2009). It is also used in skin and hair care.

Despite the numerous benefits and economic importance of this crop in Nigeria, its cucumber production and utilization have not been a viable option for farmers (Olawuyi *et al.;* 2011). Due to intensive cropping, imbalance use of inorganic fertilizers and continuous cropping among other factors has led to the soil nutrients depletion (Mahmood *et al.;* 2004). This necessitated the use of fertilizer input to supplement the low plant nutrient content of the soil. The yield of cucumber was increased through inorganic fertilizer application (Agba and Enya, 2005).Chemical fertilizers promote plant growth and give high yields. It is however, scarce, beyond the reach of the resource-challenged farmers and pollute the environment. This calls for development of alternative nutrient source to the use of inorganic fertilizer (Olawuyi *et al.;* 2011).This necessitated the use of organic fertilizers in the production of cucumbers. Organic fertilizers are readily available, affordable by the resource poor farmer and eco-friendly. In addition, organic fertilizers mineralize slowly, so a subsequent successive crop can as well benefit from its incorporation into the soil. Hence, the main aim of this study is to evaluate the effect of organic manures on the productivity of cucumbers with the soil sustainable for continuous production.

MATERIALS AND METHODS

Site Description

The field experiment was carried out during the early growing seasons of 2015 and 2016(February to June) at the Experimental Farm, Faculty of Agriculture, University of Benin, Benin City, Edo State, Nigeria. This is located between latitude 6°, 44' N and 7°,34' N and longitude 5° 40' E and 6° 43'E. The elevation of the site is about 162 m above sea level. Benin City is in the rain forest agro-ecological zone of Nigeria.

The study area lies within rain forest which has now degraded to secondary forest as a result of shifting cultivation. The dominant soil of the site was ultisols of Benin formation (Smith and Montgomery, 1962). Long term weather data (1970 - 2005) at the Nigerian Institute for Oil Palm Research (NIFOR), Benin City indicated an annual rainfall of 2000mm. The rainy season occurs from March to October and maximum rain is received in the month of June, July and August. The minimum, maximum and air temperature is 22.5, 32 and 23.6 ^{0}C, respectively. The dominant plant species at the site were*Panicum maximum* and *Mimosa* spp.

Experimental Design

The field trial comprised two organic fertilizer types (cattle and poultry manures) as the main plots and four rates of application (0, 5, 10 and 15t ha^{-1}) as sub-plots laid out in a 2 x 4split plot arrangement fitted into a randomized complete block design (RCBD) with three replications, giving rise to 24 experimental units (plots). Each plot has a dimension of 1.5 x 3 m with 0.5 m space between plots and one metre between blocks.

Soil Sample Collection, Organic Fertilizers and Laboratory Analysis

Soil sampling was done at two intervals: pre-planting and post-harvest. For the pre-planting, soil samples were collected from the experimental site at a depth of 0-15 cm using auger and bulked together to form composite soil sample for the pre-planting physical and chemical analysis. After harvest, soil samples were randomly collected from each plot and analyzed for its post-harvest chemical properties. The samples were air-dried and sieved with 2 mm sieve and used for soil chemical properties determination. Before application of organic fertilizers (poultry and cattle manures), they were sub-sampled and air-dried for five days under shade and analyzed for chemical properties.

Particle size distribution was determined by the hydrometer Method (Day, 1965), soil pH at 1:1 soil to water ratio was determined using digital glass electrode pH meter. Organic carbon was determined by wet dichromate acid oxidation Method (Page, 1982). Total nitrogen was determined by Micro-Kjedahl Method (Bremmer and Mulvancy, 1982). Available phosphorus was extracted using calcium chloride extraction Method (Houba *et al.*, 2000) Exchangeable bases (Ca, K, Mg and Na) were extracted using 1N ammonium acetate solution at pH 7.0. Calcium and magnesium contents were determined volumetrically by ethylenediaminetetra-acetic acid (EDTA) titration procedure (Black, 1965) while potassium and sodium were by flame photometer. Exchangeable acidity was determined by titration Method (Anderson and Ingram, 1993).

Organic fertilizers (poultry and cattle manures) were analyzed for pH, organic carbon, total nitrogen, phosphorus, potassium, calcium, magnesium and exchangeable acidity using similar procedures as for soil sample.

Cultural Practices

The site was cleared manually using hoe and cutlasses, debris packed and mapped out to blocks. Spade was used to make beds and organic manures incorporated into the designated plots at a rate of 0, 5, 10 and 15t ha^{-1} at four weeks prior to planting to allow for equilibration. Each individual plot was mulched with dried grass (*Panicum maximum*) after incorporation of organic fertilizer.

Seeds were sown in holes two cm deep at the rate of three seeds per hole on the 28th of March, 2015 and 2016 in the field at a spacing of 70 cm within row and 90 cm between rows giving rise to 12 plants per treatment. The plants were later thinned to one plant per stand at the two - leaf stage precisely 14 days after germination. The experiment plot was weeded manually when necessary. The experimental plot was watered every other day, except when it rained. No disease infestation was recorded during the course of the experiment. Minor incidence of cucumber beetles and Lepidoptheral larval attack were controlled by spraying with 60 ml of Upper Cott insecticide per 10 liters of water 14 days after germination.

Data Collection

Within the net plot, four plants were randomly selected and tagged for data collection. Data were collected on vine length, vine girth, number of branches and leaves and leaf area index (LAI). Vine length was measured in cm from the base of the vine to tip of the vine. Vine girth was taken at 5 cm above the base on each sample plant with the use of vernier caliper. Number of branches and leaves referred to the total count of branches and leaves per plant. The leaf area (LA) involved measurement of the length and width for randomly selected leaves of each of the four sampled plants. The means were calculated and used to estimate the LA using Flavio and Marcos (2003) formula and thus:

LA = (0.859 (LW) + 2.7) x number of leaves

Where: LA = leaf area, L = length, W = width.

From the LA, LAI was computed using Remison (1997) formula:

$$LAI = \frac{Leaf\ area}{Land\ area}$$

Days to 50 % flowering were the number of days from sowing to the time of the first six plants in each plot produced visible flowers.

At harvest, data were collected on number of fruits per plants, fruit girth, fruit length, fruit size, fruit weight per plant and fruit yield. Number of fruits per plant was the total count harvested fruits per plant of all pre-tagged plants and average computed. Fruit girth referred to the average width at the widest point in the middle portion of all harvested fruits of pre-tagged plants measured with vernier caliper. Fruit length referred to fruit measurement using tape rule from the bottom to the top of all harvested fruits of pre-tagged plants and average computed. Fruit weight per plant was estimated by the summation of weight of all harvested fruits of pre-tagged plants divided by four. Fruit size was obtained by dividing the fruit weight per plant by number of harvest fruits per plant. Fruit yield was computed based on the weight of harvested fruits per plot and converted into hectare and expressed in tonnes and thus:

$$Fruit\ yield = [\frac{Area\ of\ an\ hectare}{Area\ of\ plot} \times wt\ of\ HA] \times \frac{1}{1000} t\ ha^{-1}$$

Where HA = weight of harvested fruits (kg ha^{-1})

Data Analysis

Data collected were subjected to analysis of variance (ANOVA) after finding the mean of the data collected between the two cropping seasons using GENSTAT statistical package. Differences among treatment means were separated using the Least Significance Difference (LSD) test at 0.05 level of probability.

RESULTS

Pre-planting physical and chemical properties of the soil and composition of organic fertilizers

The result of the soil analysis is shown in Table 1. The soil textural class was Sandy loam, slightly acidic and low in total nitrogen, phosphorus, and organic carbon. The chemical properties of the organic manures used for the experiment is shown in Table 2. Chemical analysis showed that poultry manure was neutral (7.20), while cattle manure was alkaline (8.50). Both manures had the high organic carbon content. The results also show that all the manures used contained moderate content of total Nitrogen, Phosphorus, Potassium and Magnesium that can support cucumber production.

Table 1. Physical and chemical properties of soil of the experimental site prior to cropping with cucumber

Soil property	Value
Sand (g kg^{-1})	886.00
Silt (g kg^{-1})	50.40
Clay (g kg^{-1})	63.60
Textural Class	Sandy loam
pH	6.30
Organic carbon (g kg^{-1})	18.60
Total nitrogen (g kg^{-1})	0.86
Available phosphorus (mg kg^{-1})	8.15
Exchangeable cations (cmol kg^{-1})	
Calcium	1.27
Magnesium	0.44
Potassium	0.34
Sodium	0.28
Hydrogen	0.41
Aluminum	0.03

Growth of cucumber

Table 3, shows the effect of organic fertilizer application on vine length, vine girth, number of leaves, number of branches, leaf area index on days to 50 % flowering. There was no significant difference among the organic fertilizer types on the growth parameters except on vine girth where poultry manure had thicker vines. At 50 % flowering, all treated plants except days to 50 % flowering were significantly different from the control plants. Vine length, vine girth, number of leaves and leaf area index (LAI) increased as fertilizer rate increased (P<0.05). Unfertilized plots had the least values for all the parameters except days to 50 % flowering. Longest vine (120.75 cm) was observed at the application rate of 15t ha^{-1} but statistically comparable with 10 t ha^{-1} (117.58 cm) and 5t ha^{-1} (115.67 cm). This distribution pattern was repeated for number of branches and LAI. The thickest vines were observed on plots fertilized with 15t ha^{-1} (1.81 cm) and 10 t ha^{-1}(1.78 cm).This trend was repeated for number of leaves.

Table 2. Chemical composition of organic fertilizers

Parameter	Organic fertilizer	
	Poultry manure	Cattle manure
pH	7.20	8.50
Organic C (g kg^{-1})	34.40	28.73
Available P (mg kg^{-1})	9.70	6.45
Total N (g kg^{-1})	1.85	1.23
Exchangeable cations (cmol kg^{-1})		
Ca^{2+}	2.84	3.53
Mg^{2+}	0.76	0.92
K$^+$	0.72	0.42
Na$^+$	0.35	0.38
H$^+$	0.06	0.04
Al^{3+}	0.04	0.03

Table 3. Effects of different level of organic fertilizer on the growth of cucumber

Treatment	Vine length (cm)	Vine girth (cm)	No. of leaves	No. of branches	LAI	Days to 50 % flowering
Organic fertilizer type						
Cattle manure	115.92	1.61	14.42	4.07	1.28	32
Poultry manure	116.79	1.68	13.92	4.08	1.23	34
LSD$_{(0.05)}$	ns	0.054	ns	ns	ns	ns
Rate of application						
0	111.42	1.41	13.17	3.47	1.06	34
5	115.67	1.59	14.08	3.92	1.30	33
10	117.58	1.78	14.50	4.33	1.32	34
15	120.75	1.81	14.92	4.58	1.34	33
LSD$_{(0.05)}$	5.959	0.216	0.795	0.679	0.131	ns
F x R	ns	ns	ns	ns	ns	ns

Fruit yield and yield components

Table 4, shows the effect of organic fertilizer type and rate of application on number of fruits, fruit girth, fruit length, average fruit weight and fruit yield. Results showed that there was no significant difference on the fertilizer type on yield parameters. Rate of fertilizer application had significant on growth parameters except fruit size. Number of fruits per plant was highest in plots fertilized with 15t ha^{-1} (9.90) and this was statistically similar to 10 t ha^{-1} (8.10). Similar trend was observed in fruit weight and fruit yield. Fruit girth and length values observed with plots fertilized with 5, 10 and 15t ha^{-1} were statistically comparable (P>0.05) but significantly (P<0.05) than 0 t ha^{-1} (unfertilized plots). There was no significant interaction effect of fertilizer type and rate of application on the yield of cucumber.

Post-harvest soil chemical properties after cropping with cucumber

The post-harvest effect of organic fertilizer application on soil chemical properties is presented in Table 5. Organic fertilizer type had no significant effect on post-harvest soil chemical properties except on exchangeable Mg which was higher in poultry manure treated plots. The rate of application of organic fertilizer had significant effect only on organic carbon, total N, available P and exchangeable Mg .These chemical properties increased with increase in rate of fertilizer application. However, 5 – 15 t ha^{-1},were not significantly different for total N and available P. Plot fertilized with 15 t ha^{-1}organic fertilizer had the highest organic carbon (21.21 g kg^{-1}) and exchangeable Mg^{2+} (4.01). There was no significant interaction effect of organic fertilizer and rate of fertilizer application on any of the post-harvest soil chemical property.

Table 4. Effects of different level of organic fertilizer on the yield and yield components of cucumber

Treatment	No of fruits plant^{-1}	Fruit girth (cm)	Fruit length (cm)	Fruit size (g)	Fruit weight kg plant^{-1}	Fruit yield t ha^{-1}
Organic fertilizer type						
Cattle manure	7.62	16.49	16.48	181.10	1.40	3.13
Poultry manure	7.10	15.48	15.36	197.40	1.35	3.03
LSD$_{(0.05)}$	ns	ns	ns	ns	ns	ns
Rate of application						
0	4.93	15.25	14.75	183.40	0.86	1.93
5	6.50	15.94	16.79	185.80	1.20	2.69
10	8.10	16.47	16.53	205.20	1.65	3.66
15	9.90	16.29	15.62	182.40	1.80	3.99
LSD$_{(0.05)}$	2.606	0.628	1.295	ns	0.366	0.790
F x R	ns	ns	ns	ns	ns	ns

DISCUSSION

The general fertility status of the soil was low since it contained low amount of total N, available P and organic Carbon. This could be due to leaching, erosion and loss of organic matter (Duran and Smith, 1987).This observation corresponded with Agboola (1982), who reported that soils in Nigeria are generally low in inherent fertility. Consequently, the use of organic fertilizers is most effective means in improving soil fertility in order to boost crop productivity.

Cattle and poultry manure on analysis contained essential plant nutrients required for proper growth of

plants. This is in agreement with Moral et al (2005), who reported that organic fertilizers were as good as inorganic fertilizer for improving crop production. They contained high amount of organic carbon. This emphasized the manures can improve the soil physical, chemical and biological properties. Since organic carbon is a reservoir of plant nutrients, on mineralization, it would release more nutrients to the soil. These nutritious qualities of the manures may then be utilized for the growth and development of the crop. The tested crop benefited from organic fertilizer application as its growth was enhanced compared to unfertilized plants, through their superior growth. This is in agreement with Agba and Enya (2005) and Eifediyi and Remison (2009), who had reported increase in growth and yield components of cucumber in response to fertilizer application. Similarly, Ndaeyo et al (2005) also reported the crop response to fertilizer application. The response to fertilizer application is affected by nutrient reserve in the soil. According to Ndaeyo et al (2005), crops respond more to fertilizer application in soils with low nutrient reserve, than soils with high nutrient reserve. Superior performance of poultry manure in thickening vine girth can be attributed to high total N and available P in poultry manures compared to cattle manure. This observation is in agreement with Ewulo (2008) who reported that poultry manure contain high percentage of N and P for healthy growth of plants. This also suggests that more assimilates were produced in poultry manure treated plants which enhanced thicker vine production. Since Mg played an active in photosynthesis, higher Mg content available to poultry manure treated plants is an indication of higher production of assimilates resulting in secondary growth of the plant, hence thicker vines.

Increase in the number of leaves of fertilized plants compared to unfertilized plants indicated that increased nutrients in the soil increase number of leaves per plant. This was similar to findings by Law-Ogbomo and Remison (2009), who observed highest number of leaves in yam plants treated with the highest rate of fertilizer, while the least number was recorded in those treated with 0 t ha^{-1} (control). Higher number of leaves corresponding give rise to

higher leaf area index (LAI) due to adequate supply of nutrients from applied manures, led to better utilization of nutrients for the production of assimilates resulted in higher values for these characters. Increased LAI was more pronounced in plants treated with fertilizers. This could be due to higher leaf area and number of leaves observed with fertilized plants. High LAI associated with fertilized plants signify greater leaf product rates, leaf area expansion and leaf area duration (Law-Ogbomo and Remison, 2009).

After cropping with cucumber, fertility status of the soil was improved. The higher exchangeable Mg observed in poultry manure treated plots implies higher rate of mineralization of Mg. The higher organic content of the poultry manure could have probably accounted for the residual exchangeable Mg content of the soil The reduction in the pH level of the soil could be attributed to crop removal. The rate of fertilizer application had significant effect on organic Carbon, total N available P and exchangeable Mg^{2+} and increased up to300 kg Nha^{-1}. This agreed with Dinesh et al (2012) who observed significant increase in soil properties in the short term following application of organic and bio-fertilizers. Similarly, Mbagwu and Ekwealor (1990) reported that organic fertilizers apart from releasing nutrient elements to the soil, improves other soil chemical and physical properties, which enhances crop growth and development.

Table 5. Some post-harvest soil chemical properties after cropping with cucumber as influenced by different level of organic fertilizer application

Treatment	pH	Organic C (g kg^{-1})	Available P (mg kg^{-1})	Total N (g kg^{-1})	Exchangeable cations (cmol kg^{-1})					
					Ca^{2+}	Mg^{2+}	K^+	Na^+	H^+	Al^{3+}
Organic fertilizer type										
Cattle manure	5.44	18.49	8.45	0.81	0.63	0.24	0.29	0.20	0.44	0.06
Poultry manure	5.17	18.38	8.50	0.84	0.51	2.08	0.29	0.14	0.48	0.07
LSD$_{(0.05)}$	ns	ns	ns	ns	ns	1.356	ns	ns	ns	ns
Rate of application (t ha^{-1})										
0	5.22	16.32	7.90	0.81	0.51	0.19	0.28	0.16	0.50	0.07
5	5.23	16.84	8.44	0.82	0.52	0.23	0.29	0.16	0.49	0.06
10	5.37	19.37	8.64	0.83	0.59	0.22	0.29	0.18	0.44	0.06
15	5.40	21.21	8.91	0.84	0.68	4.01	0.29	0.18	0.42	0.06
LSD$_{(0.05)}$	ns	0.883	0.565	0.023	ns	1.901	ns	ns	ns	ns
F x R	ns	ns	ns	ns	ns	ns	ns	ns	ns	ns

CONCLUSION

Application of cattle and poultry manure enhanced the growth and productivity of cucumber. However, poultry manure treated plants had thicker vines. The highest yield was obtained from 15 t ha^{-1}, which was statistically comparable to 10 t ha^{-1}. Soil chemical properties were enhanced by organic fertilizer application. But poultry manure left more exchangeable Mg to the soil and also contained more organic carbon, total N and available P than cattle manure. Therefore, it is suggested that farmers in the ultisols location use 10 t ha^{-1} of poultry manure in place of inorganic fertilizer to boost production, based on convenience and cost.

REFERENCES

Agba, O.A. and Enya, V.E. 2005.Response of cucumber (*Cucumis sativus* L.) to Nitrogen in Obubra, Cross River State. Global journal of agricultural science, 4 (2):165 – 167.

Agboola, A. A. 1982. Organic manuring in tropical Africa. Symposium paper 1, International conference soil science. New Delhi. 8 – 16 July, 1982.

Anderson, F. and Ingram, I. 1993. Tropical soil biology and fertility. A handbook of methods. 2nd Edition CAB international p. 231.

Black, C. A. 1965. Methods of soil analysis. American Society of Agronomy. Monograph Madison, Wisconsin.

Bremmer, J. M. and Mulvancy, C. S. 1982. *Total nitrogen*. In: Page, A. L. (ed) Methods of soil analysis. Part 2. Chemical and microbial properties, 2nd Agronomy series p.9 Madison, W1, USA, ASA, SSSA.

Day, P. R. 1965. Particle fractionating and particle size analysis. IN: Methods of soil analysis. Part 1. Agronomy, 9: 549 – 552.

Dinesh, R., srinivasan, V., Hamza, S. and Marjusta, A. 2010. Short-term incorporation of organic manures and biofertilizers influences biochemical and microbial characteristics of soils under an annual crop [turmeric (*curcuma longa* L.)]. Biosource technology, 101: 4097 – 4702.

Duran, J. W. and Smith, M. S. 1987. Organic matter management and utilization of soil and fertilizer nutrients in soil fertility and organic matter as critical components of production systems. Soil science society of america. Special publication, (19): 101 – 121.

Efeidiyi, E. K., and Remison, S. U. 2009 .The Effects of inorganic fertilizer 101 – 121on the yield of two varieties of cucumber (*Cucumis sativus* L.), Report and opinion, 1 (5):74 - 80.

Ernestina P. 2001. Cucumber. http:www.earthnotestripod.com/cucumber. Html.

Ewulo, B. S., Ojeniyi, S. O. and Akanni, D. A. 2008.Effect of poultry manure on selected soil physical and chemical properties, growth, yield and nutrient status of tomato. African journal of agricultural research, 3(9): 612 – 616.

Grubben, G. J. H. and Denton, O. A. 2004. Plant resources of tropical Africa. 2. Vegetable PROTA foundation, Netherland. 668pp.

Law-Ogbomo, K. E. and Remison, S. U. 2009. Growth and Yield of Maize as Influenced by Sowing Date and Poultry Manure Application. Notulae BotanicaeHorti-AgrobotaniciCluj – Napoca, 37(1): 199 – 203.

Mahmood, N., Nirjhar, S., Mark, S. and Jeff, V. 2004. Evapotranspiration of two vegetation covers in a shallow water table environment. Soil science society of american journal, 69(2): 492 – 499.

Mbagwu, J.S.C. and Ekwealor, G. C. 1990. Agronomic potentials of brewer spent grains. Boil wastes, 34: 335 – 347.

Moral, R., Moreno-Caselles, J., Perez, M.N., Espinosa, P., Paredes,C. and Sosa, F. 2005. Influence of fresh and composted solid fractions of swine slurry on yield of cucumber (*Cucumis sativus* L.) Communications in soil science and plant analyses, 36(416): 517-524.

Ndaeyo, N.U., Ukpong, E.S. and John, N.M. 2005.Performance of okra as affected by organic and inorganic fertilizers on an Ultisols. Proceedings of the 38th Conference of the agriculture society of Nigeria. pp 206 - 209

Olawuyi, O. J., Babatunde, F. E., Akinbode, O. A., Odebode, A. S. and Olakoja, S. A. 2011. Influence of arbsucular mycorrhiza and NPK fertilizer on the productivity of cucumber (*Cucumis sativus* L). International journal of organic agricultural research and development, 3: 22 – 29.

Page, A. L. 1982. Methods of soil analysis. Agronomy No 9. Part 2 American Society of Agronomy, Madison, W1.

Smith, A. J. and Montgomery, R. E. 1962. Soil and land use in central western Nigeria. Ministry of Agriculture and Natural Resources, Western Nigeria. p.265.

9

Disinfection of Cowpea (*Vigna Unguiculata*) Seeds on Seedlings Development

Oscar José Smiderle[1], Manoel Luiz Silva Neto[1], Aline das Graças Souza[2*],
Jerri Edson Zilli[1] and Krisle Silva[1]

[1]*Empresa Brasileira de Pesquisa Agropecuária Roraima, Depto. de Pesquisa de
Sementes, Caixa Postal 133, 69301-970, Boa Vista, RR, Brasil.*
[2]*Universidade Federal de Pelotas, Instituto de Biologia, Depto de Botânica,
Campus Universitário s/n. Capão do Leão. CEP: 96010-900, Pelotas, RS. Brasil.
E-mail alineufla@hotmail.com
Corresponding author

SUMMARY

The fast expansion of annual crops in recent decades with little phytosanitary management has led to the dissemination of several pathogens in the Brazilian cowpea-producing regions. Although, seed treatment is a common practice, there are no registered products for the cowpea and farmers apply the fungicides recommended for soybean cultivation. This study aimed to evaluate the influence of sodium hypochlorite and fungicidal treatments on the germination and initial development of seedlings of five cowpea cultivars under greenhouse conditions. Two experiments were established, one with seeds of five cultivars stored during 120 d, and another with newly collected seeds, disinfected or not with sodium hypochlorite, and treated with five fungicides. Emergence in sand, germination speed, and dry mass of seedlings were evaluated. Germination in sand, germination speed and dry mass of seedlings obtained from seeds of cowpea cultivars were negatively influenced by sodium hypochlorite disinfection. Fungicides used negatively influenced cowpea seed vigor for BRS Cauamé and BRS Novaera, even in the absence of sodium hypochlorite. Seed germination of five cowpea cultivars was not affected by the fungicides fludioxonil, carbendazim, carbendazim + thiram and carboxin + thiram.
Key words: *Vigna unguiculata;* seedling emergence; seedling vigor.

INTRODUCTION

In Brazil, the cowpea contributes to around 30% of the planted area and 15% of the total production of beans (cowpea + common beans) (Smiderle *et al.*, 2016). Historically, the cowpea [*Vigna unguiculata* (L.) Walp.] production is concentrated in the Northeast (1.2 million ha) and North (55.8 thousand ha) regions of the country (Smiderle *et al.*, 2017a). However, the crop is gaining production areas in the Central-West region due to cultivars with traits favorable to mechanized cultivation (Oliveira *et al.*, 2015; Filgueiras *et al.*, 2009). Cowpea is a staple food for low income populations of the Brazilian Northeast region and are grown mainly under subsistence conditions (Torres *et al.*, 2015; Teófilo *et al.*, 2008; Amaral *et al.*, 2005). Its traits include a short cycle, low water requirement and hardiness, which in addition to the development in low fertility soils, makes cowpea an attractive crop (Almeida *et al.*, 2014; Xavier *et al.*, 2008).

Seed is one of the essential components for agricultural production and its genetic quality associated to the physical, sanitary and physiological conditions, directly influence the production potential of the crop (Oliveira *et al.*, 2016; Smiderle *et al.*, 2017b). Thus, to obtain high crop yields seedling emergence and the number of established seedlings per unit of area must be guaranteed (Rodrigues *et al.*, 2015). To obtain plant uniformity in the field, fungicides treatment of seed has been used intensively by farmers in large crops, such a soybean and corn; this provide immediate and long-run benefits (Ceccon *et al.*, 2013).

In the Amazonia region, studies have shown increased grain yield with the inoculation of bacteria of the genus *Bradyrhizobium* in cowpea seeds (Lacerda *et al.*, 2004; Zilli *et al.*, 2006, 2009). Positive effects of inoculation of the cowpea in different edaphoclimatic regions enables increased grain yield of the crop and provides an alternative to nitrogen fertilization (Franco *et al.*, 2002). Fungicides are chemical compounds acting as protectants, with healing and systemic action in plants (Juliatti, 2017). Chemical treatments are able to control seed pathogens present in soil and storage, and early leaf pathogens (Sallis *et al.*, 2001; Torres and Bringel, 2005; Menten and Moraes, 2010). In areas where planting is annual, seed treatment with fungicides is necessary to avoid the incidence of diseases present in the soil affecting the crop. The fast expansion of crops in recent decades, often with little phytosanitary management, has led to the dissemination of many pathogens in the cowpea-producing regions, as the seed is the main carrier of dissemination and introduction into new growing areas (Henning, 2004; Torres and Bringel, 2005). Although seed treatment is a common practice, there are no products registered for cowpea, and many farmers apply fungicides recommended for soybean (Concenço *et al.*, 2015). Nevertheless, the effect of fungicides on the germination and initial development of cowpea seedlings is unknown. In addition, initial tests for selection of nitrogen-fixing bacteria employ seed-disinfection methods, including sodium hypochlorite, to eliminate bacteria and other microorganisms that can influence the performance of new strains (Xavier *et al.*, 2006). Therefore, this study aimed to verify the influence of sodium hypochlorite and fungicide treatment on seed emergence and initial development of seedlings of five cowpea cultivars under greenhouse conditions.

MATERIAL AND METHODS

From December 2010 to January 2011, two experiments were conducted under controlled conditions (28 °C; 65 ± 5% of air relative humidity) in a greenhouse in the station Roraima, EMBRAPA, located at BR174, Km 08, municipality of Boa Vista–RR, Brazil. In the first experiment, seeds stored for 120 d were sown. In the second experiment, recently-collected seeds were utilized to verify the influence of the treatments on seeds with germination differences. Cowpea cultivars utilized in the experiments were: 'BRS Guariba', 'BRS Novaera', 'BRS Tumucumaque', 'BRS Cauamé' and 'BRS Xiquexique'. Seeds of the five cultivars were disinfected or not in a 2.5% solution of sodium hypochlorite (commercial NaClO) for 5 min, followed by five successive washings with distilled water and later drying at room temperature. Then, seed were treated with four different fungicide treatments, alone or in combination (fludioxonil, carbendazim, carbendazim + thiram, carboxin + thiram) plus the control. Fungicides were applied according to the dose recommended by the manufacturer for utilization in soybean seeds (Table 1) and dried at room temperature in the laboratory. Both experiments were conducted in a randomized block design with three replicates in a factorial scheme 5 x 5 x 2.

For each cultivar, recently collected seeds presented a mean of 95% in germination, and 85% for stored seeds. Agronomic traits such as cropping cycle in days, growth habit of the plants and 100-seed mass of the five cultivars utilized in this study are shown in Table 2.

Table 1. Concentration of active ingredients of fungicides, and dose utilized in the treatment of cowpea seeds.

Fungicide	Active ingredient	Commercial concentration	Dose of commercial product (mL kg^{-1} of seeds*)
Maxim	fludioxonil	25 g L^{-1}	2
Derosal 500sc	carbendazim	500 g L^{-1}	1
Derosal plus	carbendazim + thiram	150 +350 g L^{-1}	2
Vitavax+thiram	carboxin + thiram	200 + 200 g L^{-1}	30

*Recommended dose for soybean crop.

Table 2. Cropping cycle, growth habit, and 100-seed mass of the cultivars BRS Guariba, BRS Novaera, BRS Tumucumaque, BRS Cauamé and BRS Xiquexique, used in this research.

Cultivars	Cropping cycle (d)	Growth habit	100-seed mass (g)
BRS Guariba	65-70	Semi-erect	19.5
BRS Novaera	65-70	Semi-erect	20.0
BRS Tumucumaque	65-70	Semi-erect	20.5
BRS Cauamé	65-70	Semi-erect	17.2
BRS Xiquexique	65-75	Semi-prostrate	16.5

Each fungicidal treatment consisted of 10 g of seeds of each cowpea cultivar, arranged in 50 mL plastic cups. Following the fungicide treatment, seeds were sown in plastic pots containing 2 kg of medium-textured sand as substrate, with 10 seeds of each cultivar per pot. Three replicates were performed for each cultivar and each of the five treatments. Humidity in the pots was maintained close to 70%, by adding distilled water according to weight. After sowing, the following trials were evaluated: a) Sand Emergence: counts of normal emerged seedlings were performed from the start of emergence until 10 days, with results expressed in percentage (Brasil, 2009). b) Emergence Speed: daily counts of the number of emerged seedlings until the 10th day after planting (dap). At the end of the test, the index was calculated through the sum of the number of emerged seedlings on each day, divided by the number of days elapsed between sowing and emergence (Maguire, 1962). Dry Matter Mass: all emerged seedlings were collected after the last count on day seven. Seedlings were washed in running water to eliminate residues of the substrate and placed into paper bags for drying in an air circulation oven at 60 °C until they reached constant weight. Next, mass was determined on samples on a 0.001 g precision balance. Data on each experiment was submitted to analysis of variance and means were compared by the Tukey test (P = 0.05) using the program Sisvar (Ferreira, 2011).

RESULTS

There were no significant differences in the five evaluated cultivars nor between the different fungicide treatments in the absence of sodium hypochlorite (Table 3). Values of seed germination obtained were in accordance of accepted values for commercialization of large crops, such as cowpea (Brasil, 2009), which should be over 80 to 85%.

Mean germination speed did not differ between cultivars treated with all five fungicides and the control (Table 4) in the absence of sodium hypochlorite. Fungicide treatments did not influence dry mass for most of the cultivars, regardless of treatment with sodium hypochlorite (Table 5).

For the second experiment (Tables 6, 7 and 8), using recently collected seeds, values of emergence percentage and germination speed index obtained from the evaluated cultivars did not obtain significant difference when treated with the fungicides applied, with the exception of the cultivar BRS Cauamé, that obtained significant difference for the fungicide carbendazim + thiram, when the seeds were disinfected with sodium hypochlorite.

DISCUSSION

The application of sodium hypochlorite followed by fungicide treatments led to reduced emergence, compared to untreated seeds, with differences between the cultivars but not the fungicides treatment (Table 3) which was expected considering the genetic differences among cultivars (Xavier et al., 2005). For the cultivar BRS Novaera, germination failed (7%) even in the control seeds receiving no fungicides, with no germination on the other treatments. Sodium hypochlorite (along with the fungicide) may have caused the death of the seeds. In common beans, Jauer et al. (2002) observed germination over 80%. According to Oliveira (1997), carioca bean seeds contaminated with Rizoctonia solani, treated with carboxin + thiram, presented good indices of disease control and germination of normal seedlings, similar to that found in the absolute control, showing that the non-germination of the seeds was not due to contamination by pathogens.

Table 3. Mean* values of sand emergence obtained from seeds stored for 120 days of five cowpea cultivars treated with fungicides, plus the control, either with or without sodium hypochlorite.

Fungicides	BRS Guariba		BRS Tumucumaque		BRS Cauamé		BRS Novaera		BRS Xiquexique	
Sand emergence	%		%		%		%		%	
with NaClO										
carbendazim + thiram	40	aA	20	aAB	27	aAB	0	aB	33	aA
carbendazim	33	aA	20	aAB	33	aA	0	aB	27	aAB
carboxin + thiram	40	aA	33	aA	40	aA	0	aB	20	aAB
fludioxonil	40	aA	40	aA	40	aA	0	aB	27	aAB
Control	46	aAB	27	aABC	53	aA	7	aC	20	aBC
Mean	40	A	28	BC	39	AB	2	D	26	C
without NaClO										
carbendazim + thiram	100	a	93	A	87	a	87	a	87	a
carbendazim	93	a	87	A	87	a	93	a	87	a
carboxin + thiram	100	a	93	A	80	a	93	a	87	a
fludioxonil	100	a	87	A	87	a	93	a	87	a
Control	93	a	87	A	100	a	93	a	93	a
Mean	97	A	89	A	88	A	92	A	88	A

*Means followed by equal small letters in the column, and capital letters in the line, do not differ from one another by Tukey test (P = 0.05).

Table 4. Mean* values of germination speed (index) obtained from seeds stored for 120 days of the five cowpea cultivars treated with fungicides, plus the control, either with or without sodium hypochlorite.

Fungicides	BRS Guariba		BRS Tumucumaque		BRS Cauamé		BRS Novaera		BRS Xiquexique	
with NaClO										
carbendazim + thiram	12.7	abA	5.5	aB	6.0	aAB	0.0	aB	8.3	aA
carbendazim	5.4	bAB	6.1	aB	9.2	aA	0.0	aB	6.0	aAB
carboxin + thiram	10.1	abA	9.2	a	10.0	aA	0.0	aB	5.0	aAB
fludioxonil	10.8	abA	10.4	a	9.7	aA	0.0	aB	7.3	aAB
Control	14.7	aA	5.8	aB	14.0	aA	1.3	aB	7.3	aAB
Mean	10.7	A	7.4	AB	9.8	AB	0.27	C	6.77	B
without NaClO										
carbendazim + thiram	32.8	aA	24.2	aB	20.0	aB	21.2	aB	24.0	aB
carbendazim	30.0	aA	25.6	aAB	21.7	aB	25.8	aAB	21.2	aB
carboxin + thiram	31.7	aA	27.8	aAB	22.3	aB	20.4	aB	24.1	aAB
fludioxonil	30.9	aA	24.3	aAB	23.1	aAB	28.3	aAB	21.5	aB
Control	25.3	aA	26.4	aA	25.9	aA	28.5	aA	24.6	aA
Mean	30.1	A	25.7	B	22.6	B	24.9	B	23.1	B

*Means followed by equal small letters in the column, and capital letters in the line, do not differ from one another by Tukey test (P = 0.05).

Table 5. Mean[*] seedlings dry matter mass (g) of seeds stored for 120 d of five cowpea cultivars treated with fungicides, plus the control, either with or without sodium hypochlorite.

Fungicides	BRS Guariba		BRS Tumucumaque		BRS Cauamé		BRS Novaera		BRS Xiquexique	
Seedling dry matter mass	g		g		g		g		g	
with NaClO										
carbendazim + thiram	11.7	aA	4.3	aAB	6.2	aAB	0.0	aB	8.3	aAB
carbendazim	7.7	aAB	6.6	aAB	10.1	aA	0.0	aB	7.7	aAB
carboxin + thiram	8.9	aAB	10.4	aA	7.7	aAB	0.0	aB	5.2	aAB
fludioxonil	11.5	aA	11.7	aA	10.7	aA	0.0	aB	9.7	aA
Control	14.1	aA	7.1	aABC	12.1	aAB	2.7	aB	4.5	aAB
Mean	10.8	A	8.0	A	9.4	A	0.5	B	7.1	A
without NaClO										
carbendazim + thiram	36.8	aA	24.3	aB	21.5	aB	22.2	aB	19.8	aB
carbendazim	25.1	bAB	28.7	aA	20.5	aAB	26.3	aAB	19.7	aB
carboxin + thiram	31.7	aA	30.1	aAB	22.7	aBC	23.6	aABC	20.7	aC
fludioxonil	31.5	aA	29.7	aAB	21.9	aBC	30.7	aAB	20.1	aC
Control	24.1	bAB	30.4	aA	22.0	aAB	25.0	aAB	20.1	aB
Mean	29.8	A	28.6	AB	21.7	BC	25.6	BC	20.1	C

[*]Means followed by equal small letters in the column, and capital letters in the line, do not differ from one another by Tukey test (P = 0.05).

Table 6. Mean[*] values of germination in sand (%), using recently collected seeds of five cowpea cultivars treated with fungicides, plus the control, with and without sodium hypochlorite.

Fungicides	BRS Guariba		BRS Tumucumaque		BRS Cauamé		BRS Novaera		BRS Xiquexique	
Sand emergence	%		%		%		%		%	
with NaClO										
carbendazim + thiram	37	aA	47	aA	27	bAB	0	aB	10	aAB
carbendazim	27	aBC	40	aAB	50	aA	0	aD	13	aCD
carboxin + thiram	33	aAB	37	aA	40	abA	0	aC	13	aBC
fludioxonil	37	aA	47	aA	50	aA	0	aB	3	aB
Control	33	aA	37	aA	47	abA	3	aB	3	aB
Mean	33	A	41	A	43	A	1	B	9	B
without NaClO										
carbendazim + thiram	100	a	90	a	90	a	100	a	97	a
carbendazim	97	a	87	a	83	a	90	a	87	a
carboxin + thiram	100	a	100	a	93	a	93	a	87	a
fludioxonil	97	a	90	a	97	a	100	a	87	a
Controle	93	a	93	a	97	a	83	a	90	a
Mean	97	A	92	A	92	A	93	A	92	A

[*]Means followed by equal small letters in the column, and capital letters in the line, do not differ from one another by Tukey test (P = 0.05).

Table 7. Mean[*] germination speed (index) obtained for recently collected seeds of five cowpea cultivars treated with fungicides, plus the control with and without sodium hypochlorite.

Fungicides	BRS Guariba		BRS Tumucumaque		BRS Cauamé		BRS Novaera		BRS Xiquexique	
with NaClO										
carbendazim + thiram	9.2	aA	11.0	aA	10.4	aA	0.0	aB	4.3	aAB
carbendazim	7.2	aBC	12.7	aAB	16.2	aA	0.0	aC	3.7	aC
carboxin + thiram	10.4	aAB	14.4	aA	13.4	aA	0.0	aC	4.3	aBC
fludioxonil	9.5	aA	16.1	aA	15.2	aA	0.0	aB	0.8	aB
Control	12.0	aA	15.6	aA	12.9	aA	2.3	aB	1.1	aB
Mean	9.7	B	14.0	A	13.6	A	0.47	C	2.9	C
without NaClO										
carbendazim + thiram	40.1	aA	34.7	aAB	28.6	aB	40.3	aA	33.2	aAB
carbendazim	40.2	aA	32.8	aAB	29.5	aB	34.4	aAB	28.3	aB
carboxin + thiram	38.0	aA	36.4	aA	30.7	aAB	32.9	aAB	27.2	aB
fludioxonil	42.8	aA	28.7	aC	32.4	aBC	37.5	aAB	31.3	aBC
Control	39.7	aA	35.7	aA	33.6	aA	34.7	aA	31.7	aA
Mean	40.1	A	33.7	BC	31.0	C	35.9	B	30.3	C

[*]Means followed by equal small letters in the column, and capital letters in the line, do not differ from one another by Tukey test (P = 0.05).

Table 8. Mean[*] dry matter mass of seedlings (g) obtained for recently collected seeds of the five cowpea cultivars treated with fungicides, plus the control, with or without sodium hypochlorite.

Fungicides	BRS Guariba		BRS Tumucumaque		BRS Cauamé		BRS Novaera		BRS Xiquexique	
Seedling dry matter mass	g		g		g		g		g	
with NaClO										
carbendazim + thiram	9.0	aAB	13.1	aA	7.9	aAB	0.0	aC	4.0	aBC
carbendazim	7.6	aAB	12.2	aA	12.9	aA	0.0	aC	3.6	aBC
carboxin + thiram	10.9	aAB	12.2	aA	12.3	aA	0.0	aC	4.0	aBC
fludioxonil	8.9	aA	13.2	aA	14.2	aA	0.0	aB	0.7	aB
Control	7.3	aAB	10.3	aA	7.9	aA	0.9	aB	0.9	aB
Mean	8.7	B	12.2	A	11.8	AB	0.17	C	2.6	C
without NaClO										
carbendazim + thiram	29.1	aB	26.9	aBC	24.0	cBC	38.0	aA	20.5	aC
carbendazim	29.7	aA	31.1	aA	29.9	bcCA	32.8	abA	19.9	aB
carboxin + thiram	27.3	aAB	26.5	aAB	26.5	cAB	30.6	bA	20.3	aB
fludioxonil	33.5	aAB	30.4	aB	38.2	aA	32.1	abAB	20.8	aC
Control	31.5	aAB	28.3	aB	35.7	abA	31.5	abAB	19.9	aC
Mean	30.2	AB	28.6	B	30.9	AB	33.0	A	20.3	B

[*]Means followed by equal small letters in the column, and capital letters in the line, do not differ from one another by Tukey test (P = 0.05).

On the other hand, the application of sodium hypochlorite reduced the indices of the five cultivars and in a distinguishing way, fungicides did not influence this process, except on BRS Guariba, in which there was a reduction in the indices obtained with the application of carbendazim (Table 4). BRS Novaera was found to be the most susceptible cowpea variety to sodium hypochlorite, with a mean index reduction of 24.9 to 0.27.

Germination speed (Table 4) of the cultivar BRS Guariba was higher regardless of the treatment in the absence of sodium hypochlorite, which is desirable, since the seedlings remain vulnerable to the adverse conditions of the medium for a shorter time during the early developmental stages (Santos and Correa, 2011). In the cultivar BRS Guariba, an effect of the fungicide carbendazim was found in the absence of sodium hypochlorite, but values were similar to those of the control. The dry mass of seedlings receiving sodium hypochlorite was lower for the five cultivars, when compared to those not receiving sodium hypochlorite; BRS Novaera presented a 50-fold reduction on mass, from 25.6 g to 0.5 g. Dry mass of seedlings was determined, considering that samples with the highest values were the most vigorous, since the most vigorous seeds produce seedlings with increased growth. Due to the greater translocation of storage tissue reserves for the growth of the embryonic axis. Souza et al. (2016), indicated that reserves of nutrients stored by plants in seeds are supplied to the embryo for development and establishment of the new plant. Disinfection with sodium hypochlorite negatively affected the three variables evaluated. These results are important, considering the need to understand the effects of sodium hypochlorite application on cowpea seeds stored for 120 d.

The application of sodium hypochlorite on the seeds of the five cowpea cultivars resulted in a reduction of the mean emergence values, from 92% or greater, obtained without application, to 43% or less, regardless of the fungicidal treatment (Table 6). These values remained below 51%, and in some cases there was no germination at all, for some fungicide treatments in the seeds of BRS Novaera. On the other hand, in the absence of sodium hypochlorite, no significant differences were found between cultivars and fungicides, with means above 82%. These values are higher than the minimum necessary for seed commercialization in Brazil.

Percentage of germination means of the recently collected seeds were above that accepted for commercialization in Brazil for the five cultivars, with BRS Guariba standing out with a mean of 97% emergence. Germination means obtained in this experiment were above the 80% obtained by Teixeira et al. (2010) and 90% obtained by Santos and Correa (2011), both of which used rolls of germitest towel paper, moistened with water 2.5 times the weight of the dry substrate and placed to germinate at 25 °C. In addition, the cultivar BRS Guariba (Table 7) also stood out with a mean germination speed index of 40.1, higher than that obtained by Santos and Correa (2011), who obtained germination speed indices below 11.09 for the evaluated cultivars. It was also higher than that obtained for the same cultivar when using seeds stored for 120 d. The good performance was probably due to the use of recently collected seeds and the greater vigor of this cultivar.

Results obtained in relation to sodium hypochlorite disinfection were quite pronounced for the three variables evaluated in the two experiments. Mean germination reduction was above 60% in the two experiments and germination speed indices were reduced from more than 25 to less than 10 in both experiments.

Overall, results for germination in the two experiments (Tables 3 and 6) were higher than those obtained by Santos and Correa (2011), who investigated seeds of the cowpea cultivars BRS Tumucumaque, BRS Cauamé, BRS Itaim and BRS Guariba and lines MNC03-737F-5-1, MNC03-737F-5-4, MNC03-737F-5-9 and MNC03-737F-5-1, sown on germitest type paper towels (moistened with water equivalent to 2.5 times the weight of the dry substrate and placed to germinate at 25 °C, regarded as adequate conditions to verify the germination progress of seeds). Germination conditions in this study were not restrictive at all. According to Silva et al. (2013), efficient methodologies are important to optimize laboratory procedures and enable the analysis of further samples with reduced cost. Similarly, germination speed was lower on seeds with application of sodium hypochlorite, but without differences between fungicides groups; this also occurred in seeds without application of sodium hypochlorite. However, differences were found between cultivars, both in the presence and absence of sodium hypochlorite (Table 7). Heights showed were pronounced differences. In the presence of sodium hypochloride indices were below 16.3 and those without sodium hypochlorite reached 42.8, and never lower than 27.2.

For seedling dry matter, BRS Cauamé presented a lower mean than the control when the seeds were treated with fungicides carbendazim + thiram, carbendazim and carboxin + thiram (Table 8), showing a significant reduction in seedling development. For the cultivar BRS Novaera, seed treatment with carboxin + thiram negatively influenced seedling development, presenting the lowest indices (Table 8).

Among cultivars, significant differences were found in seedling dry mass. In the presence of sodium hypochlorite, poorer values were obtained; for cultivars BRS Tumucumaque and BRS Cauamé, the best values were obtained as a function of fungicide treatment. In seeds without sodium hypochlorite, cultivars BRS Cauamé, BRS Novaera and BRS Guariba were superior in some treatments (Table 8).

As for seedling dry mass in the second experiment, means followed the trends for nutrient storage in the grain, with BRS Cauamé standing out with 100-seed mass lower than those of cultivars BRS Guariba, BRS Tumucumaque and BRS Novaera, but producing a mean dry mass similar to these cultivars (Table 8).

Regardless of whether seeds were recently collected or had been stored for 120 d, the application of sodium hypochlorite to disinfect seeds of the five evaluated cultivars led to decreased germination and vigor of the seeds; also, cultivars BRS Cauamé and BRS Novaera presented the greatest reductions.

CONCLUSIONS

Sand germination, germination speed and dry mass of seedlings obtained from seeds of the five cowpea cultivars were negatively influenced by disinfection with sodium hypochlorite. Fungicides used negatively influenced the vigor of cowpeas seeds BRS Cauamé and BRS Novaera, even in the absence of sodium hypochlorite. Seed germination of the five cowpea cultivars is not influenced by the fungicides fludioxonil, carbendazim, carbendazim + thiram, carboxin + thiram.

REFERENCES

Almeida, W. S., Fernandes, F. R. B., Teófilo, E. M., and Bertini, C. H. C. M. 2014. Correlation and path analysis in components of grain yield of cowpea genotypes. Revista Ciência Agronômica, 45 (3): 726-736.

Amaral, J. A. B., Beltrão, N. E. M., and Silva, M. T. 2005. Zoneamento Agrícola do feijão-caupi no Nordeste Brasileiro Safra 2005/2006 - Estado da Paraíba. Embrapa Algodão. Embrapa Algodão. Comunicado Técnico, 253. Campina Grande, Brazil. 9 p.

Brasil. 2009. Ministério da Agricultura, Pecuária e Abastecimento. Regras para análises de sementes. Ministério da Agricultura, Pecuária e Abastecimento. Secretaria de Defesa Agropecuária. Brasília, DF: Mapa/ACS, 399 p.

Ceccon, G., Staut, L. A., Sagrilo, E., Machado, L. A. Z., Nunes, D. P., and Alves, V. B. 2013. Legumes and forage species sole or intercropped with corn in soybean-corn succession in Midwestern Brazil. Revista Brasileira de Ciência do Solo. 37 (1): 204-212.

Concenço, G., Ceccon, G., Marques, R. F., Marschall, R., Alves, M. E. S., Palharini, W. G., and Galon, L. 2015. Cultivos de outono-inverno na supressão de plantas daninhas em soja. Revista Brasileira de Ciência Agrárias. 10(2): 205-210.

Ferreira, D. F. 2011. Sisvar: a computer statistical analysis system. Ciência e Agrotecnologia. 35 (6): 1039-1042.

Filgueiras, G. C., Santos, M. A. S. dos., Homma, A. K. O., Rebello, F. K., and Cravo, M. da S. Aspectos socioeconômicos. In Zilli, J. E., Vilarinho, A. A., and Alves, J. M. A. 2009. A cultura do feijão-caupi na Amazônia brasileira. Boa Vista, RR: Embrapa Roraima, Brazil. p. 185-221.

Franco, M. C., Cassini, S. T. A., Oliveira, V. R., Vieira, C., and Tsai, S. M. 2002. Nodulação em cultivares de feijão dos conjuntos gênicos andino e meso-americano. Pesquisa Agropecuária Brasileira. 37: 1145-1150.

Henning, A. A. 2004. Patologia e tratamento de sementes: Noções gerais. Embrapa Soja. Documentos, 235. Londrina, Brazil. 51 p.

Jauer, A., Menezes, N. L., and Garcia, D. C. 2002. Tamanho de sementes na qualidade fisiológica de cultivares de feijoeiro comum. Revista da Faculdade de Zootecnia, Veterinária e Agronomia. 9(1): 121-127.

Juliatti, F. C. Modo de ação dos fungicidas sobre plantas e fungos. http://ppi-ppic-ipi.org/ppiweb/pbrazil.nsf. (Consulted 05 Mar. 2017).

Lacerda, A. M., Moreira, F. M. S., Andrade, M. J. B., and Soares, A. L. L. 2004. Yield and nodulation of cowpea inoculated with selected strains. Revista Ceres. 51(293): 67-82. doi: 10.1590/S1806-66902011000100001.

Maguire, J. D. 1962. Speed of germination-aid selection and evaluation for seedling emergence and vigor. Crop Science. 2(2): 176-177. http://dx.doi.org/10.2135/cropsci1962.001118 3X000200020033x.

Menten, J. O. M., and Moraes, M. H. D. 2010. Tratamento de sementes: histórico, tipos, características e benefícios. Informativo Abrates. 20(3): 52-53 http://www.abrates.org.br/images/stories/infor mativos/v20n3/minicurso03.pdf.

Oliveira, I. J., Fontes, J. R. A., and Rocha, M. M. (2015). Seleção de genótipos de feijão-caupi para adaptabilidade e estabilidade produtiva no Estado do Amazonas. Revista de Ciências Agrárias /Amazonian Journal of Agricultural and Environmental Sciences. 58(3): 292-300.

Oliveira, J. A., Andrade, M. J. B., and Fraga, A. C. 1997. Eficiência de fungicidas no tratamento de sementes de feijão (Phaseolus vulgaris L.) para o controle da podridão radicular causada por Rhizoctonia solani kuhn. Revista Brasileira de Sementes. 19(1): 91-95.

Oliveira, D. L., Smiderle, O. J., Paulino, P. P. S., and Souza, A. G. 2016. Water absorption and method improvement concerning electrical conductivity testing of Acacia mangium (Fabaceae) seeds. Revista Biologia Tropical. 64(5): 1651-1660. http://dx.doi.org/10.15517/rbt.v64i4.21944.

Rodrigues, A. P. M. S., Júnior, A. F. M., Torres, S. B., Nogueira, N. W., and Freitas, R. M. O. 2015. Tetrazolium test for evaluation of the physiological quality of seeds of *Vigna unguiculata* (L.) Walp. Revista Ciência Agronômica. 46(3): 638-644.

Sallis, M. G. V., Lucca-Filho, O. A., and Maia, M. S. 2001. Fungos associados às sementes de feijão-miúdo (V*igna unguiculata* (L.) Walp.) produzidas no município de São José do Norte, RS. Revista Brasileira de Sementes. 23(1): 36-39.

Santos, A., and Correa, A. M. 2011. Avaliação da qualidade fisiológica de sementes de feijão-caupi de porte ereto e semi-ereto produzidas em Aquidauana-MS. Anais do ENIC. 9 (1): 115-118.

Silva, R. C., Grzybowski, C. R. S., França - Neto, J. B., and Panobianco, M. 2013. Adaptação do teste de tetrazólio para avaliação da viabilidade e do vigor de sementes de girassol. Pesquisa Agropecuária Brasileira. 48 (1): 105-113.

Smiderle, O. J., Souza, A. G., Alves, J. M. A., and Barbosa, C. Z. R. 2017a. Physiological quality of cowpea seeds for different periods of storage. Revista Ciência Agronômica. 48 (5): 817-823.

Smiderle, O. J., Lima-Primo, H. E., Barbosa, H. D., and Souza, A.G. 2017b. Effect of defoliation on production components at different growth stages of cowpeas. Revista Ciência Agronômica. 48 (5): 840-847.

Smiderle, O. J., Souza, A. G., Campos, L. S., and Souza, A. A. 2016. Qualidade fisiológica de sementes de feijão-caupi obtidas em residual alternativos de adubações. Revista Congrega. 2(1): 217-224.

Souza, A. G., Smiderle, O. J., Spinelli, V. M., Souza, R. O., and Bianchi, V. J. 2016. Correlation of biometrical characteristics of fruit and seed with twinning and vigor of *Prunus persica* rootstocks. Journal of Seed Science. 38 (4): 322-328. http://dx.doi.org/10.1590/2317-1545v38n4164650.

Teixeira, I. R., Silva, G. C., and Oliveira, J. P. R.

2010. Desempenho agronômico e qualidade de sementes de cultivares de feijão-caupi na região do cerrado. Revista Ciência Agronômica. 41(2): 300-307.

Teófilo, E. M., Dutra, A. S., Pitombeira, J. B., Dias, F. T. C., and Barbosa, F. S. 2008. Potencial fisiológico de sementes de feijão-caupi produzidas em duas regiões do estado do Ceará. Revista Ciência Agronômica. 39(3): 443-448.

Torres, F. E., Sagrilo, E., Teodoro, P. E., Ribeiro, L. P., and Filho, A. C. 2015. Número de repetições para avaliação de caracteres em genótipos de feijão-caupi. Bragantia. 74(2): 161-168.

Torres, S. B. and Bringel, J. M. M. 2005. Avaliação da qualidade sanitária e fisiológica de sementes de feijão macassar. Caatinga. 18(2): 88-92.

Xavier, G. R., Martins, L. M. V., Rumjanek, N. G., and Freire filho, F. R. 2005. Variabilidade genética em acessos de caupi baseada em marcadores RAPD. Pesquisa Agropecuária Brasileira. 40(2): 353- 359.

Xavier, G. R., Martins, L. M. V., Ribeiro, J. R. A., and Rumjanek, N. G. 2006. Especificidade simbiótica entre rizóbios e acessos de feijão-caupi de diferentes nacionalidades. Caatinga. 19(2): 25-33.

Xavier, T. F., Araújo, A. S. F., Santos, V. B., and Campos, F. L. 2008. Inoculação e adubação nitrogenada sobre a nodulação e a produtividade de grãos de feijão-caupi. Ciência Rural. 38(7): 2037-2041.

Zilli, J. E., Marson, L. C., Marson, B. F., Rumjanek, N. G., and Xavier, G. R. 2009. Contribuição de estirpes de rizóbio para o desenvolvimento e produtividade de grãos de feijão-caupi em Roraima. Acta Amazônica. 39(4): 749-758.

Zilli, J. E., Valicheski, R. R., Rumjanek, N. G., Simões-Araújo, J. L., Freire Filho, F. R., and Neves, M. C. P. 2006. Eficiência de Simbiótica de Estirpes de *Bradyrhizobium* isoladas de solo do Cerrado em Caupi. Pesquisa Agropecuária Brasileira. 41(5): 201-210.

10

Nutritional Composition and Bioactive Components of Mashua (*Tropaeolum Tuberosum* Ruiz and Pavón)

D. Guevara-Freire*, L. Valle-Velástegui, M. Barros-Rodríguez, Carlos Vásquez, H. Zurita-Vásquez, J. Dobronski-Arcos and P. Pomboza-Tamaquiza

Facultad de Ciencias Agropecuarias, Universidad Técnica de Ambato, Carretera Cevallos-Quero, 180350 Cevallos, Tungurahua, Ecuador.
Email: da.guevara@uta.edu.ec
*Corresponding author

SUMMARY

The objective of this revision is to collect information about the chemical and nutritional composition and the bioactive components of mashua; an economically important tuber in the Andean region. This tuber has a high content of bioactive compounds (total phenols, Flavin 3-ols, anthocyanins, proanthocyanins, carotenoids, triterpenes, steroids, flavones and leucoanthocyanidins) which confers it with therapeutic and medicinal properties that have allowed it to be used since pre-Hispanic times, which has elicited a peak in scientific interest in recent years. Recent studies have reported the efficient use of mashua in treating benign prostatic hyperplasia, cancer, as well as chelating properties in metallic ions to remove peroxyl radicals. However, other studies find evidence of negative effects caused by the tuber due to the presence of thiocyanates that when releasing hydrocyanic block oxygen transportation in red blood cells, whilst tannins diminish voluntary feeding of poly-gastric animals, reducing digestibility and productivity. On the other hand, the bioactive components and nutritional composition of the tuber are high in relation to various Andean tubers and certain fruit that have demonstrated an important scientific contribution in medicine and industry

Keywords: antioxidant capacity, *Tropaeolum tuberosum*, phenols, glucosinolates, anthocyanins, flavonoids.

INTRODUCTION

In the inter-Andean region the domestication of roots and tubers such as *Tropaeolum tuberosum* began in the Andes, and though the exact date is unknown it is estimated to be around 5500 AC (Flannery, 1973 as cited by King and Gershoff, 1987). The tuber originated in Peru and Bolivia (León, 1964), but its cultivation has extended to other countries of the plateau, such as Ecuador, Venezuela, Colombia, (Chirinos et al., 2007), the north of Argentina, and for many years regions of New Zealand, Canada (Manrique et al., 2013), United States and England (National Research Council, 1989). In Andean countries the tuber is sown at an altitude range of between 2400 and 4300 masl and at cool temperatures that vary between 8 and 11°C (Grau et al., 2003). It is a crop that adapts to poor soils that don't need fertilization, and being a rustic plant it is resistant to plagues, nematodes and insects (Manrique et al., 2013). In Ecuador and Cuzco mashua can yield over 70000 kg/ha in experimental plots (Arbizu and Tapia, 1992; Hermann, 1992 citado por Barrera et al., 2004) in Bolivia fields have yielded between 30 and 60 t/ha and in Peru values range between 4 and 7 t/ha (Cadima et al., 2003b citado por Cadima, 2006).

Of all Andean crops mashua has the least importance in Peru, Ecuador and Bolivia (Cadima, 2006). In Ecuador production of the tuber is decreasing, as well as its genetic variability as is the case of yellow and black mashua; the reasons for this being the low culinary status of the tuber (rusticity and poverty), nescience of the beneficial properties and methods of preparation, as well as the bitter and slightly spicy flavor (Barrera et al., 2004). On the other hand in Colombian markets mashua is one of the most commercial tubers sold in large quantities, while in Peru and Bolivia it is in low demand (Surco, 2004). Mashua contains a significant value of bioactive compounds which has evidenced its role in the prevention of various diseases, these beneficial effects have been linked to the existence of certain metabolites like glucosinolates, polyphenols, anthocyanins, among others. The glucosinolates in the plant act as a natural defense against plagues and diseases; different studies have reported more than a hundred types of glucosinolates some of which have are beneficial to the health as anticarcinogens, but there are others that could act as antinutrients (Roca et al., 2005). These metabolites are popular in traditional medicine due to their antibiotic, nematicide and diuretic properties (Johns et al., 1982), some authors have linked their effect to kidney problems, liver pains, urinary disorders, anaphrodisiac properties(Hodge, 1951 as cited by Pissard et al., 2008) and anticarcinogen effects in cancerous cells of the colon and prostate (Norato et al., 2004 as cited by Roca et al., 2005).

For this reason, the objective of this study is to analyze the scientific evidence available through literature regarding the role of the bioactive compounds of mashua in human health and to supply a contribution that enriches the current knowledge about the nutritional composition of the tuber.

Characterization and morphology of mashua

Mashua is a rustic tuber with a cultivation cycle of 6 to 9 months, it is an herbaceous plant with a glabrous that is 20-30 cm tall, the tuber is a schizocarp separated into three mericarps with a rugged surface; some Ecuadorian morphotypes have 2 to 5 mericarps when ripe. The form can be conic, lengthened conic and lengthened (Cardenas, 1989; Grau et al., 2003; Barrera et al., 2004; Cadima, 2006 as cited in Cadima 2006), with erect growth, that later varies to semi prostrate when ripe (Tapia and Fries, 2007), with leaves that are solitarily distributed along the branch separated by internodes that are 1-8 cm with a 2-30 cm long petiole (Cardenas, 1989). In mashua collections from Ecuador, Peru, and Bolivia pigmented nerves have been found on the underside of the leaves (Cadima, 2006). The flowers are solitary, zygomorphs that grow from the axils of the leaves (Tapia and Fries, 2007). The surface of the tuber has a variety of colors of which prevail a yellow orange, deep yellow, light yellow, yellow, yellowish white, brown, black, intense grayish red and a secondary surface of a purple, grayish purple, yellow, grayish red, deep grayish red, light red (pink), or predominantly orange yellow with orange or blackish eyes and bands or irregular spots on tuberizations (Manrique et al., 2013).

Nutritional Composition

Mashua is a food that is highly nutritional (Table 1), that is characterized by containing a high level of protein of an elevated biological value with an ideal balance of essential amino acids. To the proteins add a high level of energy content, fiber, vitamin B1 y B2. According to Barrera et al. (2004) his tuber stands out from the other Andean roots (miso, jicama, oca, white carrot, melloco and achira) because of its content of vitamin C and elevated quantity of provitamin A (equivalents of Retinol) especially in the varieties ECU-1128 y ECU-1089. Among the minerals, those of highest concentration are phosphorus and magnesium followed by zinc and iron. In addition, its low contribution in Na and high content in K makes this tuber ideal for hypertensive people. Mashua contains various metabolites, including phenolic compounds, antioxidants, anthocyanins, isothiocyanates, and glucosinolates (National Research Council, 1989; Barrera et al., 2004; Campos et al., 2006; Manrique et al., 2013)

Table 1. Nutritional Composition of mashua (*Tropaeolum tuberosum* Ruiz y Pavón)

Parameter	WBR	WBA	DBR	DBA	References
Humid (%)	n/i	n/i	78.3 – 92.4	84.5	Gross et al. (1989); King and Gershoff (1987)
Proteins (%)	n/i	n/i	6.9 – 15.7	7.7	Gross et al. (1989); King and Gershoff (1987)
Carbohydrates (%)	n/i	n/i	69.7 – 79.5	85.8	Gross et al. (1989); King and Gershoff (1987)
Fat material (%)	n/i	n/i	– 0.4	1.0	Gross et al. (1989); King and Gershoff (1987)
Raw fiber (%)	n/i	n/i	7.8 – 8.6	0.7	Gross et al. (1989); King and Gershoff (1987)
Ash (%)	n/i	n/i	4.0 – 6.5	4.8	Gross et al. (1989); King and Gershoff (1987)
Ca (%)	n/i	n/i	n/i	0.006	Espín et al. (2001)
P (%)	n/i	n/i	n/i	0.32	Espín et al. (2001)
Mg (%)	n/i	n/i	n/i	0.11	Espín et al. (2001)
Na (%)	n/i	n/i	n/i	0.044	Espín et al. (2001)
K (%)	n/i	n/i	n/i	1.99	Espín et al. (2001)
Cu (ppm)	n/i	n/i	n/i	9.0	Espín et al. (2001)
Fe (ppm)	n/i	n/i	n/i	42.0	Espín et al. (2001)
Mn (ppm)	n/i	n/i	n/i	7.0	Espín et al. (2001)
Zn (ppm)	n/i	n/i	n/i	48.0	Espín et al. (2001)
Energy (Kcal/g)	n/i	n/i	4.19 – 4.64	4.41	Espín et al. (2001)
Starches (%)	n/i	n/i	20.01 – 79.46[a]	48.31[a]	Espín et al. (2001)
Total Sugars (%)	n/i	n/i	6.77 – 55.23[b]	28.42[b]	Espín et al. (2001)
Reducing sugars (%)	n/i	n/i	6.41 – 45.29[b]	23.65[b]	Espín et al. (2001)
Vitamin A ß-carotene (µg/100 g)	n/i	10	n/i	n/i	Collazos et al. (1996) as cited by Grau et al. (2003)
Provitamine A (Eq. Retinol /100 g)	n/i	73.56	n/i	n/i	Espín et al. (2001)
Vitamine C (mg/100 g)	n/i	77.37	n/i	n/i	Espín et al. (2001)
Lysine (mg/g of protein)	35 – 69	n/i	n/i	n/i	Grau et al. (2003); King and Gershoff (1987)
Threonine (mg/g of protein)	22 – 46	n/i	n/i	n/i	Grau et al. (2003); King and Gershoff (1987)
Valine (mg/g of protein)	25 – 88	n/i	n/i	n/i	Grau et al. (2003); King and Gershoff (1987)
Isoleucine (mg/g of protein)	25 – 44	n/i	n/i	n/i	Grau et al. (2003); King and Gershoff (1987)
Leucine (mg/g of protein)	35 – 56	n/i	n/i	n/i	Grau et al. (2003); King and Gershoff (1987)
Tyrosine (mg/g of protein)	13 – 62	n/i	n/i	n/i	Grau et al. (2003); King and Gershoff (1987)
Tryptophan (mg/g of protein)	5 – 12	n/i	n/i	n/i	Gross et al. (1989); Grau et al. (2003)
Cysteine (mg/g of protein)	1.4 – 29	n/i	n/i	n/i	Gross et al. (1989); Grau et al. (2003)

[a] Sample free from sugars and pigments. [b] Values obtained from sugars such as % Glucose. n/i = no available information. WBR: Wet Base Range. WBA: Wet Base Average. DBR: Dry Base Range. DBA: Dry Base Average.

The fraction of carbohydrate consists mainly of starch, in the form of granules, whose components are amylose 27% and amylopectin 73%, and 85% digestible starch with a fraction of 15% that is not absorbed at the level of the gastrointestinal tract (Villacrés and Espín, 1999 cite by Barrera et al., 2004). The granules of starch are smaller than in other tubers such as oca and ulluco with oval, spherical, and truncated sizes between 4.39 and 16.29 μm in length and 4.07 and 13.09 μm in diameter (Valcárcel-Yamani et al., 2013). Composed of 2-3 grains that are not very similar (Surco, 2004); which are stable in heat but with the tendency of gelification, with a lower water solubility index 0.62 ± 0.05, higher water absorption index 1.95 ± 0.04 and low swelling power 1.95 ± 0.02 (Villacrés and Espín, 1999 as cited by Barrera et al., 2004).

Bioactive Components

Bioactive compounds or phytochemicals are very important components of plants due to their therapeutic contribution to health. Studies in Mashua have shown a high content of bioactive compounds when compared to other Andean crops (Campos et al., 2006). Researchers have identified total phenols, Flavin 3-ols, anthocyanins, proanthocyanins, carotenoids (Chirinos et al., 2007), triterpenes, steroids, flavones and leucoanthocyanidins as shown in Table 2 (Barrera et al., 2004).

Polyphenols

Phenolic compounds or polyphenols are part of the plant secondary metabolites or phytochemical. Its basic structure is a molecule of phenol linked to one or more hydroxyl groups and an aromatic ring. These compounds are widely distributed in different plants, for example fruits, vegetables, roots and cereals (Peñarrieta et al., 2014). Studies with Andean tubers such as mashua, native potato, oca and ulluco have shown significant differences in the content of these metabolites, of which Mashua has the highest phenolic content (Campos et al., 2006). Polyphenols accumulate in higher concentration in purple genotypes unlike yellow varieties. The genotype ARB-5241 is the most remarkable with a total value of 3.37 mg eq. chlorogenic acid/g fresh weight.

Another important subgroup are anthocyanins that are partly responsible for the organoleptic quality of food (Tomás-Barberán, 2003). According to Campos et al. (2006) the purple mashua (standing out the genotype DP-0224) has demonstrated a higher anthocyanin content with respect to tubers such as the native oca and potato. The anthocyanin content of the tuber ranges from 0.5 to 2.05 mg. eq. cyanidin3-glycoside/g fresh weight. A detailed study by Chirinos et al. (2006) identified in extracts of mashua in 11 sub fractions of anthocyanins. These metabolites are derivatives of delphinidin (delphinidin-3-glycoside-5-acetyl rhamnoside and delphinidin-3-sophoroside-5-acetyl rhamnoside, other pigments were delphinidin-3-glucoside-5-rhamnoside, delphinidin-3-sophoroside-5-ramnoside, delphinidin-3-glycoside, cyanidin-3-sophoroside, cyanidin-3-sophoroside-5-rhamnoside, cyanidin-3-glycoside, cyanidin-3-

rutinoside, pelargonidin 3-sophoroside and pelargonidin 3-sophoroside-5-rhamnoside).

Moreover, Chirinos et al. (2008 b) characterized non-anthocyanin phenolic compounds in purple mashua genotypes. The sub-fractions identified corresponded to Flavan-3ol (gallocatechin, epigallocatechin, epicatechin and its derivatives), procyanidin B2, hydroxycinnamic acids (derivatives of o-coumaric acid and p-coumaric acid) and p-hydroxybenzoic acid (derivatives of gallic acid, hydroxybenzoic acid and protocatequico), proanthocyanidins and anthocyanidins (delphinidin). Another important metabolite in the mashua tuber and with antioxidant properties are the carotenoids. The same study by the researchers Campos et al. (2006) shows a content that varies between 1 to 25 mg carotene/g in fresh weight in yellow genotypes (the varieties ARB-5576, M6COL2C and DP-0207 stand out). Total carotene values are higher than those registered in commercial potatoes, native potatoes and papaya, but lower compared to foods such as tomato, mango and carrot as shown in Table 3.

Glucosinolates

Eight types of glucosinolates have been identified in the Tropaeolaceae family such as glucotropaeoline, glucolepidiin, glucosinalbine, glucoconringiina, glucoaubrietina, glucoputranjivina, glucococleorina and glucorafenina all derived from amino acids (Fahey et al., 2001; Bayer and Appel, 2003 ; Ramallo et al., 2004).

The identification of the principle glucoaubrietin (4-methoxybenzyglucosinolate) in the Tropaeolum tuberosum species by Ruíz and Pavón. Later three other aromatic glucosinolates were identified by Ortega et al. (2006): 4-hydroxybenzyl glucosinolate (OHB, Glucosinalbine), benzyl glucosinolate (B, Glucotropaeoline), m-methoxybenzyl glucosinolate (MOB, Glucolimnatin). The glucosinolates are the precursors of the isothiocyanates distributed in sixteen families of dicotyledonous angiosperms and domestic species (Fahey et al., 2001).

The cultivated species Tropaeolum tuberosum sub-specie tuberosum (seeds tuber, leaves and flowers) present p-methoxybenzyl isothiocyanate and N, N-di (methoxy-4-benzyl) thiourea and in the wild species Tropaeolum tuberosum subspecies wild were identified benzyl-isothiocyanate, 2-propyl-isothiocyanate, 2-butyl-isothiocyanate and N, N- di (4-methoxybenzyl) thiourea (Johns and Neil, 1981). Numerous studies point out the medicinal properties of glucosinolates and their derivatives, among them Benzyl-isothiocyanate as responsible for acting against tumor cells and which is associated as an anticancer agent and p-methoxy-benzyl glucosinolate

Table 2. Bioactive components and reference values in dry and wet bases

Components	Units	Wet base	Dry base	References
Total Phenolics	mg eq. Chlorogenic acid/g		9 – 21[a]	Chirinos et al., (2007)
	mg eq. Chlorogenic acid/g	0.92 – 3.37[d]		Campos et al., (2006)
	mg eq. Gallic acid/100g	174.9 – 374.4[g]		Chirinos et al., (2006)
	mg eq. Gallic acid/100g	311– 343[h]		Aro and Tipacti (2016)
Flavan–3 oles	mg eq. catechin /g		0.2 – 5.3[a]	Chirinos et al., (2007)
Anthocyanins	mg. eq. Cyanidin3-glycoside/g		2.4-5.7[a]	Chirinos et al., (2007)
	mg. eq. Cyanidin3-glycoside/100g	45.5 – 131.9[g]		Chirinos et al., (2006)
	mg. eq. Cyanidin3-glycoside/g	0.5 – 2.05[c]		Campos et al., (2006)
Carotenoids	µg eq. β carotene/g		70-132[a]	Chirinos et al., (2007)
	µg eq. β carotene/g	1 – 25[b]		Campos et al., (2006)
Antioxidant Capacity	µmol equivalent of Trolox/g ABTS		80 – 378[a]	Chirinos et al., (2007)
	µmol equivalent of Trolox/g ABTS	16.2 – 45.7[g]		Chirinos et al., (2006)
	µmol equivalent of Trolox/g ABTS	35 – 92[h]		Calsin et al., (2016)
	µmol equivalent of Trolox/g ORAC)		59 – 389[a]	Chirinos et al., (2007)
Hydrophilic antioxidant capacity	µg equivalent of Trolox/g ABTS	955 – 9800[e]		Campos et al., (2006)
Lipophilic antioxidant capacity	µg equivalent of Trolox/g ABTS	93 – 279[f]		Campos et al., (2006)
Glucosinolates (antinutrient)	µmol/g		36.5 – 90.0[i]	Ramallo et al., (2004)
	µmol/g		0.27 – 50.74	Ortega et al., (2006)
Linolenic acid	%		48.70[j]	Ramallo (2004)
α-linolenic acid	%		22.13[j]	Ramallo (2004)
Palmitic acid	%		21.2[j]	Ramallo (2004)
Oleic acid	%		3.96[j]	Ramallo (2004)
Stearic acid	%		1.47[j]	Ramallo (2004)
Vacnic acid	%		1.30[j]	Ramallo (2004)

[a] Genotypes ARB 5241, DP 0224 y AGM 5109. [b] Genotypes ARB- 5576, M6COL2C y DP-0207 (highest values). [c] Genotypes colored. [d] Genotypes ARB-5241, DP-0224 y AGM-5109 (highest values). [e] Genotypes ARB-5241, DP- 0224 y ARV-5366 (highest values). [f] Genotypes M6COL2C, ARV-5366 y DP-02-03 (highest values). [g] Genotypes ARB-5241, DP-0224 y AGM-5109. [h] Genotypes ARB-5241, DP-5366 y AGM-0224. [i] Variety Kulli kisaño, Kellu anaranjado, Kellu cheschi, Kellu A, Zapallo, Kellu B. [j] Variety Kulli kisaño, Kellu anaranjado, Kellu cheschi, Kellu A, Zapallo, Kellu B.

which is related to reproductive function (Johns et al., 1982; Hasegawa et al., 1992; Pintao et al., 1995 cite by Quispe et al., 2015).

A recent study developed in male Holtzman rats demonstrated the efficacy of mashua with benign prostatic hyperplasia (BPH). The researchers used three doses of which the second dose (500mg/kg/rat) and third dose (800mg/kg/rat). Rats treated with the tuber obtained the best results in echo-graphic and histopathological studies, despite this, the effects of mashua was not superior to finasteride (antiandrogenic drug used for BPH); concluding that the two methods were efficient against BPH (Aire-Artezano et al., 2013).

Fatty Acids

Ramallo (2004) investigated fatty acids in the mashua flour via gas phase chromatography. The author pointed out a high content of polyunsaturated fatty acids (70.8%), among six varieties of mashua; highlighting the presence of linoleic acid n-6 (48.70%), α-linolenic n-3 (22.13%), palmitic (21.2%), and with less importance oleic acid n-9 (3.96%), stearic (1.47%) and Cis-vacénico (1.30%). The ratio between linoleic/α-linolenic acids for mashua was 2.2, compared with other vegetables such as rapeseed is 2.49; walnut is 4.64. It's value is optimal according to the nutritional recommendations that establishes a ratio less than 5. The study of polyunsaturated fatty acids of the series: n-3, n-6 and n-9 are of interest due to their therapeutic properties. Polyunsaturated fatty acids (PUFAs) n-3 and n-6 are necessary during pregnancy, specifically for the child's fetal and cognitive brain development (Rodríguez-Cruz et al., 2005). α-linolenic acid (n-3) acts in the prevention of coronary diseases, inhibiting angiogenesis, possessing cytotoxic activity on tumor cells and reducing blood cholesterol levels (Serrano et al., 2006). Linoleic acid (n-6) reduces cholesterol levels. Deficiencies of linoleic acid (n-6) involves inflammatory processes at the level of animal and human skin. Excessive consumption of n-6 PUFAs suggests a tendency to acquire some type of cancer, mitochondrial DNA damage and cardiac disorder (Rodríguez-Cruz et al., 2005; Urango et al., 2008).

Flavonoids

Chirinos et al. (2008) have shown that phenolic compounds such as anthocyanidins, Flavan monomers 3 oles, flavonols and proanthocyanins contribute to the antioxidant capacity of mashua, the latter metabolite being the most significant. The author and his collaborators have reported a range of 34.7 - 39.2% of the condensed tannin.

Proanthocyanins are complex structures derived from flavan-3-oles (Peñarrieta et al., 2014), and are important for their health benefits. These metabolites are attributed to have antioxidant, anti-carcinogenic, cardioprotective, antimicrobial and neuroprotective properties (Navarro et al., 2013). Studies in animals show an inhibitory effect against the oxidation of tissues and low density lipoproteins are preventing the formation of thrombi in the circulatory system, in addition they kidnap and neutralize free radicals (Vázquez- Flores et al., 2012).

The content of flavan-3-oles in the mashua tuber varies between 0.2 - 5.3 mg eq. catechin / g (Chirinos et al., 2007). Flavan-3-oles, flavanols or flavanols, are considered minor flavonoids chemically formed by a C ring and an OH group in the 3-position of the heterocyclic 2-phenylbenzopyran (Peñarrieta et al., 2014; Soriano, 2003). Studies on the benefits of flavan 3 oles mention its ability to regulate the synthesis of glutathione known for its antioxidant power and stabilizes or deactivates free radicals that attack cells and damages them (Moskaug et al., 2005). Regarding the anti-cancer activity, Torres et al. (2002) report that flavan-3-ols decrease the increase of cancer cells, especially those that come from the human colon. An unwanted effect due to the high intake of these flavonoids can lead to a leukemia in the case of infants. In addition to inhibiting certain enzymes such as topoisomerases responsible for acting on the DNA topology (Moskaug et al., 2005).

Other metabolites reported in mashua by Chirinos et al. (2006) are the anthocyanins that vary between 45.5-131.9 mg. eq. cyanidin3-glycoside / 100g. They are an extensive group of water soluble pigments responsible for a high range of colors and are present in the plant kingdom (Fenema, 1993 as cited by Ortíz et al., 2011).

Anthocyanins are considered a subclass of flavonoids. Approximately 300 anthocyanins have been identified, however, the most important are: Pelargonidin, delphinidin, cyanidin, petunidin, peonidin and malvidin known as anthocyanidins because of the combination of these with different sugars. Its coloration depends on factors such as: ph effect, union with other compounds and ions, effects of enzymes, acids, presence of oxygen, the increase of hydroxyl phenolic ring in its chemical composition generates blue color and an increase in methoxyl generates red color (Badui, 2006). The importance of anthocyanins lies mainly in the benefits for human health (De Pascual-Teresa and Sanchez-Ballesta, 2008) due to it's farmalogical and therapeutic conditions (Astrid, 2008).

The therapeutic relationship of anthocyanins is given by its antioxidant activity (Ghiselli et al., 1998). Investigations carried out live and in vitro demonstrate that cyanine extracted from red beans that are more rhamnose; Red soy which has glucose play an important role in immunological effects, especially in the suppression or prevention of tumors (Koide et al., 1997). Cyanidin presents greater antimutagenic activity than peonidin (Yoshimoto et al., 2001). The flavonoids present in some fruits and vegetables are important in the human diet (Yoshimoto et al., 2001). For example, the content of flavonoids present in strawberries and spinach, avoided effects of neural and cognitive aging (Joseph et al., 1998). According to varios studies, three types of pigments have been identified (antioxidants, flavonoids, anthocyanins and carotenoids) that help with the reduction of degenerative diseases in humans, such as cancer and coronary diseases (Badui, 2006).

Table 3. Bioactive components and referential values for different vegetables

Component	Units	Range	References
Antioxidant Capacity			
Sweet potato	(µg equivalent of Trolox/g Fresh weight DPPH)	12 409 ± 1024	Cevallos-Casals and Cisneros-Zevallos (2003)
Purple Corn	(µg equivalent of Trolox/g Fresh weight DPPH)	21 351 ± 121	Cevallos-Casals and Cisneros-Zevallos (2003)
Cranberries	(µg equivalent of Trolox/g Fresh weight DPPH)	5646 ± 389	Cevallos-Casals and Cisneros-Zevallos (2003)
Hydrophilic antioxidant capacity			
Native Potatoes	(µg equivalent of Trolox/g Fresh weight ABTS)	860 – 3780	Campos et al. (2006)
Oca	(µg equivalent of Trolox/g Fresh weight ABTS)	1637 – 4771	Campos et al. (2006)
Ulluco	(µg equivalent of Trolox/g Fresh weight ABTS)	483 – 1524	Campos et al. (2006)
Cranberry	(µg equivalent of Trolox/g Fresh weight ABTS)	6900 – 9572	Roca et al. (2007)
Lipophilic antioxidant capacity			
Native Potatoes	(µg equivalent of Trolox/g Fresh weight ABTS)	115 – 361	Campos et al. (2006)
Oca	(µg equivalent of Trolox/g Fresh weight ABTS)	69 – 320	Campos et al. (2006)
Total phenolics			
Ulluco	(mg eq. Chlorogenic acid/g Fresh weight)	0.41 – 0.77	Campos et al. (2006)
Oca	(mg eq. Chlorogenic acid/g Fresh weight)	0.71 – 1.32	Campos et al. (2006)
Potato	(mg eq. Chlorogenic acid/g Fresh weight)	0.64 – 2.32	Campos et al. (2006)
Blackberries	(mg eq. Gallic acid/100g Fresh weight)	226 ± 4.1	Wang Shiow and Lin (2000)
Black raspberries	(mg eq. Gallic acid/100g Fresh weight)	267 ± 4.3	Wang Shiow and Lin (2000)
Raspberries	(mg eq. Gallic acid/100g Fresh weight)	234 ± 5.1	Wang Shiow and Lin (2000)
Strawberry	(mg eq. Gallic acid/100g Fresh weight)	103 ± 2.0	Wang Shiow and Lin (2000)
Sweet potato	(mg eq. Chlorogenic acid/100g Fresh weight)	1756 ± 64	Cevallos-Casals and Cisneros-Zevallos (2003)
Purple corn	(mg eq. Chlorogenic acid/100g Fresh weight)	945 ± 82	Cevallos-Casals and Cisneros-Zevallos (2003)
Cranberries	(mg eq. Chlorogenic acid/100g Fresh weight)	574 ± 35	Cevallos-Casals and Cisneros-Zevallos (2003)
Anthocyanin			
Ulluco	(mg. eq. Cyanidin3-glycoside/g Fresh weight)	0.41 – 0.77	Campos et al. (2006)
Oca	(mg. eq. Cyanidin3-glycoside/g Fresh weight)	0.14 – 1.3	Campos et al. (2006)

Table 3. Continuation

Component	Units	Range	References
Potato	(mg. eq. Cyanidin3-glycoside/g Fresh weight)	0.08 – 0.8	Campos et al. (2006)
Blackberries	(mg. Eq. Cyanidin3-glycoside/100 g Fresh weight)	152.8 ± 8.0	Wang Shiow and Lin (2000)
Black raspberries	(mg. Eq. Cyanidin3-glycoside/100 g Fresh weight)	197.2 ± 8.5	Wang Shiow and Lin (2000)
Raspberries	(mg. Eq. Cyanidin3-glycoside/100 g Fresh weight)	68.0 ± 3.0	Wang Shiow and Lin (2000)
Strawberry	mg. Eq. Pelargonidin 3 glycoside/100g Fresh weight)	31.9 ± 4.1	Wang Shiow and Lin (2000)
Sweet potato	(mg. Eq. Cyanidin3-glycoside/100 g Fresh weight)	1642 ± 92	Cevallos-Casals and Cisneros-Zevallos (2003)
Purple corn	(mg. Eq. Cyanidin3-glycoside/100 g Fresh weight)	182 ± 2	Cevallos-Casals and Cisneros-Zevallos (2003)
Cranberries	(mg. Eq. Cyanidin3-glycoside/100 g Fresh weight)	276 ± 25	Cevallos-Casals and Cisneros-Zevallos (2003)
Carotenoids			
Oca	(μg eq. β-carotene/g Fresh weight)	2 – 25	Campos et al. (2006)
Tomato	(μg β-carotene/g Fresh weight)	56 – 210	Gross (1987)
Papaya	(μg β-carotene/g Fresh weight)	4.08	Gross (1987)
Mango	(μg β-carotene/g Fresh weight)	74.3	Gross (1987)

Tannins

In the plants there is a diversity of secondary compounds that when consumed create a toxic effect as a defensive response to attack by predators. Tannins and certain phenolic compounds constitute the chemical protection of the plant (Chaves, 1994; Sepúlveda et al., 2003). Different varieties of mashua contain condensed tannins at a value of 2210 mg.eq. catechin / 100 g (López, 2001 as cited by Urresta, 2010). Condensed tannins are derivatives of flavan-3-oles; they are found in leaves, fruits and barks (Peñarrieta et al., 2014). The negative biological effect of these metabolites on health is to precipitate salivary proteins, decrease the digestion and absorption of dietary protein. In addition to preventing the absorption of metals such as non-heme iron and carbohydrates, thus slowing the growth process. Studies in animals, show a dose of 0.5 to 2 g / kg / day (5% of the diet) without acute toxic effects, however it shows problems during growth (Vázquez-Flores et al., 2012).

On the other hand, tannins can have negative effects on the nutritional content of forages of ruminant feeders. At concentrations of 6-10% of the dry matter (DM) decreases the voluntary feeding of the animal, reducing digestibility and productivity. While at a moderate and low concentration, ie 2-4% of the DM, its effect is positive where digestion is not depressed

there is greater absorption of essential amino acids etc. (Barry et al., 1986, Reed et al., 1990, Barry and MacNabb 1999, as cited by Otero e Hidalgo, 2004).

Anitoxidate Capacity of mashua

Several investigations relate the antioxidant potential of the bioactive compounds (polyphenols, flavan-3-oles anthocyanins and carotenoids) of the Mashua with respect to its hereditary characters and processes of plant development.From the available studies, a significant relationship of antioxidant activity and genetic variety is indicated. The study of Campos et al. (2006) affirms a high hydrophilic antioxidant capacity of the varieties ARB-5241, DP-02-24 and ARV-5366 compared with Andean crops such as the oca, native potatoes and ulluco. While the genotypes M6COL2C, ARV-5366 and DP-02-03 present a similar lipophilic antioxidant capacity among tubers (mashua, Native Potatoes and oca). The lipophilic fraction contributes between 2-19% of the total antioxidant capacity. On the other hand, in an analysis between genotypes, several researchers claim that purple mashua is eight to ten times richer in antioxidant activity than the yellow variety (Chirinos et al., 2007 as cited by Chirinos et al., 2008).

The maturation of the Mashua is associated with changes in its physical and chemical structure; in such a way, that the content of some bioactive components and the antioxidant capacity varies according to the stage of development. Chirinos et al. (2007) found in the genotypes ARB 5241, DP 0224 and AGM 5109 purple color, collected after 7.5 months of planting, a high antioxidant capacity (271-446 µmol equivalent of Trolox/g dry matter ORAC), a high content of phenols (14-24 mg eq. chlorogenic acid/g fresh weight) and an increase in anthocyanin content (3.7 - 8.7 mg eq cyanidin3-glycoside/100g fresh weight). While the carotenoids were only evidenced in yellow mashuas of the genotypes DP 0207, AVM 5562 and ARB 5576 with a range of 8.9 - 14.4 (µg eq. B Carotene/g Fresh weight). During the post harvest, the mashua appeared to increase their bioactive components with the exception of the carotenoids that remained stable. The genotypes DP 0224, ARB 5241 and AGM 5109 stand out as those with the highest phenolic content and antioxidant power (Chirinos et al., 2007).

Antioxidants prevent the oxidation of the cell membrane by forming stable complexes. (Goodam, 1998; Halliwell, 1990 as cited by Avello and Suwalsky, 2006). The interest of antioxidant capacity has intensified due to its pharmacological and therapeutic benefits. Some of the positive effects of antioxidant capacity are related in the treatment of strokes and neurodegenerative diseases (Castañeda et al., 2008). Studies in animals with metastatic lung cancer have shown inhibition of the tumor neovascularization process (Monte et al., 1994 as cited by Castañeda et al., 2008). An important aspect about the study of mashua is to know the contributions of the total antioxidant capacity of the tuber and its incidence in health. Studies with phenolic compounds (flavonoids, anthocyanins and phenols) and antioxidant capacity from purified mashua extracts have demonstrated the ability to prevent oxidative damage of LDL and inhibit erythrocyte hemolysis. These results suggest that mashua phenolic compounds have the capacity to chelate metal ions as well as eliminate peroxyl radicals (Chirinos et al., 2008).

Correlation between antioxidant properties and phenolic compounds

The measurement of the antioxidant capacity in food is considered as a relevant contribution in recent years, since the oxidation resistance of a product is determined (Zulueta et al., 2009). Antioxidant activity is evaluated by different methods either in vitro or live; several studies coincide with the use of chemical techniques (in vitro) such as DPPH (2.2-Diphenyl-1-picrilhydracil), ORAC (Absorption Capacity of Oxygen Radicals) and ABTS (2.2'-azino-bis (3-ethylbenzthiazoline- 6-sulfonic) of which the most commonly used methods are ABTS and DPPH.

The study done by Chirinos et al. (2007) found a high correlation between the ABTS and ORAC methods (0.794 <r <0.953, P <0.01) for eight mashua genotypes, both techniques describe the same oxidative mechanism. The researchers highlighted a significant association of the content of Flavan 3 oles (FA), anthocyanins (AT) and total phenols (FT) with respect to the ABTS antioxidant capacity with a coefficient of r = 0.897, P <0.01; r = 0.891, P <0.05; r = 0.879, P <0.01, respectively. The genotypes analyzed were ARB 5241 (FA and AT) and the DP 0224 (FT) genotype. The study concludes a significant difference between genotypes.

Chirinos et al. (2006) determined a high correlation between the antioxidant capacity (ABTS) and the phenolic content with a positive coefficient (r= 0.9873). However, when the comparison between the antioxidant capacity (ABTS) and the fraction rich in anthocyanins is made, the coefficient decreases (r= 0.637). The researchers suggest that the phenolic compounds were directly responsible for antioxidant activity (ABTS).

Campos et al. (2006) agrees with the data obtained by Chirinos et al. (2006). Different phenolic profiles of mashua tubers influenced the correlation index between hydrophilic antioxidant capacity (ABTS), total anthocyanins and total phenol content. The researchers found a low correlation of anthocyanins (r2 = 0.48, P = 0.11) and a significant correlation with respect to the phenolic content (r2 = 0.84, P = 0.00). Similar results were evidenced when comparing the content of phenolic compounds and the ORAC antioxidant activity in four purified phenolic extracts of mashua (fractions I, II, III and IV); the correlation coefficients were positive (r = 0.9983, 0.9801, 0.9891 and 0.9976). The study deduces that the phenolic compounds were responsible for the antioxidant activity (ORAC) in the four fractions. (Chirinos et al., 2008b).

Nutritional Properties of mashua

In traditional medicine, mashua is used as a digestive cleanser and cicatrizant. In Ecuador is taken as an infusion mixed with caballochupa (*Equisetum* sp.) and corn silk (*Zea mays*) to treat kidney problems. It is also cooked with brown sugar (panela) for prostatitis and gonorrhea (Cárdenas 1989; Espinoza et al., 1996 citado por Cadima, 2006). It can be used as a treatment of skin ulcers and to kill lice (Oblitas, 1969 as cited by Johns et al., 1982).

Around 600 carotenoids are distributed in the form of red, orange and yellow liposoluble pigments in fruits and vegetables, however, only 20 have been related to certain biological properties (García-Casal et al., 2013). The β Carotene is the most remarkable metabolite because of its potential health benefit. Like other components it has the capacity to become a source of vitamin A (α-carotene, β-carotene and β-cryptoxanthin) (Vitale et al., 2010; García-Casal et al., 2013). Carotenoids have been investigated for their possible preventive properties of diseases such as cancer, arteriosclerosis, cataracts, macular degeneration, premature aging among others (Sánchez et al., 1999). Studies with Andean crops such as mashua suggest that carotenoids and other phenolic compounds found in the tuber can be considered an important source of antioxidants for health. (Chirinos et al., 2007). The vitamin A content in the tuber corresponds to 10, expressed in β-carotene as shown in Table 1.

Liebler et al. (1997) have demonstrated the antioxidant capacity of β-carotene when incorporated in the carotenoid in liposomes. The result showed a correct inhibition of lipid peroxidation that was induced by AAPH, however, the effect was ineffective when the carotenoid was added to preformed liposomes. In the same way Nagler et al. (2003) as cited by Krinsky and Johnson (2005), stated that β-Carotene effectively inhibits the oxidation of linoleic acid, but not that of α-linoleic acid. In general, β-Carotene has been linked to its protective effect against photooxidation and radical-mediated peroxidation, reduction of precancerous lesions in the cervix and oral cavity, and, moreover, to prevent sunburn (Krinsky and Johnson, 2005; Taylor, 1996; Sánchez et al., 1999). Nonetheless, there exists disadvantages to the β-Carotene supplementation, it is that a high dose could inhibit the intestinal absorption of other nutrients which are relevant to the prevention of carcinogenic diseases (Mobarhan et al., 1994; Sánchez et al., 1999). Other studies suggestion that supplementation with beta carotene could cause a risk of cardiovascular disease and lung cancer (Taylor, 1996)

Table 4. Recommend daily intake reference values and bio-active components

Components	Units	Values	Intake	References
Total polyphenols	mg/day/person	2803	Spain	Saura-Calixto (2007)
	mg/day/person	2500 – 3000	Spain	Roldan and Carbajal (2012)
	mg/day/person	64	Japan	Martínez-Valverde et al. (2000)
Flavonoids	mg/day/person	6	Finlandia	Martínez-Valverde et al. (2000)
	mg/day/person	25	Spain	Roldan and Carbajal (2012)
Proanthocyanins	mg/day/person	240 – 450	Spain	García-Casal et al. (2013)
Anthocyanidins	mg/day/person	185 – 215	n/i	Wu et.al (2006) citado por Beas et al. (2011)
Carotenoids	mg/day/person	1–2 with reports of up to 10	United States and United Kingdom	(García-Casal et al., 2013)
	mg/day/person	3 – 4.3	Spain	Roldan and Carbajal (2012)
Antioxidant Capacity	μmol equivalent of Trolox/person/day.	3528	Spain	Saura-Calixto (2007)
Glucosinolate	mg/k of body mass (toxic dose in HCN)	1.5	n/i	Kermanshai et al. 200 as cited by (Rinc, 2014)
	mg/day	6.5	Spain	Roldan and Carbajal (2012)
Polyunsaturated fatty acids (AGPIs)	Proportion	n–6: n–3 of 5–10: 1/day.	FAO/ OMS	Rodríguez-Cruz et al., (2005)
	AGPIs n-3 (g/ day)	1.1 a 1.5	FAO/ OMS	Rodríguez-Cruz et al. (2005)

n/i= not available information

Bio-active compounds and recommended intake

In recent years, researchers have associated the consumption of foods rich in antioxidants as a source of benefits for health. For this reason, Medical and Traditional publications of the last decades have been examined, and the levels of intake per day/person of polyphenols, anthocyanins, carotenoids, antioxidant capacity and glucosinolates have been identified as shown in Table 4. This is prior to a comparison between the actual content of these metabolites and their comparison with other foods, these values are reported in Table 2 and Table 3 respectively.

The content of polyphenol in mashua oscillates in a range of 174.9-374 mg eq. gallic acid/100g higher than those reported for blackberry, black raspberry, red raspberry and strawberry (226, 267, 234, 103 mg eq. gallic acid/100g respectively). If this phytonutrient is analyzed in the context of the recommended daily intake (RDI) in Spain, 2500-3000 mg/day/person is required. The levels of anthocyanins reported in the tuber vary between 45.5-131.9 (mg. eq. cyanidin3-glycoside/100g) which is an inferior content when compared to the blackberry, black raspberry, sweet potato, purple corn and cranberries (152.8, 197.2, 1642, 182, 276 mg. eq. cyanidin3-glycoside/100g respectively). The daily intake requirements in values of mg/day/person of these metabolites is of 185-215, higher than those found in the tuber.

Other metabolites of importance are the carotenoids, in mashua a content of 1-25 µg eq. β-carotene/g, whose value is lower than the tomato and mango (56 – 210; 74.3 µg β-carotene/g fresh weight respectively) and equal to oca (2 – 25 µg eq. β-carotene/g) and higher than the papaya (4.08 µg eq. β Carotene/g) have been identified. Spain has established a recommended intake of 3-4.3 mg/day/person. The ABTS value found for the antioxidant capacity of the mashua is of 16.2-92 µmol equivalent to Trolox/g ABTS, which is inferior to those found in sweet potato, purple corn and cranberries (12409, 21351, 5646 µg equivalent to Trolox/g fresh weight DPPH respectively). Nevertheless, the daily requirement per person in Spain is higher, reporting a value of 3528 µmol equivalent to Trolox/person/day. The authors have indicated a content of 0.27-90.0 µmol/g glucosinolates in mashua. This metabolite is very controversial because of its properties have been recommended in Spain with an ingestion doses of 6.5 mg/day.

Secondary effects of the mashua on health

The Tropaeolum tuberosum like all other members of the Tropaeolaceae family have isothiocyanates such as glucosinolates. These metabolites could be responsible for suppressing of the sexual appetite and the lowering of the reproductive potential of the Incas during the military operations of the XVI century, according to popular belief (Johns et al., 1982). Taking into consideration this background Johns et al. (1982) demonstrated an anti-aphrodisiac action in the mashua (m-methoxybenzyl glucosinolates) in rats that were feed the tuber. The

rats with the treatment and the control group had the same capacity to impregnate the females, however, the rats that were feed the tuber showed a lowering of 45% in the levels of testosterone/dihydrotestosterone in their blood. Current studies like that of Cárdenas-Valencia et al. (2008) demonstrated the effect of the mashua (benzyl-glucosinolates) in the reduction of the testicular function in a test done with male Holtzman rats for 42 days. The results indicated a reduction in sperm for 12 to 42 days of the treatment. Moreover, there was a delay in the transit of spermatozoids by the epididymis for 7 days and with an accelerated movement after day 12 up to day 21. The results did not find any significant differences between the levels of testosterone between the treatment rats and control group.

Vásquez et al. (2012) suggested that mashua lowers sperm parameters in the male reproductive system. The author and collaborators used an extract of a hydro-alcoholic Tropaeolum tuberosum in laboratory rats. The effect was a lowering of the sperm in transit and an increment of the number of spermatozoids after 21 days of the treatment. One of the disadvantages of consuming mashua is the presence of thiocyanates. The Ecuadorian genotypes of fresh mashua presents thiocyanates with values of 23-33 mg / 100 g (Dolores and Espín 1997; Grau et al., 2003; as cited by Grau et al., 2003). These metabolize upon being hydrolyzed liberate molecules hydrogen cyanide; the INIAP reported that the values fluctuate between 33.55 y 23.11 mg of cyanide per 100 g of the plant (INIAP, 1996 as cited by Urresta, 2010). These values are low when compared to other food like sorghum with 250 mg, Cassava leaves 104 mg and Cassava root 53 mg of cyanide per 100 g of the plant.

The process of cooking, washing and sifting, exposure to sunlight or in combination lowers the toxicity of the cyanide to values of 9.2 to 9.4 mg per hundred. Through exposure to light, cooking and fermentation with yeasts, the levels can be dropped to 0.36 mg which is below the limit of toxicity. The larger, yellow reddish varieties of Mashua have a higher content of cyanides (INIAP, 1996 as cited by Urresta, 2010). Cyanide block the transportation of oxygen to the red blood cells. The ingestion of 60 mg de cyanide (dry weight) for a person who weighs 60 kg is considered a fatal dose (relationship of 1 mg per kilo of body weight), which could be one of the possible causes of polyneuropathies (Zaninovic´, 2003).

CONCLUSION

In conclusion, the antioxidant capacity and the polyphenol content of the mashua are higher than that which is reported for many of the fruits and tubulars of high consumption in the Americas. The scientific investigation consolidates the preventive role that the bioactive components of the mashua have in the prevention of cardiovascular illness, cancer, and other neurodegenerative illnesses. Moreover, it contributes in the food industry by avoiding the lipid oxidation of vegetable oils and pork

REFERENCES

Aire-Artezano, G., Charaja-Vildoso, R., Cruz-Santiago H., Guillermo-Sánchez B., Gutarra-Vela, M., Huamaní-Charagua, P., Pari-Ñaña, R. 2013. Efecto de Tropaeolum tuberosum frente a la hiperplasia prostática benigna inducida en ratas Holtzman. CIMEL Ciencia E Investigación Médica Estudiantil Latinoamericana. 18: 1–13.

Astrid, G. 2008. Las Antocianinas Como Colorantes Naturales Y Compuestos Bioactivos : Revisión. Acta Biológica Colombiana. 13: 27–36.

Avello, M., Suwalsky, M. 2006. Radicales libres, antioxidantes naturales y mecanismos de protección. Atenea (Concepción). 494: 161–172. https://doi.org/10.4067/S0718-04622006000200010

Badui, D. 2006. Química de los Alimentos (Pearson ed). México.

Barrera, V. H., Tapia, C., Monteros, A. 2004. Raíces y Tubérculos Andinos: Alternativas para la conservación y uso sostenible en el Ecuador. Serie: Conservación y uso de la biodiversidad de raíces y tubérculos andinos: Una década de investigación para el desarrollo (1993-2003). (Instituto Nacional Autónomo de Investigaciones Agropecuarias, Ed.). Quito, Ecuador - Lima, Perú: Centro Internacional de la Papa, Agencia Suiza para el Desarrollo y la Cooperación.

Bayer, C., Appel, O. 2003. Tropaeolaceae, 683: 400–404.

Cadima, X. 2006. Tubérculos. Botánica Económica de Los Andes Centrales, 347–369.

Calsin, M., Aro, J., Tipacti, Z. 2016. Evaluación de la eficacia de antioxidantes de IsañoEvaluación de la eficacia de antioxidantes de Isaño (Tropaeolum tuberosum Ruiz&Pavón) en la oxidación de aceite de soya. Revista de Investigación Altoandin. 18: 143–150.

Campos, D., Noratto, G., Chirinos, R., Arbizu, C., Roca, W., Luis, C.-Z. 2006. Antioxidant capacity and secondary metabolites in four species of Andean tuber crops: native potato (Solanum sp.), mashua (Tropaeolum tuberosum Ruiz & Pav´on), Oca (Oxalis tuberosa Molina) and ulluco (Ullucus tuberosus Caldas). Journal of the Science of Food and Agriculture: 86: 1481–1488. https://doi.org/10.1002/jsfa

Cárdenas-Valencia, I., Nieto, J., Gasco, M., Gonzales, C., Rubio, J., Portella, J., Gonzales, G. F. 2008. Tropaeolum tuberosum (Mashua) reduces testicular function: effect of different treatment times. Andrologia, 40: 352–357. https://doi.org/10.1111/j.14390272.2008.00868.x

Cardenas, M. 1989. Manual de plantas económicas de Bolivia. (2nd ed.). La Paz y Cochabamba: Los Amigos del Libro.

Castañeda, B. C., Ramos, L. E., Ibáñez, V. L. 2008. Evaluación de la capacidad antioxidante de siete plantas medicinales peruanas. Revista Horizonte Medico. 8: 56–72.

Cevallos-Casals, B. A., Cisneros-Zevallos, L. 2003. Stoichiometric and kinetic studies of phenolic antioxidants from Andean purple corn and red-fleshed sweetpotato. Journal of Agricultural and Food Chemistry. 51: 3313–3319. https://doi.org/10.1021/jf034109c

Chaves, S. 1994. Contenido de taninos y digestibilidad in vitro de algunos forrajes troicales. Agroforesteía En Las Américas, 10–13.

Chirinos, R., Campos, D., Arbizu, C. 2007. Effect of genotype, maturity stage and post-harvest storage on phenolic compounds, carotenoid content and antioxidant capacity, of Andean mashua tubers Journal of the. Retrieved from http://onlinelibrary.wiley.com/doi/10.1002/jsfa.2719/full

Chirinos, R., Campos, D., Arbizu, C., Rogez, H., Rees, J.-F., Larondelle, Y., Luis, C.-Z. 2007. Effect of genotype, maturity stage and post-harvest storage on phenolic compounds, carotenoid content and antioxidant capacity, of Andean mashua tubers (Tropaeolum tuberosum Ruiz & Pavón). Journal of the Science of Food and Agriculture, 87, 437–446. https://doi.org/10.1002/jsfa

Chirinos, R., Campos, D., Betalleluz, I., Giusti, M., Schwartz, S., Tian, Q., Larondelle, Y. 2006. High-Performance Liquid Chromatography with Photodiode Array Detection (HPLC − DAD)/ HPLC − Mass Spectrometry (MS) Profiling of Anthocyanins from Andean Mashua Tubers ´ z and Pavo ´ n) and Their (*Tropaeolum tuberosum* Ruız Contribution to the Overall. Journal Agricultural Food Chemistry 54: 7089−7097. https://doi.org/10.1021/jf0614140 CCC:

Chirinos, R., Campos, D., Costa, N., Arbizu, C., Pedreschi, R., Larondelle, Y. 2008b. Phenolic profiles of andean mashua (Tropaeolum tuberosum Ruíz & Pavón) tubers: Identification by HPLC-DAD and evaluation of their antioxidant activity. Food Chemistry, 106: 1285–1298. https://doi.org/10.1016/j.foodchem.2007.07.024

Chirinos, R., Campos, D., Warnier, M., Pedreschi, R., Rees, J. F., Larondelle, Y. 2008 a. Antioxidant properties of mashua (*Tropaeolum tuberosum*) phenolic extracts against oxidative damage using biological *in vitro* assays. Food Chemistry. 111: 98–105. https://doi.org/10.1016/j.foodchem.2008.03.038

De Pascual-Teresa, S., Sanchez-Ballesta, M. T. 2008. Anthocyanins: From plant to health. Phytochemistry Reviews. 7: 281–299. https://doi.org/10.1007/s11101-007-9074-0

Espín, S., Brito, B., Villacrés, E., Rubidio, A., Nieto, C., Grijalva, J. 2001. Composición química, valor nutricional y usos potenciales de siete especies de raices y tubérculos Andinos. Acta Cientifica Ecuatoriana. 7(1).

F. A. Tomás-Barberán. 2003. Los polifenoles de los alimentos y la salud. Alimentacion. Nutricion y Salud. 10: 41–53

F.Saura-Calixto, I.G.J.S. 2007. Caracterización de los alimentos tradicionales de la dieta española: alegaciones nutricionales y alegaciones en salud. 1–32.

Fahey, J. W., Zalcmann, A. T., Talalay, P. 2001. The chemical diversity and distribution of glucosinolates and isothiocyanates among plants. Phytochemistry. 56: 5–51.

Fahey, J. W., Zalcmann, A. T., Talalay, P. 2001. The chemical diversity and distribution of glucosinolates and isothiocyanates among plants. Phytochemistry. 56: 5–51.

Flannery, K. V. 1973. The origins of agriculture. Annual Review Anthropology. 2: 271–310.

García-Casal, M. N., Landaeta, M., De Baptista, G. A., Murillo, C., Rincón, M., Rached, L. B., … Peña-Rosas, J. P. 2013. Valores de referencia de hierro, yodo, zinc, selenio, cobre, molibdeno, vitamina C, vitamina E, vitamina K, carotenoides y polifenoles para la población venezolana. Archivos Latinoamericanos de Nutricion. 63: 338–361.

Ghiselli, A., Nardini, M., Baldi, A., Scaccini, C. 1998. Antioxidant Activity of Different Phenolic Fractions Separated from an Italian Red Wine. Journal of Agricultural and Food Chemistry. 46: 361–367. https://doi.org/10.1021/jf970486b

Grau, A., Ortega, R., Nieto, C., Hermann, M. 2003. Mashua *Tropaeolum tuberosum* Ruíz & Pav. Promoting the conservation and use of underutilized and neglected crops. (International Potato Center, Ed.). Lima, Peru: International Plant Genetic Resources Institute, Rome, Italy.

Gross, J. 1987. Pigments in Fruits. Academic Press. London.

Gross, R., Koch, F., Malaga, I., Miranda, A.F De, Schoeneberger, H., Trugo, L.C. 1989. Chemical Composition and protein quality of some local Andean Food Sources. Food Chemistry. 34: 25–34.

Hodge, W. 1951. Three native tuber foods of the high Andes. Economic Botany. 5: 185–201.

Johns, T., Kitts, W. D., Newsome, F., Towers, G. H. N. 1982. Anti-reproductive and other medicinal effects of *Tropaeolum tuberos* um. Journal of Ethnopharmacology, 5: 149–161. https://doi.org/10.1016/0378-8741(82)90040-X

Johns, T., Neil Towers, G. H. 1981. Isothiocyanates and thioureas in enzyme hydrolysates of *Tropaeolum tuberosum*. Phytochemistry. 20: 2687–2689. https://doi.org/10.1016/0031-9422(81)85268-5

Joseph, J.A., Shukitt-Hale, B., Denisova, N.A., Prior, R.L., Cao, G., Martin, A., Bickford, P.C. 1998. Long-term dietary strawberry, spinach, or vitamin E supplementation retards the onset of age-related neuronal signal-transduction and cognitive behavioral deficits. The Journal of Neuroscience: The Official Journal of the Society for Neuroscience. 18: 8047–8055.

King, S. R., Gershoff, S. N. 1987. Nutritional evaluation of three underexploited andean tubers: *Oxalis tuberosa* (Oxalidaceae), *Ullucus tuberosus* (Basellaceae), and *Tropaeolum tuberosum* (Tropaeolaceae). Economic Botany. 41: 503–511. https://doi.org/10.1007/BF02908144

Koide, T., Hashimoto, U., Kamei, H., Kojima, T., Hasegawa, M., Terabe, K. 1997. Antitumor effect of anthocyanin fractions extracted from red soybeans and red beans *in vitro* and *in vivo*. Cancer Biotherapy and Radiopharmaceuticals. 12: 277–280. https://doi.org/10.1089/cbr.1997.12.277

Krinsky, N.I., Johnson, E. J. 2005. Carotenoid actions and their relation to health and disease. Molecular Aspects of Medicine. 26: 459–516. doi.org/10.1016/j.mam.2005.10.001

León, J. 1964. Plantas Alimenticias Andinas Boletın Técnico No. 6. Lima, Peru.

Liebler, D.C., Stratton, S.P., Kaysen, K.L. 1997. Antioxidant Actions of β-Carotene in Liposomal and Microsomal Membranes: Role of Carotenoid-Membrane Incorporation and α-Tocopherol. Archives of Biochemistry and Biophysics. 338: 244–250. https://doi.org/10.1006/abbi.1996.9822

Manrique, I., Arbizu, C., Vivanco, F., Gonzales, R., Ramírez, C., Chávez, O., Ellis, D. 2013. *Tropaeolum tuberosum* Ruíz and Pav. Colección de germoplasma de mashua conservada en el Centro Internacional de la Papa (CIP) (Primera). Lima, Perú: Centro Internacional de la Papa.

Martínez- Valverde, I., Periago, M., Ros, G. 2000. Significado nutricional de los compuestos fenólicos de la dieta. Archivos Latinoamericanos de Nutricion. 50: 5–14.

Martínez Roldan, C., Carbajal Azcona, Á. 2012. Componentes bioactivos de los alimentos. Manual Práctico de Nutrición y Salud.

Mobarhan, S., Shiau, A., Grande, A., Kolli, S., Stacewicz-Sapuntzakis, M., Oldham, T., Frommel, T. 1994. 3-Carotene Supplementation Results in an Increased Serum and Colonic Mucosal Concentration of n-Carotene and a Decrease in a-Tocopherol Concentration in Patients with Colonic Neoplasia. Cancer Epidemiology, Biomarkers and Prevention. 3: 1–505.

Moskaug, J.Ø., Carlsen, H., Myhrstad, M. C., Blomhoff, R. 2005. Polyphenols and glutathione synthesis regulation. American Society for Clinical Nutrition. 81: 277S–83S.

National Research Council. 1989. Roots and tubers, in Lost Crops of the Incas: Little Known Plants of the Andes with Promise for Worldwide Cultivation. National Academy Press. Washington, DC.

Navarro, M., Monagas, M., Quesada, S., Murillo, R., Bartolomé, B., Sánchez-Patán, F., Garrido, I. 2013. Extractos fenólicos de *Uncaria tomentosa* l. (uña de gato) costarricense: composición estructural y bioactividad. In SILAE XXII.

Ortega, O., Kliebenstein, D., Arbizu, C., Ortega, R., Quiros, C. 2006. Glucosinolate survey of cultivated and feral mashua (Tropaeolum Tuberosum Ruiz y Pavón) in the Cuzco Region of Perú.

Ortíz, M.A., Reza, C., Gerardo, R., Madinaveitia, C., Ciencias, F. De, Universidad, Q., Artículo, A. 2011. Propiedades funcionales de las antocianinas. Biotecnia. 13: 16–22.

Otero, M. J., Hidalgo, L. 2004. Taninos condensados en especies forrajeras de clima templado: efectos sobre la productividad de rumiantes afectados por parasitosis gastrointestinales (una revisión). Livestock Research for Rural Development. 16: Retrieved from http://www.lrrd.cipav.org.co/lrrd16/2/oter16 02.htm

Peñarrieta, M., Tejeda, L., Mollinedo, P., Vila, J., Bravo, J. 2014. Phenolic compounds in food. Revista Boliviana de Química. 31: 68–81. https://doi.org/10.1007/s00394-008-2002-2

Pissard, A., Arbizu, C., Ghislain, M., Bertin, P. 2008. Influence of Geographical Provenance on the Genetic Structure and Diversity of the Vegetatively Propagated Andean Tuber Crop, Mashua (*Tropaeolum tuberosum*), Highlighted by Intersimple Sequence Repeat Markers and Multivariate Analysis Methods. International Journal of Plant Sciences. 169: 1248–1260. https://doi.org/10.1086/591979

Quispe, C., Mansanilla, R., Chacón, A., Blas, R. 2015. Análisis de la variabilidad morfológica del "Añu" *Tropaeolum tuberosum* Ruiz & Pavón procedente de nueve distritos de la región cusco. Ecología Aplicada. 14: 1–12.

Ramallo, R. 2004. Análisis exploratorio de los ácidos grasos del isaño (*Tropaeolum tuberosum*). Investigación y Desarrollo. 77: 71–77.

Ramallo, R., Wathelet, J. P., Le Boulengé, E., Torres, E., Marlier, M., Ledent, J. F., Larondelle, Y. 2004. Glucosinolates in isaño (Tropaeolum tuberosum) tubers: Qualitative and quantitative content and changes after maturity. Journal of the Science of Food and Agriculture. 84: 701–706. https://doi.org/10.1002/jsfa.1691

Rinc, A. 2014. Biosintesis De Los Glucosinolatos E Importancia Nutricional Humana Y Funciones De Protección a Las Plantas. Revista Alimentos Hoy. 22: 64–80.

Roca, W. M., Ynouye, C., Manrique, I., Arbizu, C., Gomez, R. 2007. Indigenous Andean Root and Tuber Crops: new sources of food for the new millennium Indigenous Andean Root and Tuber Crops: New Foods for the New Millennium ANDEAN ROOT AND TUBER. Chronica Horticulturae. 47: 1–19.

Roca, W., Manrique, I. 2005. Tubérculos Andinos Para La Nutrición y La Salud. Agrociencia. IX: 195–201.

Rodríguez-Cruz, M., Tovar, A., Del Prado, M., Torres, N. 2005. Mecanismos moleculares de acción de los ácidos grasos poliinsaturados y sus beneficios en la salud. Revista de Investigación Clínica, 57: 457–472.

Rosalba Beas, F., Guadalupe Loarca, P., Salvador Horacio Guzmán, M., Rodriguez, M. G., Nora Lilia Vasco, M., Fidel Guevara, L. 2011. Potencial nutraceutico de componentes bioactivos presentes en huitlacoche de la zona centro de México. Revista Mexicana de Ciencias Farmaceuticas. 42: 36–44.

Sánchez, A., L, F.-C., Langley, E., Martín, R., G, M., Sánchez, S. 1999. Carotenoides: Estructura, Función, Biosíntesis, Regulación y Aplicaciones. Revista Latinoamericana de Microbiología. 41: 175–191.

Sepúlveda Jiménez, G., Porta Ducoing, H., Rocha Sosa, M., Sepúlveda-Jiménez, G., Porta-Ducoing Mario Rocha-Sosa, H. 2003. La Participación de los metabolitos secundarios en la defensa de las plantas. Revista Mexicana de Fitopatologí. 21: 355-363

Serrano, M.E.D., López, M.L., Espuñes, T.D.R. S. 2006. Componentes bioactivos de alimentos funcionales de origen vegetal. Revista Mexicana de Ciencias Farmaceuticas: 37: 58–68.

Soriano, E. 2003. Los metabolitos de las plantas y las células cancerosas I. los flavonoides. REB. Revista de Educación Bioquímica. 22: 191–197.

Surco, F. 2004. Caracterización de almidones aislados de tuberculos andinos: mashua (*Tropaeolum tuberosum*), Oca (*oxalis tuberosa*), Olluco (*Ullucus tuberosus*) para su aplicación tecnologica. 1–55. Retrieved from http://cybertesis.unmsm.edu.pe/handle/cybert esis/2588

Tapia, M. E., Fries, A. M. 2007. Guia de campo de los cultivos andinos. Fao; Anpe-Perú.

Taylor, S. 1996. Beta-carotene, carotenoids, and disease prevention in humans. Faseb Journal. 10: 690–701.

Torres, J. L., Lozano, C., Julià, L., Sánchez-Baeza, F. J., Anglada, J. M., Centelles, J. J., Cascante, M. 2002. Cysteinyl-flavan-3-ol conjugates from grape procyanidins. Antioxidant and antiproliferative properties. Bioorganic and Medicinal Chemistry. 10: 2497–2509. https://doi.org/10.1016/S09680896(02)00127 -X

Urango, L., Montoya, G., Cuadros, M., Henao, D., Zapata, P., López, L., Gómez, B. 2008. Efecto de los compuestos bioactivos de algunos alimentos en la salud. Perspectivas En Nutrición Humana. 11: 124–4108.

Urresta, B. 2010. Evaluación del valor nutricional de la harina de mashua (Tropaeolum Tuberosum) en dietas para pollos de engorde. Escuela Politécnica Nacional.

Valcárcel-Yamani, B., Rondán-Sanabria, G. G., Finardi-Filho, F. 2013. The physical, chemical and functional characterization of starches from andean tubers: Oca (*Oxalis tuberosa* molina), olluco (*Ullucus tuberosus* caldas) and mashua (*Tropaeolum tuberosum* ruiz & pavón). Brazilian Journal of Pharmaceutical Sciences. 49: 453–464.

https://doi.org/10.1590/S1984825020130003
00007

Vásquez, J.H., Gonzales, J.M., Pino, J.L. 2012. Decrease in spermatic parameters of mice treated with hydroalcoholic extract *Tropaeolum tuberosum* "mashua" Revista Peruana de Biología. 19: 89–93.

Vázquez- Flores, A., Alvarez- Parrilla, E., Lopez-Díaz, J., Wall- Medrano, A., De la Rosa, L. 2012. Taninos hidrolizables y condensados: naturaleza química, ventajas y desventajas de su consumo. Nutricion Hospitalaria. 6: 84–93. doi.org/10.3305/nh.2015.31.1.7699

Villacrés, E., Espín, S. 1999. Evaluación y rendimiento, características y propiedades del almidón de algunas raíces y tubérculos andinos. raíces y tubérculos andinos. (Centro Internacional de la Papa, Ed.), Avances de la Investigación Tomo 1. (Vol. 1). Lima Perú.

Vitale, A. A., Bernatene, E. A. y Pomilio, A. B. 2010. Carotenoides en quimioprevención : Licopeno Carotenoids in chemoprevention : Lycopene. Acta Bioquímica Clínica. 44: 195–238.

Wang, Shiow Y., Lin, H.-S. 2000. Antioxidant Activity in Fruits and Leaves of Blackberry, Raspberry, and Strawberry Varies with Cultivar and Developmental Stage. Journal of Agricultural and Food Chemistry. 48: 140–146.

Yoshimoto, M., Okuno, S., Yamaguchi, M., Yamakawa, O. 2001. Antimutagenicity of Deacylated Anthocyanins in Purple-fleshed Sweetpotato. Bioscience, Biotechnology, and Biochemistry: 65: 1652–1655. https://doi.org/10.1271/bbb.65.1652

Zaninovic´, V. 2003. Posible asociación de algunas enfermedades neurológicas con el consumo excesivo de la yuca mal procesada y de otros vegetales neurotóxicos. Colombia Médica. 34: 82–91.

Zulueta, A., Esteve, M. J., Frígola, A. 2009. ORAC and TEAC assays comparison to measure the antioxidant capacity of food products. Food Chemistry. 114: 310–316. https://doi.org/10.1016/j.foodchem.2008.09.033

11

Use of Alternative Substrates for Broccoli Seedling Production under Greenhouse Conditions

Hernán Zurita-Vásquez*, Luciano Valle, Marcia Buenaño, Deysi Guevara,
Gonzalo Mena and Carlos Vásquez

*Facultad de Ciencias Agropecuarias, Universidad Técnica de Ambato, Carretera
Cevallos-Quero, 180350 Cevallos, Tungurahua, Ecuador.
Email: hernanzurita@yahoo.es
Corresponding author

SUMMARY

In this study, the effect of two alternative growing substrates (corncob and *Azolla anabaena*) on some vegetative parameters (days to emergence, stem diameter, plant height and root volume) in seedlings of broccoli hybrid Coronado were evaluated. Both materials were ground and used according to the following treatments: 100% corncob (T1), 100% *Azolla* (T2), 50% corncob + *Azolla* 50% (T3), corncob 75% + *Azolla* 25% (T4), 25% corncob + *Azolla* 75% (T5) and then compared to a commercial substrate (BM2) (T6). Substrates were uniformly mixed, deposited in germination trays and watered at field capacity. Additionally, physic-chemical characteristics were determined in the different substrates used for broccoli seedling production. Stem diameter, height plant and root volume showed to be statistically higher in seedlings grown in commercial substrate (0.18 cm, 5.27 cm and 0.61 cm^3, respectively) followed by those seedlings grown in 100% *Azolla* (0.17 cm, 4.92 cm and 0.49 cm^3, respectively) ($p < 0.001$). Seedlings growing in substrates with higher corncob proportion showed lower values in these vegetative parameters. Based on our results, *Azolla* showed potential to be used as a seed substrate for production of broccoli seedlings, thus decreasing the use of peat and consequently the production costs in nurseries.

Keywords: *Brassica*, substrates, *Azolla*, maize residues, organic agriculture.

INTRODUCTION

Broccoli (*Brassica oleracea* var. *italica*) is an annual vegetable of importance for human nutrition and vegetable oil production (Lopes *et al.*, 2012), however, its yield and quality can be affected by several factors including fertilization regime (Feller and Fink, 2005), as well as, for the date and method of sowing (transplanting or direct sowing), among others (Reta *et al.*, 2004). In relation to the sowing method, transplant is preferred because it ensures to obtain more vigorous seedling from a few number of seeds. In addition, it allows to get an early harvest (Laviola *et al.*, 2006) and to increase productivity (Andreoli *et al.*, 2002).

Additionally, quality of seedling depends on the quality of seeds and the type of substrate (Lopes *et al.*, 2012). Substrates must show good water retention capacity, porosity to promote oxygen diffusion, to be free of pathogens and provide nutrients for the plant (Silva *et al.*, 2001, Smiderle and Minami, 2001). The overexploitation of some types of substrates to obtain commercial seedlings can cause an ecological impact in places where they are extracted from, for example: in peat exploitation sites in northern Europe (Maroto, 2000). Therefore, there is a growing concern to use alternative substrates, which include by-products from either wood industry (pine bark compost) or agricultural production (coconut fiber, some plant straw fibers) (Maroto, 2000).

In recent years, the use of *Azolla* has gained popularity as a sowing substrate because it has shown to confer positive effects on agriculture productivity since its symbiosis with a nitrogen fixer seaweed, *Anabaena azollae* (Petruccelli *et al.*, 2015). Although the *Azolla-Anabaena* complex has been mainly used as fertilizer in rice plantings, it has a wide potential to be used as bio fertilizer in other crops (Pabby *et al.*, 2003). Previous works that included the use of *Azolla* mixed with other types of substrates have proved to be easy handling and to have an acceptable water content to be used either as fertilizer or as a substrate in nurseries in different crops (Ríos, 2014). Thus, the use of *Azolla* in combination to kekilla allowed to obtain good quality broccoli plants (Gavilanez, 2015).

On the other hand, corncob is also a viable alternative for use as sowing substrate since of the total hemicellulose (34%), approximately 94% corresponds to xylan, making it attractive for the development of nitrogen fertilizers with prolonged or slow action (Córdoba *et al.*, 2013). In consideration of the above, in this study the effect of corncob and *Azolla anabaena* as substrate for sowing on the vegetative parameters of the *B. oleracea* seedlings were evaluated

MATERIALS AND METHODS

Location and substrate collection

Study was conducted in Parroquia Montalvo, Canton Ambato, Province of Tungurahua (01°24'00 "S, 78°23'00" W, 2600 masl). Corncob was obtained from the waste of the grain harvest while *A. anabaena* was collected from the water reservoir of the Experimental Farm at the Faculty of Agricultural Sciences (FCAGP), Technical University of Ambato (UTA) in Querochaca, Province of Tungurahua. Previuos to be used, *Azolla* was shade dried at room temperature for five days, as described by Petruccelli *et al.* (2015).

Preparation of substrates and treatments

Each substrate was ground in a one-HP electric mill (Fritsch™; model Pulverisette 15; Germany) to obtain 1 mm particles. Firstly, all substrates were disinfected using Vitavax Flow™ (2 mL/20 L of water) and uniformly mixed, according to the following treatments: T1; corncob 100%, T2; *Azolla anabaena* 100%, T3; corncob 50% + *Azolla anabaena* 50%, T4; corncob 75% + *Azolla anabaena* 25%, T5; corncob 25% + *Azolla anabaena* 75%, T6; BM2 commercial substrate. After that, substrates were placed on germination trays and watered to field capacity. Previously, a chemical analysis of each substrate was made in order to determine the nutrient supply in each case (Table 1). This research was conducted in a plastic-covered greenhouse providing 35% shade.

Variable considered

The number of days to emergence was determined taking into account the days from sowing until 50% of the seedlings had emerged in each treatment. On the other hand, stem diameter, plant height and root volume were measured 28 days after sowing (before being transplanted).

Physical and chemical analysis

Electrical conductivity (EC) and pH were determined according to Violante and Adamo (2000). Total nitrogen and carbon content were determined by the Dumas method, using the CHN LECO 628 elemental analyzer (LECO Corporation). The total phosphorus content was determined by colorimetric method of the vanado-molybdate complex (Genesys 20 Spectrophotometer) (Anderson and Ingram, 1993). Potassium, calcium, magnesium, cooper, manganese and zinc contents were determined by wet digestion (AA Perkin Elmer 100 spectrophotometer) (Shirin et al., 2008), while K2O, CaO and MgO were calculated by transformation from pure elements to their respective oxide forms. All analyzes were replicated three times.

Experimental design and statistical analysis: The study was carried out in a completely randomized design with six treatments and six repetitions. All the variables were subjected to analysis of variance and mean comparison was made by using a Tukey test (p <0.05) (INFOSTAT version 2016).

Table 1. Chemical analysis to the substrates used for broccolis seedling production

Sustrate	pH	CE $\mu S\backslash cm$	N	P_2O_5	K_2O	CaO (%)	MgO	C	C:N	Cu	Mn ppm	Zn
T1	6,05	1275	1,14	0,87	0,34	2,96	0,67	45,58	39,38	19	38	58
T2	6,35	1948	1,51	1,38	0,16	2,86	0,54	35,18	23,30	19	353	37
T3	6,33	1888	1,39	0,90	0,10	2,50	1,28	42,98	30,92	20	159	20
T4	6,33	1493	1,25	0,76	0,10	1,37	0,28	43,65	34,92	39	96	39
T5	6,29	2004	1,50	0,85	0,26	2,70	0,47	39,12	26,08	20	223	20
T6	4,96	720	0,71	1,35	0,09	6,12	3,28	35,24	49,63	20	79	40

T1: corncob 100%; T2: *Azolla* 100%; T3: corncob 50% + *Azolla* 50%; T4: corncob 75% + *Azolla* 25%; T5: corncob 25% + *Azolla* 75%; T6: BM2 (commercial)

RESULTS AND DISCUSSION

Agronomic parameters

Days to emergence: the type of substrate used affected the number of days to the emergence of broccoli seedlings (p <0.001) (Table 2). Lower number of days (from sowing up to 50% of the seedlings emerged) was observed in the T6 treatment (commercial substrate) with an average of 5.0 days, while the longest time was observed in T3, in the which seeds were delayed up to 5.5 days as compared to T6. The rest of the treatments showed intermediate values. In general, broccoli seeds have shown to require between 8 and 15 days for total emergence, depending on the depth of sowing and covering of seed (Corpocauca, 2007), soil aeration conditions and the speed of plant growth (Quesada and Méndez, 2005). Additionally, seed germination also depends on the substrate used (Oliveira et al., 2015). Thus, higher germination rate in *Tabebuia heptaphylla* and *Terminalia argentea* was observed in substrates with predominance of clay particles compared to sandy soils or vermiculite (Bocchese et al., 2008, Oliveira and Farias, 2009)

Stem diameter: significant differences by effect of the substrate type were detected in stem diameter of broccoli seedlings (p <0.001). Maximum values were observed in seedlings grown on substrate containing 100% *Azolla* (T2) and on the commercial substrate (T6) (Table 2). Contrarily, lower values were shown in seedlings planted in substrates with higher content of ground corncob (T1, T3, T4 and T5). Petruccelli et al. (2015) found that olive plants caused showed better growth and accumulation of biomass when grown in substrates with 50% *Azolla*. These authors suggested that *Azolla* could be an excellent

component of the substrates used in nursery conditions for plants production.

Plant height: the type of substrate used also caused significant differences in plant height (p <0.001) (Table 2). Higher height was observed in plants growing or in the substrate containing 100% *Azolla* (T2) or on the commercial substrate (T6). Similar to that observed with stem diameter, plants that grew on those substrates containing higher proportion of corncob showed lower height, being 2.38 to 2.55 times smaller when 100% was used of corncob (T1) in comparison to T2 and T6, respectively, whereas when the proportion of corncob used in the substrate decreased to 25% (T5) this difference was smaller, being 1.32 and 1.41 times lower in comparison with the same treatments (T2 and T6, respectively).

Probably higher values obtained in plants grown on the substrate with 100% *Azolla* (T2) could be because nitrogen, phosphorus and potassium contents in *Azolla* are similar to those in the commercial substrate BM2 (T6). On the other hand, lower values in plant height in T1 (100% corncob) could be explained by low phosphorus as well as high zinc content since phosphorous can cause general stunting (Bertsch, 2009) whereas zinc provokes competition for P_2O_5 and Mg absorption, which are important for stem growth (Table 1).

Root volume: In general, plants grown on the commercial substrate (T6) showed the maximum root volumes at 28 days after sowing, followed by those grown on 100% *Azolla* (T2) substrate in which root volume was only 20% lower (Table 2). Conversely, plants cultivated with substrates containing higher corncob percentage showed root volumes 68.9 and 57.4% lower when sown in 75% corncob + 25%

Azolla (T4) and 100% corncob (T1), respectively. Although corncob plays a role in the conservation of soil, water and soil carbon, it is still unknown what its contribution of nutrients (Wienhold *et al.*, 2011), so that more detailed studies are still required to measure its impact on other crops. Similar to that observed by Petruccelli *et al.* (2015), both the pH and EC values tended to increase as *Azolla* content increased (Table 1), however, this does not seem to have negatively affected growth of broccoli plants since they show acceptable development in conditions close to neutrality and these particular salinity conditions (Sánchez-Monedero *et al.*, 1997).

Table 2. Variation in the agronomic parameters (days to emergence, stem diameter, plant height and root volume) in Coronado hybrid broccoli seedlings grown on different substrates

Substrate	Days to emergence	Stem diameter	Plant height	Root volume
T1	7,17±0,983ab	0,12±0,028 c	2,06±0,396 c	0,26±0,057 c
T2	8,67±1,366ab	0,17±0,010 a	4,92±0,470 a	0,49±0,036 b
T3	9,50±2,074ab	0,12±0,011 c	2,66±0,349 c	0,26±0,048 c
T4	10,50±1,975b	0,12±0,008 c	2,53±0,249 c	0,19±0,0256 c
T5	9,33±2,251ab	0,14±0,013bc	3,73±0,294 b	0,40±0,034 b
T6	5,00±2,041a	0,18±0,014 a	5,27±0,488 a	0,61±0,070 a
Valor P	0,001	0,001	0,001	0,001

Values in a column followed by the same letter did not show significant differences according to Tukey's test at p< 0,001. T1: corncob 100%; T2: *Azolla* 100%; T3: corncob 50% + *Azolla* 50%; T4: corncob 75% + *Azolla* 25%; T5: corncob 25% + *Azolla* 75%; T6: BM2 (commercial)

On the other hand, those substrates containing *Azolla* 100 or 75% showed 2.12 and 2.11 times more N in relation to the commercial substrate, which could be explained by the high concentration of this element in its dry biomass (Bhuvaneshwari and Singh, 2015). Most of the studies have indicated that broccoli is highly demanding in nitrogen, therefore high N content in T2 could explain better plant growth; however, requirements may vary as cultivation conditions and cultivars (Rincón-Sánchez *et al.*, 2001). Additionally, the presence of calcium in 100% *Azolla* (T2) could favor the activation of several plant enzyme systems, which stimulate root and leaf development (INPOFOS, 1997). Likewise, there is a positive effect between the amount of Mg that the crop absorbs and the root and aerial growth (Cakmak and Yazici, 2010), so that together with Ca and N, they could have influenced positively the growth of the seedlings.

Additionally, C/N ratio could also have influenced the best development of broccoli seedlings. On the one hand, C/N ratio influences the availability of mineral nitrogen absorbed by plants and, along with the soil humidity, temperature and aeration favors the microbial activity. According to previous studies, it is estimated that microorganisms generally use 30 parts of C for each part of N (Jhorar *et al.*, 1991). Based on this, the observed C/N ratio in the substrates with Azolla could ensure a source of mineral nitrogen available for seedling growth.

According to Bilderback *et al.* (2005), substrates used for plant production in nurseries should show physical-chemical characteristics within acceptable ranges to ensure quality of seedlings. However, these values should not be generalized, but should be adapted to certain groups of plants with similar requirements (Petruccelli *et al.*, 2015).

CONCLUSIONS

Azolla proved to be the best substrate since it allowed the seeds of broccoli to germinate in the shortest time, in addition to the seedlings showing better characteristics of stem diameter, plant height and root volume, comparable with the seedlings obtained with the commercial substrate. In this regard, the use of *Azolla* as substrate for broccoli seedlings production could be a sustainable alternative because it is a source of nutrients necessary for plant growth and development in nursery.

REFERENCES

Anderson, J., Ingram, I. 1993. Tropical Soil Biology and Fertility: A handbook of methods. Oxford: CAB International. pp: 237.

Andreoli, C., Andrade, V.R., Zamora, S.A., Gordon, M. 2002. Influência da germinação da semente e da densidade de semeadura no estabelecimento do estande e na produtividade de milho. Revista Brasileira de Sementes. 24:1-5. DOI 10.1590/S0101-31222002000100001.

Bertsch, F. 2009. Absorción de nutrimentos por los cultivos. First edition. San José de Costa Rica: Asociación Costarricense de la Ciencia del Suelo. pp: 307.

Bhuvaneshwari, K., Singh, P.K. 2015. Response of nitrogen-fixing water fern *Azolla* biofertilization to rice crop. 3 Biotech. 5:523-529. DOI: 10.1007/s13205-014-0251-8

Bilderback, T.E., Warren, Jr S.L., Owen, J.S., Albano, J.P. 2005. Healthy substrates need physicals too. HortTechnology. 15:747-751.

Bocchese, R.A., Oliveira, A.K.M., Melotto, A.M., Fernandes, V., Laura, V.A. 2008. Efeito de diferentes tipos de solos na germinação de sementes de *Tabebuia heptaphylla*, em casa telada. Cerne. 14:62-67.

Cakmak, I., Yazici, A.M. 2010. Magnesium: a forgotten element in crop production. Better Crops. 94(2):23-25.

Córdoba, J.A., Salcedo, E., Rodríguez, R., Zamora, J.F., Manríquez, R., Contreras, H., Robledo, J., Delgado, E. 2013. Caracterización y valoración química del olote: degradación hidrotérmica bajo condiciones subcríticas. Revista Latinoamericana de Química. 41:171-184.

Corpocauca (Corporación para el Desarrollo Del Cauca). 2007. Alianza productiva para el fortalecimiento a la cadena de hortalizas en brócoli (*Brassica oleracea* L.) Municipio de Pasto, Corregimiento De Gualmatan, Nariño. Ministerio De Agricultura y Desarrollo Rural. http://observatorio.misionrural.net/alianzas/productos/brocoli/pasto/preinversion_%20BRoCOLI.pdf. Consulted on March 25, 2017.

Feller, C., Fink, M. 2005. Growth and yield of broccoli as affected by the nitrogen content of transplants and the timing of nitrogen fertilization. HortScience. 40: 1320-1323.

Gavilanez, E. J. 2015. Evaluación del helecho de agua asociado con anabaena como sustrato ecológico para producción de plantas de brócoli.M. Sc. Thesis. Universidad Técnica de Ambato.

INPOFOS (Instituto de la Potasa y el Fosfato). 1997. Manual Internacional de la Fertilidad del Suelo, Quito, Ecuador.

Jhorar, B.S., Phogat, V., Malik, R.S. 1991. Kinetics of composting rice straw with glue waste at different carbon: nitrogen ratios in a semiarid environment. Arid Soil Research and Rehabilitation. 5:297-306.

Laviola, B.G., Lima, P.A., Wagner Jr., A., Mauri, A.L., Viana, R.S., Lopes, J.C. 2006. Efeito de diferentes substratos na germinação e no desenvolvimento inicial de jiloeiro (Solanum gilo Raddi), cultivar verde claro. Ciência e Agrotecnología. 30:415-421.

Lopes, J.C., Mauri, J., Ferreira, A., Alexandre, R.S., Freitas, A.R. 2012. Broccoli production depending on the seed production system and organic and mineral fertilizer. Horticultura Brasileira. 30:143-150. DOI: 10.1590/S1413-70542006000300005.

Maroto, J.M. 2000. Elementos de Horticultura General. España: Mundiprensa Libros S.A. pp: 481.

Oliveira A.K.M., Farias G.C. 2009. Efeito de diferentes substratos na germinação de sementes de *Terminalia argentea* (Combretaceae). Revista Brasileira de Biociências. 7:320-323.

Oliveira, A.K.M., Souza, S.A., Souza, J.S., Carvalho Jr., M.B. 2015. Temperature and substrate influences on seed germination and seedling formation in *Callisthene fasciculata* Mart. (Vochysiaceae) in the laboratory. Revista Árvore. 39:487-495. DOI: 10.1590/0100-67622015000300009.

Pabby, A., Prasanna, R., Singh, P.K. 2003. Azolla-Anabaena symbiosis: from traditional agriculture to biotechnology. Indian Journal of Biotechnology. 2:26-37.

Petruccelli, R., Briccoli Bati, C., Carlozzi, P., Padovani, G., Vignozzi, N., Bartolini, G. 2015. Use of *Azolla* as a growing medium component in the nursery production of olive trees. International Journal of Basic and Applied Sciences. 4:333-339. DOI: 10.14419/ijbas.v4i4.4660

Quesada, G., Méndez, C. 2005. Evaluación de sustratos para almácigos de hortalizas. Universidad de Costa Rica. Agronomía Mesoamericana. 16:171-183.

Reta, D.G., Faz, R., Moreno, L.E. 2004. Rendimiento y calidad del brócoli establecido en siembra directa y trasplante en cinco fechas. Agrofaz. 4:537-542.

Rincón-Sánchez, L., Pellicer-Botía, C., Sáez-Sironi, J., Abadía-Sánchez, A., Pérez-Crespo, A., Marín-Nartínez C. 2001. Crecimiento y absorción de nutrientes del brócoli.

Investigación Agraria: Producción y Protección Vegetal. 16:119-129.

Ríos, C.A. 2014. Determinación de los métodos y tiempos de secado de *Azolla* para obtener un sustrato orgánico en la Parroquia Pinguilí, Cantón Mocha, Provincia de Tungurahua. M. Sc. Thesis. Universidad Técnica de Ambato.

Sánchez-Monedero, M.A., Bernal, M.P., Antón, A., Noguera, P., Abad, A., Roig, A., Cegarra, J., 1997. Utilización del Compost como sustratos para semilleros de plantas hortícolas en Cepellon. p. 78-85. In Proceedings of the I Congreso Ibérico y III Nacional de Fertirrigación, Murcia, España.

Shirin, K., Naseem, S., Bashir, E., Imad, S., Shafio, S. 2008. A comparison of digestion methods for the estimation of elements in *Dodonaea viscosa*: a native flora of Wadh, Balochistan, Pakistan. Journal of the Chemical Society of Pakistan. 30:90-95.

Silva, R.P., Peixoto, J.R., Junqueira, N.T.V. 2001. Influência de diversos substratos no desenvolvimento de mudas de maracujazeiro azedo (*Passiflora edulis* Sims f. *flavicarpa* DEG). Revista Brasileira de Fruticultura. 23:377-381. DOI: 10.1590/S0100-29452001000200036

Smiderle, O.S., Minami, K. 2001. Emergência e vigor de plântulas de goiabeira em diferentes substratos. Revista Científica Rural. 6:38-45.

Violante, P., Adamo, P. 2000. Determinazione del grado di reazione (pH), metodi di analisi chimica del suolo, F. Angeli (Ed.). Roma 10-13.

Wienhold, B.J., Varvel, G.E., Jin, V.L. 2011. Corn cob residue carbon and nutrient dynamics during decomposition. Agronomy Journal. 103:1192-1197. DOI: 10.2134/agronj2011.0002

12

Ecophysiological Responses of Potato (*Solanum Tuberosum*) to Salinity and Nitrogen Fertilization in Screenhouse, Cameroon

P. T. Tabot*, S.N. Mbega and N.F.J. Tchapga

Department of Agriculture, Higher Technical Teachers' Training College Kumba, University of Buea. P.O. Box 249 Kumba, Cameroon.
email:pascal.tabot@ubuea.cm, mbegasanamanathalie@yahoo.fr, fjtchapgangandjui@yahoo.fr
*Corresponding author:

SUMMARY

Irrigated agriculture under fertilization is a practice that can ensure food security but has the unintended consequence of secondary salinization of arable land. There is the need to forecast production in this increasingly saline environment. An experiment of 4x4 factorial was carried out in screenhouse in Cameroon, to test the responses of potato to salinity under nitrogen fertilization. Plants were grown over two months and standard growth and physiological measurements taken and analysed. Results showed that in conditions of increased sediment salinity up to 12 ppt, tuber yields were greatly suppressed, and did not respond to fertilization. Plants grown under freshwater irrigation had higher root:shoot ratio and tuber development. Plants irrigated with freshwater had higher water use efficiency (WUE), irrespective of whether they were fertilized or not; WUE decreased as salinity increased. Photosynthetic efficiency dropped in all treatments. Under freshwater conditions, potato plants present a more efficient physiology that ultimately results in better growth and tuber development. This results are significant for future potato farming under increasing sediment salinity. In areas where potato is produced, measures should be taken to prevent or reduce the rate of sediment salinization.

Keywords: Secondary salinization; Water use efficiency; Harvest Index; Ecophysiology; Potato

INTRODUCTION

Potato (*Solanum tuberosum* L.) is a member of the family Solanaceae. Many members of this family including tomato, eggplant, peppers, tobacco, and the wild night-shade are important food, medicinal, or ornamental plants. In 2010, average world farm yield for potato was 17.4 tons per hectare (FAOSTAT, 2010). In Cameroon, the Northwest and West Regions are the main centres of potato production (Fontem *et* al., 2001). Significant production also occurs in the Lebialem Highlands of the South West Region. Potato produced in Cameroon is exported across the CEMAC and ECOWAS markets (Acquah and Lyonga, 1994). The crop has now assumed a cash crop status, with an annual output of 150000 tones, grown on 70000 ha of the national territory. With its status as a cash crop, potato is in high demand within the country, and production barely meets demand. There is also potential for production to meet increasing export needs. However, as with other crops increased production depends on irrigation in the off-season, coupled with the application of fertilizers. This inevitably leads to secondary salinization of arable soils. About 20% of agricultural land worldwide is salinized (Shrivastava and Kumar, 2015), while estimates of salinized crop land are more severe at 50% (Paul and Lade, 2014). Secondary salinization is further compounded by climate change, as increased temperatures increase the rate of precipitation of soluble salts on farmlands (Horie et al., 2012). As arable land becomes more saline, there is need to determine crop plant responses in a view to forecast future production. How potato responds to saline soils is therefore of particular importance. Several authors have studied responses of potato to increasing salinization. Van Hoorn et al. (1993) showed that soil salinity resulted in water stress on potato plants, but the overall water use efficiency was not affected. Water use efficiency measures the rate of conversion of irrigation water to plant biomass. Levy (1992) studied salinity effects on different potato cultivars and showed that irrespective of cultivar, growth and tuber yield of potato decreased with increasing salinity. This is the trend generally expected of glycophytes, for example Kumar and Khare (2016) on susceptible and resistant rice cultivars and Blanco et al. (2008) on corn. However, nitrogen fertilization is expected to alleviate to some degree, salinity stress in some plants. Hamed et al. (2008) showed that growth of Batis maritima and Crithmum maritimum under saline conditions was limited by restrictions on N uptake at high NaCl concentrations, but there is as yet no clear evidence that augmenting soil nitrogen concentrations would lead to more uptake by plants. At higher salinities, a higher fraction of nitrogen absorbed is diverted to the production of glycinebetaine and/or proline, compatible organic solutes associated with osmotic regulation (Storey et al., 1977, Naidoo and Naidoo, 2001). Few studies of the effects of interaction of these nutrients with salinity on crop plants exist, especially in sub Saharan Africa, with the finality of forecasting future production in current sites, and the need and

possibility of expanding production especially in areas in which potato is not traditionally produced.

The necessity for such studies is now prompted by our varying climates across the world. In the current study the ecophysiological responses of potato to salinity under nitrogen fertilization were studied through its growth and tuber formation, chlorophyll fluorescence as a marker of stress, and water use efficiency, all under the intermediation of nitrogen fertilization.

MATERIALS AND METHODS

The study site

The experiment was conducted in a screen house constructed at SOWEFCU (South West Farmers' Cooperative Union) premises in Kumba, Meme division, which is located at 04.628°58″N latitude and 009.444°98″E longitude, at an altitude of about 237 m asl. Kumba falls within the Cameroon Agro-ecological Zone IV characterized by monomodal rainfall. Mean annual temperature stands at 25 °C and the total average rainfall in the subdivision is about 2200mm (IRAD- Barombi, 2013), and the natural vegetation is equatorial forest. Agriculture is the backbone of the economy, but potato is not among the crops traditionally produced in this agroecological zone.

Green house design and management

The experiment was conducted in a screen house of dimension 16m by 4m. It was constructed with wood and the roof was covered with a transparent polythene to avoid rain. The walls were covered with a mesh, estimated to have adequate light transmission quality and ambient CO_2 levels. Inside of the screenhouse, tables of dimension 4 by 1.5m were constructed with wood at the height of 60cm, to avoid contact with soil. On these tables, experimental pots were placed. During the experiment, temperature within the screenhouse ranged from 24.5 to 41.5 °C with a relative humidity range of 49 to 80.5% at midday.

Source and characteristics of planting material

Seed potato of the Cipira variety (Solanum tuberosum L. var Cipira) were obtained from IRAD Dschang, Cameroon. This variety is characterized by oval tubers, brown flesh and is resistant to mildew and viruses and moderately resistant to bacterial swelling. The cropping cycle is 90 to 120 days, with a potential yield of 30 – 35 tons/ha. The seeds were pre-sprouted before planting (Figure 1).

Figure 1: Pre-sprouted seed potato used in the experiment

Experimental design and treatments

There were two factors in this experiment, namely fertilization and salinity. NPK 20:10:10 fertilizer was applied at four levels namely (0, 0.2, 0.4, 0.6 g/plant) representing F0 to F3 respectively. The second factor was salinity, administered at 4 levels namely 0, 4, 8 and 12 ppt. The experimental design was a 4 x 4 factorial experiment with 3 replications, which ran for two months. Salinity treatments were obtained by dilution of seawater of 35 ppt with freshwater.

Planting

Two disease- and pest free tubers of uniform size and single sprouts were planted per 2L pot filled to the brim with homogenized top soil. The soil was clay loam and slightly acidic with a pH of 5.5. The effective cation exchange capacity was 5.45 with base saturation of 15.5%. The percentage organic matter was 3.82% and carbon: nitrogen ratio of 11. The potato tubers were planted so that the buds were directed upward. Where the tuber had more than one shoot, one was retained and the others destroyed. Plants were irrigated with freshwater for two weeks to establish, before treatment application. The plants were subsequently thinned to one plant per pot.

The four rates of fertilizer (0g, 0.2g, 0.4g and 0.6g) were applied two weeks after planting. Irrigation then followed with water of the respective salinity (0, 4, 8 and 12 ppt). All irrigation was done at the rate of 1.2L/pot/week, calculated from the annual precipitation of 2200mm per year.

Weed, pest and disease control

The interior of the screenhouse was cleaned before planting. After sowing weeds were controlled manually on emergence in the pots as well as in the screenhouse entirely. Insects were controlled by the application of Cipercal 12 ec (Cypermethrin 12 g/) three days before planting, throughout the greenhouse. The soil contained in the pots was treated with Ridomil Gold Plus 66 wp (Mefenoxam(Metalaxyl-M) 6%+ CuO 60 %), then by Mancoxyl Plus 720 wp (Cypermethrin 12 g/) one week after planting and two weeks afterwards.

Data collection

Parameters measured over time

Plant height, number of leaves, number of stems, collar diameter, leaf length and width, and chlorophyll fluorescence were measured weekly, between 11 am and 1 pm. Plant height was measured from the base of the plant to the tip of the shoot using a metre rule. Collar diameter was measured using a digital caliper. Leaf and stem numbers were counted, and leaf length and width measured for use in calculating relative leaf area. Chlorophyll fluorescence was measured using a Pocket Plant Efficiency Analyzer (Hansatech Instruments Ltd., Norfolk, UK), after 30 seconds of dark adaptation. The dark adaptation time was determined a priori. Fluorescence parameters measured included the ratio of variable fluorescence and the maximum fluorescence (Fv/Fm) and the performance index (PI). These measurements were performed at midday. During each measuring session, relative humidity and room temperature were measured. For all parameters, the first measurement 2 weeks after planting represented the base measurement.

Parameters measured following harvest

The relative growth rate (RGR) was calculated from the initial and final plant heights as follows (Tabot and Adams, 2012):

$$RGR \left(\frac{cm}{wk}\right) = \frac{LN\, H2 - LN\, H1}{t2 - t1}$$

Where LN = natural logarithm, H2 = final height, H1 = initial height, t = time in weeks.

At harvest, number of tubers, tuber yield, root:shoot ratio, harvest index and water use efficiency were determined. Number of tubers were counted per treatment and tuber yield determined by weighing on a sensitive balance. The ratio of the roots to the shoot was calculated after separation of roots from shoots, and carefully cleaning both (Tabot and Adams, 2012):

$$Root: shoot\ ratio = \frac{root\ mass\ (g) + tuber\ mass\ (g)}{shoot\ mass\ (g)}$$

The harvest index was calculated as follows (Amanullah and Inamullah, 2016):

$$Harvest\ Index = \frac{Tuber\ mass\ (g)}{Total\ Plant\ mass\ (g)}$$

Water use efficiency (WUE) was calculated as the ratio between the total plants biomass (g) produced and the total amount of water (L) applied during irrigation (After Egbe *et al.*, 2014).

$$WUE \left(\frac{g}{L}\right) = \frac{total\ plants\ biomass}{Total\ amount\ of\ water\ applied}$$

Data analysis

Data were analyzed for variance using the GLM ANOVA function, after tests for normality and homogeneity of variance. Means were separated through Tukey HSD test. Because some parameters were not normally distributed, Spearman Rank Correlation analyses were done to examine relationships between parameters. All analyses were done at $\alpha = 0.05$ using Minitab® version 17 statistical package (Minitab Inc, CA, USA).

RESULTS

Table 1 shows the significance of main and interaction effects of parameters measured over time. For the growth parameters, there were statistical differences in height in response to salinity, fertilization and time. The effect of interaction of salinity and fertilization on height was also significant (p<0.05). The response of

collar diameter to treatments and interaction effects was similar to that of height (Table 1). Number of stems varied significantly under treatment with salinity, and increased with time, and the interaction of salinity and fertilizer on stem number was also significant. For the main effects, salinity significantly affected leaf area, while the effect of fertilization was not significant. Chlorophyll fluorescence (fv/fm) was not significantly affected by treatments, but varied significantly over time.

Table 2 is of the significance of responses of root:shoot ratio, water use efficiency, number of tubers, tuber yield, harvest index and relative growth rate to salinity, fertilization and the resulting interaction. At $\alpha = 0.05$, salinity significantly influenced root: shoot ratio and water use efficiency; tuber numbers, tuber mass and harvest index were significantly affected at $\alpha = 0.001$. Harvest index varied significantly with fertilization, and the interaction between salinity and fertilization significantly affected all yield parameters (Table 2). Relative growth rate of potato plants was significantly affected by salinity treatments.

Responses of parameters to salinity

Figure 2 details responses of variables to salinity. Plant height was statistically similar at 0 and 4 ppt but plants treated with 8ppt saline water were significantly taller. This is reflected in the RGR (Figure 2 B). Leaf area decreased significantly as salinity increased from 0 to 4 ppt, with a slight increase at higher salinities. Chlorophyll fluorescence decreased quantitatively as salinity increased from 4ppt to 12 ppt but this difference was not statistically significant. Root:shoot ratio was highest (1.76) in plants irrigated with freshwater, and decreased significantly as salinity increased to 12ppt. Tuber yield measured as fresh tuber mass decreased sharply with an increase in salinity. The heaviest tubers (20.36g) were produced by plants irrigated with freshwater Figure 3. Tuber yield decreased thereafter, to 0.38g in plants irrigated with water of 12 ppt salinity (Figures 2 and 3).

Table 1: Significance of main and interaction effects of salinity and fertilization on height, number of stems, leaf area and chlorophyll fluorescence of potato measured over time in screenhouse, Cameroon

Source	Height	No. of Stems	CD	Leaf area	fv/fm
Salinity (S)	***	***	***	***	ns
Fertilization (F)	*	ns	***	ns	ns
Time (T)	***	***	***	***	***
S x F	**	***	***	ns	ns
S x T	ns	ns	ns	ns	ns
F x T	ns	ns	ns	ns	ns
S x F x T	ns	ns	ns	ns	ns

CD = collar diameter, fv/fm = ratio of variable to maximum chlorophyll fluorescence, * = significant at $\alpha = 0.05$, ** = significant at $\alpha = 0.01$, *** = significant at $\alpha = 0.005$, ns = not significantly different

Table 2: Significance of main and interaction effects of salinity and fertilization on height, number of stems, leaf area and chlorophyll fluorescence of potato measured over time in screenhouse, Cameroon

Source	Root:Shoot ratio	WUE	No. tubers	Tuber yield(g)	HI	RGR
Salinity (S)	*	*	***	***	***	*
Fertilization (F)	ns	ns	ns	ns	*	ns
S x F	ns	ns	**	**	***	ns

WUE = water use efficiency, HI = Harvest Index, RGR = relative growth rate, * = significant at α = 0.05, ** = significant at α = 0.01, *** = significant at α = 0.005, ns = not significantly different

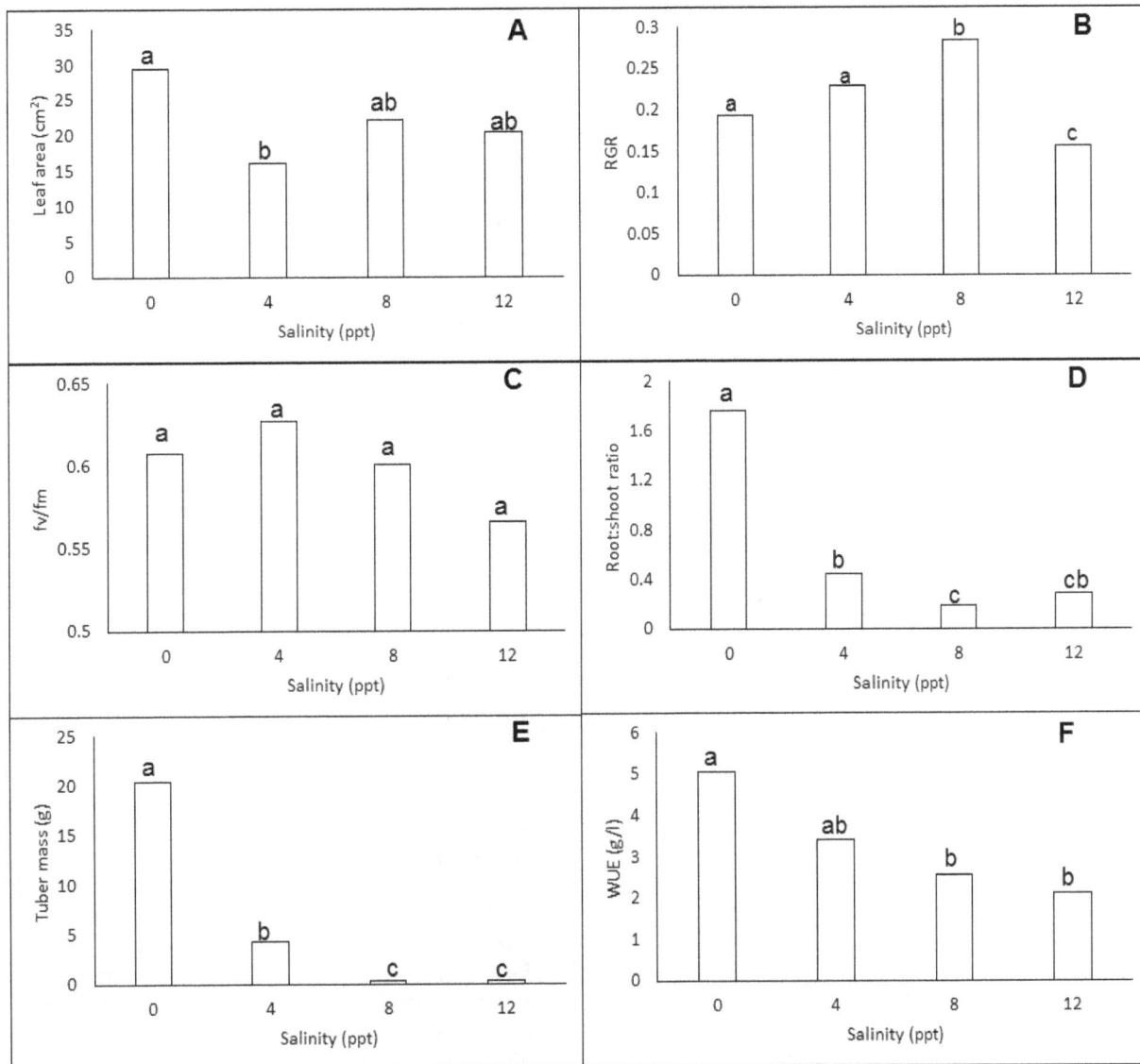

Figure 2. Responses of measured variables to salinity treatments. A =leaf area, B= plant height, C= chlorophyll fluorescence, D=root/shoot ratio, E=tuber mass and F=water use efficiency. Bars represent means. Means separated through ANOVA with Tukey pairs wise comparisons at α=0.05. Bars with the same letter for each variable are not significantly different.

Figure 3 Tubers per plant harvested after 8 weeks of salinity treatment in screenhouse. Salinity treatments ranged from 0 ppt to 12 ppt

Responses of parameters to fertilization

Plant height increased significantly with increase in fertilization, from 35.7cm in plants that were not fertilized, to 48cm in plants treated with 0.6g fertilizer. There was no significant fertilizer effect on leaf area, chlorophyll fluorescence, root:shoot ratio and water use efficiency (data not presented).

Effects of interactions between fertilizer and salinity on potato responses

The effects of the interaction between salinity and fertilization on parameters measured are presented in Figure 4. As seen from Tables 1 and 2, these interactions had a significant effect on plant height, tuber yield and Harvest index. At 0ppt, plants that were fertilized were taller than those not fertilized. At higher salinities, height of unfertilized plants was statistically comparable to those of fertilized plants. At all fertilizer levels, there was higher tuber yield from pots irrigated with freshwater and yield decreased sharply irrespective of fertilization as salinity increased. At 4ppt, the unfertilized plants had better yield than the fertilized plants. With respect to water use efficiency, plants irrigated with freshwater and supplied 0.2g fertilizer had the best WUE (8.61g/l) while the least was in unfertilized plants irrigated with water of 12 ppt

salinity (1.3g/l). Fertilized plants at this high salinity had better WUEs (Figure 4).

Figure 5 shows the interaction of plant height with chlorophyll fluorescence over time. Over time, as plant height increased, fv/fm decreased. Beyond the 5th week, plant height also began to experience a decrease. There was a mild negative correlation between plant height and fv/fm ($\rho = -0.197$, $p = 0.001$) but no correlation with RGR.

Correlations

There were significant negative correlations between WUE, number of tubers, tuber mass, biological yield and harvest index with salinity. Some of these correlations are presented in Figure 6. There were no significant correlations with fertilizer. Root shoot ratio and WUE correlated positively with related yield parameters. Growth parameters measured over time (data not shown) seem to be negatively affected by salinity, but indeed the only negative correlations were between salinity and number of stems ($\rho = -0.197$, $p = 0.001$). Plant height, number of stems and number of flowers were significantly positively related to time (($\rho = 0.488, 0.225, 0.145$ respectively, $p < 0.05$), while chlorophyll fluorescence correlated negatively with time ($\rho = -0.536$, $p = 0.000$).

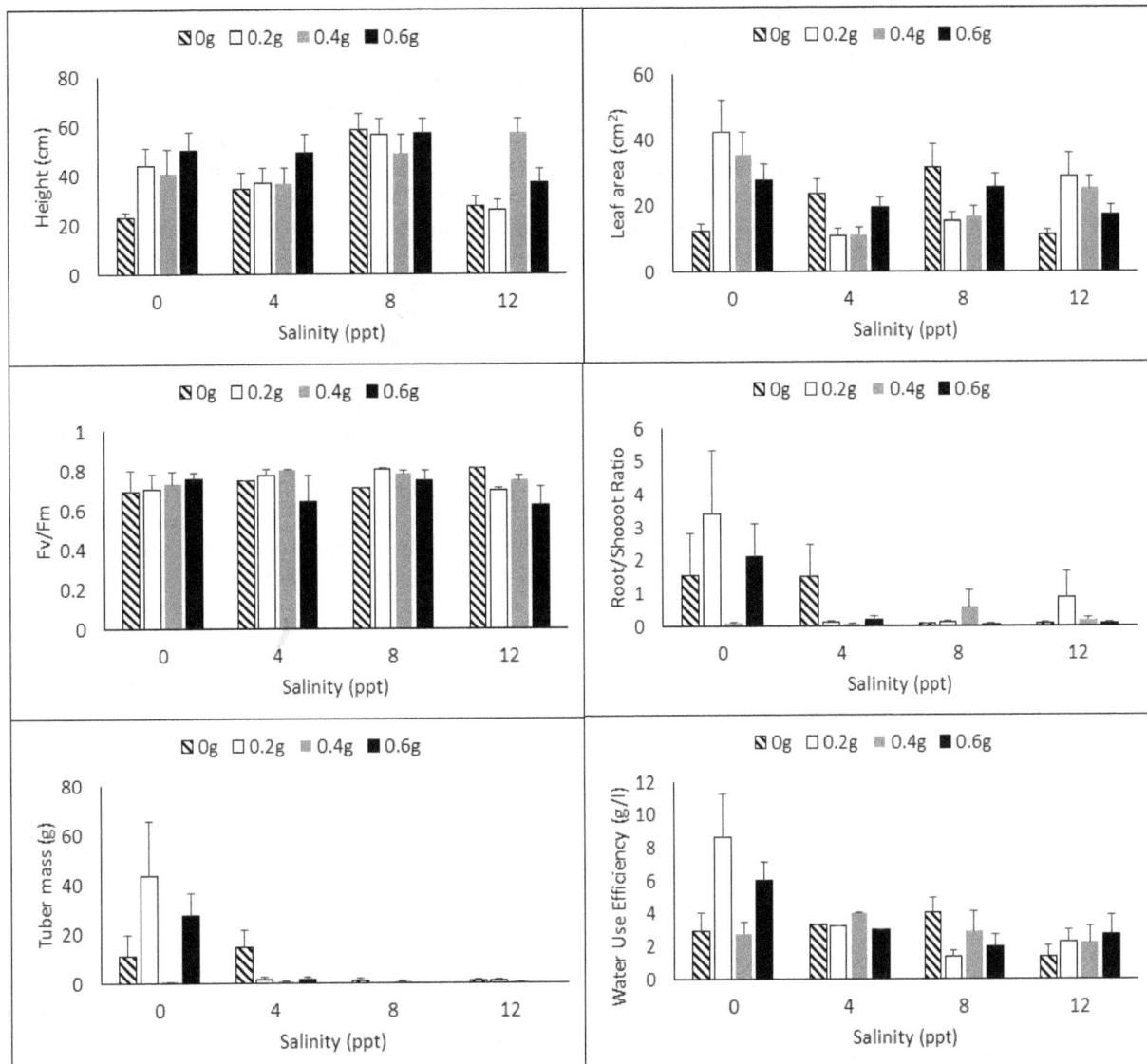

Figure 4. Responses of plant height, leaf area, chlorophyll fluorescence, root:shoot ratio, tuber yield and water use efficiency of potato to fertilization under four salinity regimes in screenhouse, Cameroon

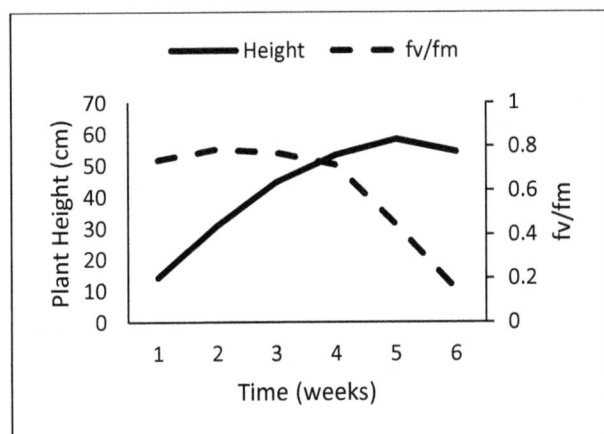

Figure 5. Relationship between fv/fm and plant height over 6 weeks in screenhouse, Cameroon

DISCUSSION

Morphological responses to salinity and fertilization

Statistically, salinity effects on growth parameters were significant, but the pattern is not clear; plant height were stimulated up to 8ppt, while leaf area decreased as salinity increased from 0ppt. For glycophytes, the expected result is a decrease in growth with salinity increase (Horie *et al.*, 2012). Glycophytes like potato lack mechanisms of tolerance to salinity like ion compartmentalization, salt secretion etc. However, the very similar heights and leaf area show that for the first six weeks of growth, potato grown in screen house was more tolerant of salinity than expected. This could be due to counteracting effects of nitrogen fertilization. It has been shown that nitrogen fertilization augments plant growth and yield (Muhammad *et al.*, 2015; Saravia *et al.*, 2016). We intended to find a role for nitrogen in mediating salinity stress. One way in which salinity stress affects plants is inhibition of nutrient uptake (Munns and Tester, 2008). Incidentally, nitrogen is a key component of most systems that tolerate salt stress. The fact that nitrogen addition did not translate to increased growth suggests that it went into systems that improve potato tolerance to salinity. Nitrogen is a key component of proline and glycinebetaine, essential compatible solutes that are responsible for salinity tolerance, as they scavenge reactive oxygen species and stabilize cell membranes (Naidoo and Naidoo, 2001; Munns and Tester, 2008; Tabot and Adams, 2012). The levels of these solutes were not assessed in this study, but other authors have already established their role in other plants (Naidoo and Naidoo, 2001; Tabot and Adams, 2012). The contribution of Nitrogen is evident in the WUE, where fertilized plants at 12ppt had better WUE than unfertilized ones. While nitrogen

fertilization might have improved plants' tolerance to salinity, tuber yield in number and mass decreased significantly as salinity increased from 0ppt; there were significant negative correlations between salinity and tuber yield, tuber number, harvest index and other parameters dependent on tuber yield, but no correlation of these parameters with fertilization. This suggests the salinity stress is predominant and only mildly moderated by Nitrogen fertilization. Salinity typically increases oxidative stress in plants including water stress and restrictions in nutrient uptake (Munns and Tester, 2008). The result is that photosynthesis efficiency is reduced (Redondo-Gómez *et al.*, 2007), and much of the photosynthate is diverted to stress tolerance mechanisms, with little left for the sinks. Tuber formation is thus suppressed, as is the harvest index. Therefore, in conditions of increased sediment salinity, it is expected that potato growth could persist if supplemented with nitrogen fertilization, but tuber yields will be greatly suppressed.

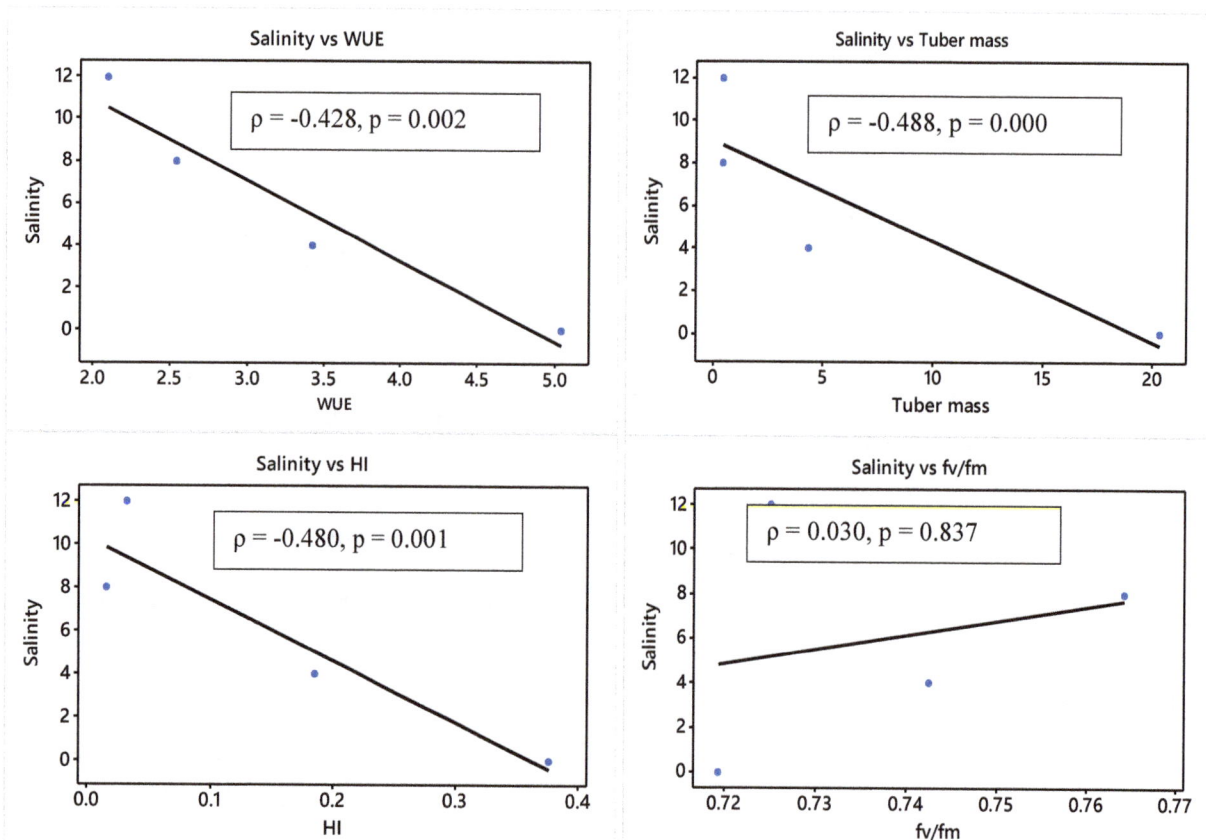

Figure 6 Correlations between salinity and water use efficiency, tuber mass, harvest index and chlorophyll fluorescence. ρ = Rho, the Spearman Rank Correlation coefficient; p = the level of significance.

Physiological responses to salinity and fertilization

Chlorophyll fluorescence measured as an index of the efficiency of photosynthesis did not vary significantly with any main effect but decreased significantly over time. The ratio fv/fm was consistently below 0.8, which is the expected threshold for healthy unstressed plants, but above 0.3, which is the value of photosynthetic efficiency beyond which plants are deemed non- viable and would most possibly die (Woo *et al.,* 2008). This shows that under salinity treatments, potato plants were stressed across the board and this stress is bound to result in reduced productivity. Similar stress levels across salinity treatments suggest that other environmental conditions affected plants in all treatments. The RH was often low and screenhouse temperatures high, which themselves contribute to stressing the plants. Salinity-stressed plants typically exhibit higher root:shoot ratios, as photosynthate is divert to improve root architecture for enhanced nutrient uptake (Tabot *et al.,* 2012). However, in the present study, potato plants grown under freshwater irrigation had higher root:shoot ratio, as the conditions were more favourable for root and tuber development; if the plant were tolerant of salinity we would expect similar root shoot ratios at the higher salinities but this decreased sharply. Similarly, WUE varied significantly with salinity but not fertilization. Plants irrigated with freshwater had higher water use efficiency, irrespective of whether they were fertilized or not. Water use efficiency gives an indication of the plants' ability to convert irrigation water into biomass (Egbe *et al.*, 2014). Since the potato was the same variety we expected similar WUE, but this decreased as salinity increased. Under freshwater conditions, potato plants present a more efficient physiology - more efficient photosynthesis, better water balance, better gaseous exchange and more efficient moisture and nutrient uptake that ultimately results in better growth and tuber development, and more efficient use of water.

CONCLUSION

Our results are significant for the future of potato farming in an environment of increasing salinity of arable land. While salinity might stimulate growth in height of plants, it drastically reduces tuber formation, and supplementation with nitrogen fertilization is not enough to counter the retarding effects of sediment salinization. Therefore, in potato producing areas, measures should be taken to prevent or reduce the rate of sediment salinization.

Acknowledgement
We thank the South West Farmers' Cooperative Union for providing premises for the screenhouse to be constructed, and the Department of Botany and Plant Physiology for providing relevant equipment.

REFERENCES

Acquah, E., Lyonga, S. 1994. Internal Review of the Tropical Root and Tubers Research Project. 78 p.

Amanullah, Inamullah. 2016. Dry Matter Partitioning and Harvest Index Differ in Rice Genotypes with Variable Rates of Phosphorus and Zinc Nutrition. Rice Science. 23(2):78-87. DOI: 10.1016/j.rsci.2016.09.006

Blanco, F.F., Folegatti, M.V., Gheyi, H.R., Fernandes, P.D. 2008. Growth and yield of corn irrigated with saline water. Scientia Agricola. 65(6): 574-580. DOI: 10.1590/S0103-90162008000600002

Paul, D., Lade, H. 2014. Plant-growth-promoting rhizobacteria to improve crop growth in saline soils: a review. Agronomy for Sustainable Development. 34 (4): 737-752. DOI: 10.1007/s13593-014-0233-6

Egbe, E.A., Forkwa, E.Y., Ayamoh, E.E. 2014. Evaluation of seedlings of three woody species under four soil moisture capacities. British Journal of Applied Science and Technology. 4(24):3455-3472. DOI: 10.9734/BJAST/2014/10854

FAOSTAT. 2011. Production-Crops, 2010 data. Food and Agriculture Organization of the United Nations.

Fontem, D.A., Demo, P., Njualem, D.K. 2001. Status of potato production, marketing and utilisation in Cameroon. Study funded by the Rockefeller Foundation. Unpublished.

Hamed, K.B., Messedi, D., Ranieri, A., Abdelli, C. 2008. Diversity in the response of two potential halophytes (*Batis maritima* and *Crithmum maritimum*) to salt stress.71-81. In: Abdelly, C., Öztürk, M., Ashraf, M., Grignon, C. (eds). Biosaline agriculture and high salinity tolerance. Berlin: Birkhäuser.

Horie, T., Karahara, I., Katsuhara, M. 2012. Salinity tolerance mechanisms in glycophytes: and overview with the central focus on rice plants. Rice. 5:11. DOI: 10.1186/1939-8433-5-11

Kumar, V., Khare, T. 2016. Differential growth and yield responses of salt-tolerant and susceptible rice cultivars to individual (Na^+ and Cl^-) and additive stress effects of NaCl. Acta Physiologiae Plantarum. 38: 170. DOI: 10.1007/s11738-016-2191-x.

Levy, D. 1992. The response of potatoes (*Solanum tuberosum L.*) to salinity: plant growth and tuber yields in the arid desert of Israel. Annals of Applied Biology. 120 (3): 547-555. DOI: 10.1111/j.1744-7348.1992.tb04914.x

Muhammad, N., Hussain, Z., Rahmndil, Ahmed, N. 2015. Effect of different doses of NPK fertilizers on the growth and tuber yield of potato. Life Science International Journal. 9 (1, 2, 3, 4): 3098-3105.

Munns, R., Tester, M. 2008. Mechanisms of salinity tolerance. Annual Review of Plant Biology. 59: 651 - 681. DOI:

10.1146/annurev.arplant.59.032607.092911

Naidoo, G., Naidoo, Y. 2001. Effects of salinity and nitrogen on growth, ion relations and proline accumulation in *Triglochin bulbosa.* Wetlands Ecology and Management. 9: 491 - 497. https://doi.org/10.1023/A:1012284712636

Qadir, M., Quillerou, E., Nangia, V., Murtaza, G., Singh, M., Thomas, R.J., Drechel, P., Noble, P. 2014. Economics of salt-induced land degradation and restoration. Natural Resources Forum. 38: 282–295. DOI: 10.1111/1477-8947.12054

Redondo-Gómez, S., Mateos-Naranjo, E., Davy, A.J., Fernandez-Munoz, F., Castellanos, E.M., Luque, T., Figueroa, M.E. 2007. Growth and photosynthetic responses to salinity of the salt-marsh shrub *Atriplex portulacoides.* Annals of Botany. 100: 555 - 563. DOI: 10.1093/aob/mcm119

Storey, R., Ahmad, N., Jones, R.G.W. 1977. Taxonomic and ecological aspects of the distribution of glycinebetaine and related compounds in plants. Oecologia. 27: 319 - 332. DOI: 10.1007/BF00345565

Shrivastava, P., Kumar, R. 2015. Soil salinity: A serious environmental issue and plant growth promoting bacteria as one of the tools for its alleviation. Saudi Journal of Biological Sciences. 22(2): 123-131. DOI: 10.1016/j.sjbs.2014.12.001.

Tabot, P.T., Adams, J.B. 2012. Morphological and Physiological responses of *Triglochin buchenaui* Kocke, Mering & Kadereit to various combinations of water and salinity: implications for resilience to climate change. Wetlands Ecology and Management. 20:373-388. DOI: 10.1007/s11273-012-9259-1

van Hoorn, J.W., Katerji, N., Hamdy, A., Mastrorilli. 1993. Effect of saline water on soil salinity and on water stress, growth, and yield of wheat and potatoes. Agricultural Water Management. 23 (3): 247-265. DOI: 10.1016/0378-3774 (93)90032-6.

Levy, D. 1992. The response of potatoes (*Solunum tuberosum* L.) to salinity: plant growth and tuber yields in the arid desert of Israel. Annals of Applied Biology. 120(3): 547-555. DOI: 10.1111/j.1744-7348.1992.tb04914.x

Woo, N.S., Badger, M. R., Pogson, B. J. 2008. A rapid, non-invasive procedure for quantitative assessment of drought survival using chlorophyll fluorescence. Plant Methods. 4 (27): 1746-4811. DOI: 10.1186/1746-4811-4-27.

13

Effect of Feeding with Dried Distillers Grains with Solubles and Rice Polishings on Scrotal Circumference and Semen Parameters of Ram Lambs

Miguel A. Domínguez-Muñoz[1], Fernando Martínez-Cordero[1],
Miguel Mellado[2], Cecilia C. Zapata-Campos[3], Claudio Arzola-Alvarez[4],
and Jaime Salinas-Chavira[3]*

[1]*Department of Reproduction, College of Veterinary Medicine and Animal Science,
Autonomous University of Tamaulipas, Mexico.*
[2]*Department of Animal Nutrition, Universidad Autónoma Agraria Antonio Narro,
Saltillo, México.*
[3]*Department of Nutrition, College of Veterinary Medicine and Animal Science,
Autonomous University of Tamaulipas, Mexico. Email: jsalinasc@hotmail.com*
[4]*Department of Animal Nutrition, College of Animal Science and Ecology, Universidad
Autónoma de Chihuahua, Chihuahua, México*
Corresponding author

SUMMARY

The effects of level of dried distillers grains with solubles (DDGS) and rice polishings (RP) on the scrotal circumference and semen parameters were examined in 22 crossbred (Pelibuey x Dorper) ram lambs (average live weight 24.84 ± 2.07 kg, approx. 5 months of age). The lambs were randomly allotted to four treatments in a 2 x 2 factorial design. The two factors were dietary levels of RP (0 and 15%) and of DDGS (0 and 30%). The experiment lasted 52 days. The animals were weighed on days 0, 34 and 52, and semen variables were measured on days 34 and 52. Semen evaluation included sperm morphology, mass and progressive motility, and sperm concentration. On day 34, sperm concentration showed an interaction of both effects (P=0.01) and was greater in rams in group 4 (both RP and DDGS) than in those in groups T2 (RP alone) or T3 (DDGS alone). Sperm morphology (% normal cells) was higher (main treatment effect) in rams fed with 30% DDGS than in those not fed with DDGS (P<0.05). The other variables showed no treatment effect. On day 52, none of the ejaculate variables varied with treatments. In conclusion, the diets did not affect body weight of ram lambs, but did affect the percentage of normal spermatozoa at 34 d of the trial, although this effect was not observed after 52 d of the trial; thus, in the long run, these diets do not alter scrotal circumference or semen parameters of young Pelibuey x Dorper ram lambs.
Key words: Distillers grains; Rice polishings; Ram lambs; Semen characteristics.

INTRODUCTION

Growth and economic viability of the sheep industry will not be profitable if not carried out in an orderly manner, organizing meat production, as some producers will focus on breeding stock, others on lamb production and others on feedlot lambs.

Dried distillers grains with solubles (DDGS) are an excellent source of energy and protein for sheep (Schauer et al., 2008). Dried distillers grains with soluble are a byproduct of the ethanol industry, which are formed after the fermentation of cereal grains, mainly corn, by Saccharomyces yeast (Klopfenstein et al., 2008). Large amounts of DDGS are obtained, providing a significant source of protein, amino acids, energy, phosphorus and other nutrients for feeding sheep and other domestic animals (Lumpkins et al., 2005; Zhang et al., 2010). They can be included in growing or finishing animals, lowering production costs compared with conventional ingredients such as sorghum grain and soybean meal (Schauer et al., 2005). Therefore, the use of high levels of DDGS can be economically justifiable.

Rice is a grain of great economic and nutritional importance because it is widely cultivated; however, most production is destined for human consumption. For animal feeding, rice polishings (RP) are used; this feed is obtained when the surface of the endocarp is removed along with rice flour residues, and it constitutes approximately 20% of the amount of rice for human consumption. To make intensive sheep meat production more profitable, the use of available feed resources should be optimized, where RP can be part of the rations for sheep fattened in feedlots (Tabeidian and Sadeghi, 2009; Salinas-Chavira et al., 2013).

Currently, both DDGS and RP are commonly used in diets for sheep but there is little information on the effect of these feeds on the reproductive efficiency of ram lambs. In a preliminary study, sheep fed with diets containing DDGS showed reduced sperm concentration (Van Emon et al., 2013), the effect attributed to high sulfur levels in DDGS that reduced copper absorption of lambs and could negatively influence spermatogenesis. According to NRC (1996), RP shows high levels of crude fat (15%), NDF (26.1%), ash (10.4%) and sulfur (0.19%) which combined with the high levels of these nutrients in DDGS (10.7% crude fat, 46% NDF, 4.6% ash and 0.44% sulfur) may affect digestive and reproductive function of lambs. The information on this subject is scarce. Therefore, the aim of this study was to assess whether there is a negative effect of DDGS and RP on scrotal circumference and semen characteristics of ram lambs.

MATERIALS AND METHODS

This study was conducted at the College of Veterinary Medicine and Animal Science at the Autonomous University of Tamaulipas, Ciudad Victoria, Mexico. The study site is located at 23 °N and 97 °W, at an altitude of 340 meters. The average annual temperature is 24.3 °C, with an average annual rainfall of 926 mm.

The animal management protocol for this study was approved by the Bioethics and Animal Welfare Committee, FMVZ, UAT. The protocol followed the ARRIVE guidelines and the EU Directive 2010/63/EU for animal experiments.

Experimental treatments

The present study consisted of four diets, prepared using two levels of dried distillers grains with solubles (DDGS; 0 and 30%) and two levels of rice polishings (RP; 0 and 15%) in a 2 x 2 factorial design. Treatments were: T1 with 0% DDGS and 0% RP (control diet), T2 with 0% DDGS and 15% RP, T3 with 30% DDGS and 0% RP, and T4 with 30% DDGS and 15% RP (Table 1). In each case, the inclusion of DDGS and/or RP replaced a portion of sorghum grain, soybean meal and/or urea, to maintain similar levels of CP and ME. All diets were balanced to contain 14.8% CP and 2.8 Mcal/kg ME, considering the nutritional requirements of feedlot sheep indicated by NRC (2007).

Animals and management

The study was conducted in April and May 2015. Twenty-four non-castrated crossbred (Pelibuey × Dorper) ram lambs were housed indoors in individual pens that were 1.20 m long, 0.70 m wide and 1 m high, with individual feeding bunks. Lambs were allowed ad libitum access to feed and water. Feed was offered twice daily at 9:00 and 16:00 h. Offered feed and orts were weighed daily. The provided feed per day was approximately 110% of the prior day's consumption. Information of the feeding management of animals was not statically considered because the study was focused in the reproductive variables of animals. The ram lambs were randomly divided into four treatments with six repetitions (sheep). During the course of the study, two animals were eliminated from the study due to causes unrelated to the study, leaving at the end a total of six repetitions in T1 and T3, and five repetitions in T2 and T4. The adaptation period was 15 days and the experiment lasted 52 days.

At the beginning of the experiment the ram lambs weighed 24.84 ± 2.07 kg and were approximately 5 months old; they received vitamins A, D and E intramuscularly (250,000 IU, 37,500 IU and 25 mg, respectively; Vigantol ADE®, Bayer, Mexico). An oral dose of 50 mg/kg of the anthelminthic Closantel (Closantel®, Panavet, Mexico) was given, with a second application 15 days later.

Reproductive variables

During the trial period, ram lambs were weighed and scrotal circumference was measured on days 0, 34 and 52; semen evaluation were performed on days 34 and 52.

Scrotal circumference (SC) was calculated with a flexible tape placed over the middle of the scrotum; the testicles were softly pushed toward the base of the scrotum and kept in that position during measurement. Daily scrotal growth for both sampling periods (34 and 52 d) was calculated.

Semen was collected and evaluated following the methods described by Evans and Maxwell (1990). One semen sample was obtained from each ram lamb on days 34 and 52 using an electroejaculator (Standard Precision, Denver, CO, USA).

Sperm motility was measured at two levels: mass motility and progressive motility. Mass motility was calculated by observing a fresh sperm preparation (at 100 ×) and scoring it on a 0-to-5 scale (0= no movement, 5= intense whirling motion). Progressive

motility was evaluated by observing (at 400 ×) the number of sperm that moved more or less linearly in the microscopic field and estimated as a percentage of total sperm.

Sperm concentration was measured using a Neubauer chamber from semen diluted at 1:300. Sperm morphology was examined using semen smears stained with Rose Bengal; 100 cells were counted, distinguishing between morphologically normal and abnormal sperm.

Scrotal circumference and semen variables were compared between both sampling periods (days 34 and 52), to show the improvement in these variables as the rams matured.

Data analysis

Data were analyzed using a general linear model in a 2 × 2 factorial arrangement for unbalanced data, with 2 levels of dried distillers grains with solubles (0 or 30%) and 2 levels of rice polishings (0 or 15%). The model considered main effects and their interaction. Tukey's test was used to compare means when the interaction was significant. Student's t-test was used to compare semen variables between the two periods. The statistical models were evaluated using the GLM procedure of SAS. A significant effect was considered at P≤0.05.

Table 1. Percentage and nutrient composition of the experimental diets (% DM).

Ingredients	0% DDGS		30% DDGS	
	0% RP (T1)	15% RP (T2)	0% RP (T3)	15% RP (T4)
Sorghum stover	10	10	10	10
Sorghum grain	66.75	53.68	48.05	34
Soybean meal	14	11.3	1.85	0
Urea	0.5	0.57	0.04	0.05
Sugarcane molasses	5	5	5	5
Mineral supplement*	1.5	1.5	1.5	1.5
Salt	0.25	0.25	0.25	0.25
DDGS	0	0	30	30
Rice polishings	0	15	0	15
Tallow	2	2.7	3.31	4.2
Total	100	100	100	100
Nutrient composition on DM basis				
CP (%)	14.86	14.86	14.84	14.84
ME (Mcal/kg)	2.80	2.80	2.80	2.80

* Premix of macro and microminerals for feedlot lambs (Ovitec 302 F ®Tenusa Thrown; Monterrey, NL, México). Contains calcium carbonate, sodium bicarbonate, common salt, molasses, ammonium sulfate, ammonium chloride, magnesium oxide, ferric oxide (pigment), vitamin A acetate, Vitamin D3, vitamin E acetate, thiamine, antioxidant, ionophore (bovatec), aureomycin, artificial flavor, compounds of: manganese, zinc, iron, iodine, selenium and cobalt. DDGS=distillers grains with solubles; RP= Rice polishings.

RESULTS

Results for live body weight and semen characteristics after day 34 of the feeding trial are shown in Table 2. Sperm concentration showed an interaction (P=0.01) between factors, where concentration was greater (P < 0.05) in rams in group T4 (30%DDGS + 15% RP) than those in T2 (15% RP) or T3 (30% DDGS). The percentage of normal sperm was higher (main effect of treatment) in animals receiving DDGS than animals with no dietary DDGS (P = 0.04). There was no treatment or interaction effect on the other variables.

The corresponding results for day 52 are shown in Table 3. There was no treatment or interaction effect on live body weight or sperm characteristics (P > 0.05). Comparisons between variables recorded on days 34 and 52 are shown in Table 4. Values for live body weight, scrotal circumference, mass and progressive motility, and sperm concentration increased with age (P < 0.01). Daily scrotal growth or sperm morphology were not influenced by day of sampling.

DISCUSSION

In the present study ram lambs showed no difference among treatments and had an adequate nutrition regardless of group. The average live weight of ram lambs was 24.84, 36.27 and 40.11 kg at 1, 34 and 52 days of the feeding trial, respectively. Based on weight, the animals in the current study reached puberty, particularly at semen sampling periods (34 and 52 d). In other studies, puberty was reached at 21.4 kg in Dorper-cross rams (Villasmil-Ontiveros et al., 2011), while in Pelibuey rams, it has been reached from 23.3 kg (Aguilar-Urquizo et al., 2013) to 32.6 kg of live weight (Valencia-Méndez et al., 2005). In the present study ram lambs at 52 days showed increased reproductive function, when the value for progressive motility approached the range of 70 to 90% described by Valdez (2013).

In agreement with the present study, Tufarelli et al. (2011) observed that scrotal circumference was not influenced by dietary treatments. In the current study, on average it was 29.09 and 31.86 cm at 34 and 52

days of feeding, respectively, which coincides with the 28.40 cm described for rams (Sarlós et al., 2013). Aguilar-Urquizo et al. (2013) reported that scrotal circumference varied from 20.8 to 21.9 cm and live weight from 19.5 to 23.3 kg at the first normal ejaculate. They also reported that the onset of puberty in Pelibuey rams was not influenced by diet consumption of phytoestrogens in foliage from M. Alba or H. rosa-sinensis.

In the current study, scrotal growth was not influenced by dietary treatments; on average it was 0.13 and 0.15 (cm/d) at 34 and 52 days of feeding, respectively; these values are in agreement with Moreno-Cáñez et al. (2013), who reported daily testicular development of 0.15 cm/d in growing lambs. In other studies, nutrition improvements have enhanced reproductive characteristics in rams (Jiménez-Severiano et al., 2010; Kheradmand et al. 2006).

The percentage of normal sperm on day 34 of the feeding trial was higher in animals receiving DDGS, while at day 52 of the feeding trial it was not influenced by dietary treatment. Overall average percentage of normal sperm in the present study was 63.91% and 66.27% at 34 and 52 days, respectively. These values exceed in 15% the minimum level required for semen used for artificial insemination (Hafez and Hafez, 2000). In part, results could be influenced by season because the study was conducted during the spring, when sheep show the highest percentages of sperm abnormalities (López et al., 2011).

Sperm production in rams is influenced by health, age, conformation, nutrition, handling and environmental factors (Ghorbankhani et al., 2015). Testicular size, sperm production and quality are highest during the fall season (Hafez and Hafez, 2000; Zamiri et al., 2010); at other times, semen quality may be lower, but values may be suitable for good fertility throughout the year (Zamiri et al., 2010). In a tropical environment, Cárdenas-Gallegos et al. (2015) reported no influence of season on libido in Pelibuey, Blackbelly, Dorper and Katahdin rams.

Table 2. Growth and reproductive variables evaluated in ram lambs fed with different levels of distillers grains with solubles (DDGS) and rice polishings (RP; 34-d period).

| | 0% DDGS | | 30% DDGS | | P value | | |
	0% RP (T1)	15% RP (T2)	0% RP (T3)	15% RP (T4)	RP	DDGS	Interac.
Initial weight (kg)	25.8 ± 0.8	25.8 ± 2.9	24.0 ± 1.2	23.3 ± 1.7			
Initial SC[1] (cm)	24.8 ± 1.7	24.7 ± 2.4	24.1 ± 2.5	24.6 ± 2.7			
Data at 34-d							
Weight (kg)	37.1 ± 2.8	36.4 ± 2.2	37.0 ± 2.3	34.3 ± 2.4	0.12	0.34	0.35
Scrotal circumference (cm)	29.0 ± 1.3	28.2 ± 1.3	29.8 ± 1.6	29.2 ± 1.3	0.24	0.14	0.90
Scrotal growth (cm/d)	0.12 ± 0.03	0.12 ± 0.04	0.15 ± 0.06	0.14 ± 0.06	0.68	0.32	0.75
Mass motility (0-5)	3.0 ± 1.8	1.6 ± 2.2	3.0 ± 1.7	4.0 ± 0.7	0.79	0.14	0.11
Progressive motility (%)	55.0 ± 24.3	40.0 ± 33.2	58.3 ± 17.2	70 ± 7.1	0.86	0.12	0.18
Sperm concentration (10^9/mL)	2.43 ± 0.99[AB]	1.50 ± 1.36[B]	1.73 ± 1.38[B]	3.73 ± 1.06[A]	0.31	0.24	0.01
Morphology (% normal cells)	60.5 ± 12.8	58.2 ± 6.2	73.2 ± 4.2	62.6 ± 10.8	0.12	0.04	0.31

[1]SC: scrotal circumference.
Within rows, values showing different superscripts differ (P < 0.05).

Table 3. Growth and reproductive variables evaluated in ram lambs fed with different levels of distillers grains with soluble (DDGS) and rice polishings (RP; 52-d period).

| | 0% DDGS | | 30% DDGS | | P value | | |
	0% RP (T1)	15% RP (T2)	0% RP (T3)	15% RP (T4)	RP	DDGS	Interaction
Weight (kg)	40.8 ± 3.1	40.7 ± 2.2	40.6 ± 2.1	38.1 ± 2.6	0.24	0.24	0.30
Scrotal circumference (cm)	31.5 ± 2.6	31.0 ± 2.2	32.8 ± 1.3	32.0 ± 1.9	0.43	0.18	0.84
Scrotal growth (cm/d)	0.14 ± 0.12	0.16 ± 0.18	0.16 ± 0.10	0.16 ± 0.10	0.96	0.79	0.81
Mass motility (0-5)	4.3 ± 0.8	3.6 ± 1.5	4.2 ± 0.4	4.2 ± 0.8	0.40	0.66	0.36
Progressive motility (%)	68.3 ± 13.3	60 ± 29.2	70 ± 0.0	72 ± 8.4	0.64	0.36	0.46
Sperm concentration (10^9/mL)	5.28 ± 6.02	2.53 ± 1.83	6.00 ± 3.28	5.80 ± 2.33	0.38	0.27	0.45
Morphology (% normal cells)	67.2 ± 7.0	67.0 ± 6.0	66.2 ± 9.3	64.6 ± 3.8	0.78	0.58	0.82

Table 4.Comparison of the variables evaluated between both evaluation periods (34 vs. 52-d)

	34-d	52-d	P value
Weight (kg)	36.27 ± 2.52	40.11 ± 2.62	$P < 0.01$
Scrotal circumference (cm)	29.11 ± 1.41	31.86 ± 1.93	$P < 0.01$
Gross motility (0-5)	2.91 ± 1.77	4.09 ± 0.92	$P < 0.01$
Progressive motility (%)	55.91 ± 23.23	67.73 ± 15.41	$P < 0.01$
Sperm concentration (10^9/mL)	2.32 ± 141	4.97 ± 3.84	$P < 0.01$
Morphology (% normal cells)	63.9 ± 10.4	66.27 ± 6.6	$P=0.19$
Scrotal growth (cm/d)	0.13 ± 0.05	0.15 ± 0.12	$P=0.20$

Average sperm concentration for the four treatments was within the $2\text{-}6 \times 10^9$/mL range described by Evans and Maxwell (1990). As mentioned in the present study, lambs at 34 days of feeding had attained puberty, which allowed an increased sperm concentration by 52 days. Van Emon *et al.* (2013) studied Suffolk x western whiteface rams beginning at 40.4 kg (90 days old) and lasted 117 days; they reported that increasing dietary DDGS levels decreased sperm concentration. Despite some differences between the latter and the present study, semen evaluations ended at comparable ages. In the present study, animals receiving the diet with both DDGS and RP for 34 days showed a higher sperm concentration than those receiving either RP or DDGS alone. However, at 52 days none of the semen variables were influenced by dietary treatments. High dietary levels of sulfur in diets with DDGS can also alter the use of selenium and copper absorption (Richter *et al.*, 2012). This may explain the reduction in the concentration of sperm in rams consuming DDGS at 34 days; however this effect was absent at 52 days, indicating that, for the current study, sperm cell concentration was not affected by DDGS or rice polishing. It is unclear whether the difference at 34 days was due to diet (RP or DDGS) or to any of the other factors discussed above. Other studies in lambs have reported negative effects of feeds such as cotton seed and their by-products on the quality of the semen, especially on progressive motility and sperm vigor (Cunha *et al.*, 2012; Paim *et al.*, 2016).

CONCLUSION

The use of 30% of dried distillers grains with solubles with or without rice polishings in diets for young Pelibuey × Dorper rams did not alter weight gain, scrotal circumference or quality of the semen after 52 days of feeding; therefore these feedstuffs may be safely supplied to these animals destined for breeding. Further studies should be undertaken for longer periods of feeding.

REFERENCES

Aguilar-Urquizo, E., Sangines-Garcia, J. R., Delgadillo, J. A., Capetillo-Leal, C. and Torres-Acosta, J. F. J. 2013. The onset of puberty of Pelibuey male hair sheep is not delayed by the short term consumption of Morusalba or Hibiscus rosa-sinensis foliage. Livestock Science. 157:378-383.

Cárdenas-Gallegos, M. A., Aké-López, J. R., Magaña-Monforte, J. G. and Centurión-Castro, F. G. 2015. Libido and serving capacity of mature hair rams under tropical environmental conditions. Archivos Medicina Veterinaria. 47:39-44.

Cunha, M., Gonzalez, C., de Carvalho, F. and Soares, A. 2012. Effect of diets containing whole cottonseed on the quality of sheep semen. Acta Scientarum. Animal Sciences. 34:305-311.
http://dx.doi.org/10.4025/actascianimsci.v34i3.12963

Evans, G. and Maxwell, W. M. C. 1990. Inseminación artificial de ovejas y cabras. 1th ed. Acribia., Zaragoza, Spain.

Ghorbankhani, F., Souri, M., Moeini, M. M. and Mirmahmoudi, R. 2015. Effect of nutritional state on semen characteristics, testicular size and serum testosterone concentration in Sanjabi ram lambs during the natural breeding season. Animal Reproduction Science. 153:22-28.

Hafez, E. S .E. and Hafez B. 2000. Reproducción e inseminación artificial en animales. 7th ed. McGraw-Hill Interamericana. Mexico City, Mexico.

Jiménez-Severiano, H., Reynoso, M. L., Roman-Ponce, S. I. and Robledo, V. M. 2010. Evaluation of mathematical models to describe testicular growth in Blackbelly ram lambs. Theriogenology. 74:1107-1114.

Kheradmand, A., Babaei, H. and Batavani, R. A. 2006. Effect of improved diet on semen quality and scrotal circumference in the ram. Veterinarski Arhiv. 76:333-341.
https://hrcak.srce.hr/5142

Klopfenstein, T. J., Erickson, G. E. and Bremer, V. R. 2008. Board-invited review: use of distillers by-products in the beef cattle feeding industry. Journal Animal Science. 86:1223-1231.

López, A., Regueiro, M., Castrillejo, A. and Pérez-Clariget, R. 2011. La época del año y el plano nutricional y su influencia sobre la morfología espermática en carneros Corriedale en pastoreo. Veterinaria (Montevideo). 47:15-21.

Lumpkins, B., Batal, A. and Dale, N. 2005. Use of distillers grains plus soluble in laying hen diets. Journal Applied Poultry Research. 14:25-31.

Moreno-Cáñez, E., Ortega-García, C., Cáñez-Carrasco, M.G. and Peñúñuri-Molina, F. 2013. Evaluación del comportamiento posdestete en corral de futuros sementales ovinos de raza Katahdin y Pelibuey en Sonora. Tecnociencia Chihuahua. 7:7-15.

National Research Council (NRC). 2000. Nutrient requirements of beef Cattle. Updated 7th rev. ed. Natl. Acad. Press, Washington, DC.

National Research Council (NRC). 2007. Nutrient Requirements of Small Ruminants. Sheep, goats, cervids, and new world camelids. National Academic Press. Washington, D.C., U.S.A.

Paim, T. D., Viana, P., Brandao, E., Amador, S., Barbosa, T., Cardoso, C., Lucci, C. M., Rodrigues de Souza, J., McManus, C., Abdallal, A.L. and Louvandini, H. 2016. Impact of feeding cottonseed coproducts on reproductive system of male sheep during peripubertal period. Scientia Agricola. 73:489-497. https://doi.org/10.1590/0103-9016-2015-0377

Richter, E. L., Drewnoski, M. E. and Hansen, S. L. 2012. Effects of increased dietary sulfur on beef steer mineral status, performance, and meat fatty acid composition. Journal Animal Science. 90:3945-3953.

Salinas-Chavira, J., Perez, J. A., Rosales, J. A., Hernandez, E. A. and La, O. 2013. Effect of increasing levels of rice polishings on ruminal dry matter degradability and productive performance of fattening sheep. Cuban Journal Agricultural Sciences. 47:375-80.

Sarlós, P., Egerszegi, I., Balogh, O., Molnár, A., Cseh, S. and Ratky, J. 2013. Seasonal changes of scrotal circumference, blood plasma testosterone concentration and semen characteristics in Racka rams. Small Ruminant Research. 111:90-95.

Schauer, C. S., Anderson, L. P., Stecher, D. M., Pearson, D. and Drolc, D. 2005. Influence of dried distillers grains on feedlot performance and carcass characteristics of finishing lambs. Western Dakota sheep and beef day. 46:31-33.

Schauer, C. S., Stamm, M. M., Maddock, T. D. and Berg, P. B. 2008. Feeding of DDGS in lamb rations. Sheep Goat Research Journal. 23:15-19.

Tabeidian, S. A. and Sadeghi, G. 2009. Effect of replacing barley with rice bran in finishing diet on productive performance and carcass characteristics of Afshari lambs. Tropical Animal Health Production. 41: 91-796.

Tufarelli, V., Lacalandra, G. M., Aiudi, G., Binetti, F. and Laudadio, V. 2011.. Influence of feeding level on live body weight and semen characteristics of Sardinian rams reared under intensive conditions. Trop. Anim. Health Prod. 43:339-345.

Valdez, J. D. 2013. Efecto del dodecil sulfato iónico adicionado a un diluyente libre de yema de huevo sobre la calidad del semen ovino congelado. MSc thesis, Universidad de Cuenca, Cuenca, Ecuador.

Valencia-Méndez, J., Trujillo-Quiroga, M. J., Espinosa-Martínez, M. A., Arroyo-Ledezma, J. and Berruecos-Villalobos, J. M. 2005. Pubertad en corderos Pelibuey nacidos de ovejas con reproducción estacional o continua. Revista Científica. FCV-LUZ. 15:437-442.

Van Emon, M. L., Vonnahme, K. A., Berg, P. T., Redden, R. R., Thompson, M. M., Kirsch, J. D. and Schauer, C. S. 2013. Influence of level of dried distillers grains with solubles on feedlot performance, carcass characteristics, serum testosterone concentrations, and spermatozoa motility and concentration of growing rams. Journal Animal Science. 91:5821-5828.

Villasmil-Ontiveros, Y., Aranguren, J., Madrid-Bury, N., González, D., Rubio, J., González-Stagnaro, C., Portillo, M. and Yañez, L. 2011. Age and body weight at puberty of crossbred ram lambs in Zulia State, Venezuela. Actas Iberoamericanas Conservación Animal. 1:419-422.

Zamiri, M. J., Khalili, B., Jafaroghli, M. and Farshad, A. 2010. Seasonal variation in seminal parameters, testicular size, and plasma testosterone concentration in Iranian Moghani rams. Small Ruminant Research. 94:132-136.

Zhang, S. Z., Penner, G. B., Yang, W. Z. and Oba, M. 2010. Effects of partially replacing barley silage or barley grain with dried distillers grains with solubles on rumen fermentation and milk production of lactating dairy cows. Journal Dairy Science. 93:3231-3242.

14

Impact of Privi-Silixol Foliar Fertilizer in Combination with Di-Ammonium Phosphate and Mycorrhiza on Performance, Npk Uptake, Disease and Pest Resistance on Selected Crops in a Greenhouse Experiment

G.N. Karuku[1]* and M. Maobe[2]

Department of Land Resources and Agricultural Technology, University of Nairobi; Email, karuku_g@yahoo.com
Department of Plant Science and Crop Protection, University of Nairobi. Email: gmoe54321@gmail.com
*Corresponding author

SUMMARY

Despite silicon not considered an essential nutrient, it is typically abundant in soils and is known to have beneficial effects when added to rice crops and several other plants. These beneficial effects include disease and pest resistance, structural fortification, and regulation of the uptake of other ions. In this study, the effect of silicic acid fertilization (Privi-Silixol) on the increased biomass, economic yields, pest and disease tolerance or resistance, NPK uptake and chlorophyll content for five crop plants (Cowpea, common beans, cabbage, maize, and rice) was evaluated. The approach was executed through a controlled greenhouse experiment using Acid Washed Sand as a neutral medium. Crops planted with Privi-Silixol alone or in combination with full or half rates of recommended inorganic fertilizer performed significantly ($P \leq 0.05$) better compared to all other treatments. Plants treated with Privi-Silixol had higher dry matter yield (DMY), chlorophyll content and NPK uptake. Maize, common bean, cowpea, cabbage and rice had better disease and pest resistance compared to the control plants. Collectively, these results indicate beneficial effects of silicon in DMY, chlorophyll content, pest and disease resistance, but additional studies are needed under farmers' conditions to conclusively understand the interactions of silicon with other interacting factors in the field.
Key words: silicon; biomass; yield; chlorophyll content; draught tolerance.

INTRODUCTION

Weathering is the main factor in the availability of silicon in soils (Song et al., 2012). As weathering increases, available silicon is generally depleted (Narayan et al., 1999). This phenomenon occurs mostly in tropical regions of the earth. At a lower pH which is common in the tropics, silicic acid (H_4SiO_4) is more soluble and less likely to dissociate (Tubaña and Heckman. 2015) and is in equilibrium with soil SiO_2 at pH 3.1 and at a concentration of 0.794mM.

Silicon amendments can be important for optimal crop yields (Alvarez et al., 1988; Korndörfer and Lepsch, 2001).The essentiality of silicon for plant growth has long been a question of interest to plant nutrition researchers. Uptake of silicon from soil increases with increasing soil water content (Hemmi, 1933; Williams and Shapter, 1955) and varies by species and by plant group (Jones and Handreck 1967; Ma et al., 2001; Richmond and Sussman, 2003). Rice plants appear to perform active uptake of silicon (Ma and Yamaji, 2006; Van Soest, 2006), at least in hydroponic solutions. Ma and Yamaji (2006) suggest that there is a gene that encodes a Si uptake transporter in rice while Cornelis et al. (2010) suggest that uptake is passive in forest trees.

The presence of Si in plants has been found to alleviate many abiotic and biotic stresses, leading to the incorporation of silicates into many fertilizers. The deposition of silica as a physical barrier to penetration and reduction in the susceptibility to enzymatic degradation by fungal pathogens has been examined (Yoshida et al., 1962). However, there is debate as to whether this increased strength is sufficient to explain the protective effects observed (Fauteux et al., 2005). Investigations on defense mechanisms by Belanger et al. (2003) and Rodriguez et al. (2003), studying wheat and rice blast, respectively indicated that these species were capable of inducing biologically active defense agents, including increased production of glycosylated phenolics and antimicrobial products such as diterpenoid and phytoalexins in the presence of silica. Experiments performed on cucumber leaves following fungal infection showed that further resistance to infection is acquired by expression of a proline-rich protein together with the presence of silica at the site of attempted penetration (Kauss et al., 2003).

Metal toxicity (e.g. Mn, Cd, Al and Zn), salinity, drought and temperature stresses can be alleviated by Si application (Liang et al., 2007).Drought tolerance brought about by the application of 'Si' may result from decreased transpiration (Epstein, 1999) and the presence of silicified structures in plants suggest a reduction of leaf heat-load, providing an effective cooling mechanism and thereby improving plant tolerance to high temperatures (Wang et al., 2005).

Resistance to salt stress is associated with the enhancement of enzymes such as superoxide dismutase (SOD) and catalase, preventing membrane oxidative damage (Zhu et al., 2004; Moussa, 2006).

Shubhodaya Mycorrhiza on the other hand is a commercial root fungi manufactured by Cosme Biotech PVT Ltd India, that enhances efficient utilization of nutrients and water in low fertility soils and under draught conditions, respectively. It contains 10^5 ineffective propagules per kilogram. It is applied in nurseries for transplanted crops while in field crops like maize, it can also be applied or injected into the soils at sowing or seeds maybe coated using juggery syrup and Shubhodaya Mycorrhiza sprinkled on them. In field and greenhouse trials conducted by Karuku et al. (2014c) in five agro-ecological zones in Kenya, Shubhodaya mycorrhiza was shown to increase root and shoot biomass, yields of field crops and branching of rose flowers. The French beans' root mass, pod weight, Stover and seed weight increased by 27.9, 43.5, 40.8 and 55.7%, respectively and were all significantly different at ($p \leq 0.05$) compared to the control. Plots treated with Shubhodaya Mycorrhiza gave higher yields in beans, maize, collianda while tomatoes recorded highest number of fruits at 99.1% increment above control. Flowering and fruiting was observed to occur earlier than in the other plots not treated with the root fungi.

Use of fertilizers such as K_4SiO_4 have a potential of making the crop resistant to attack by pests while improving nutrient status and not interfering much with the biodiversity is bound to give positive results. Si strengthens tissues and acts as an insecticide and a fungicide (Datnoff and Rodrigues, 2005). Chen et al. (2011) also found that silicon increased photosynthetic rate on a per-leaf basis. According to Lee et al. (2010), the addition of 2.5 mM Si to soybean plants "is beneficial in hydroponically grown plants as it significantly improves growth attributes and effectively mitigates the adverse effects of NaCl induced salt stress". According to Shen et al. (2010), the addition of 1.7 mM Si significantly increased soybean dry mass by 26% when subjected to -0.5 MPa of polyethylene glycol stress (PEG) stress. Sonobe et al. (2010) reported that 1.78 mM Si (SiO_2) in a 15% PEG 6000 (v/v) solution (to create -0.6 MPa) at 23 days increased shoot dry weight and second-nodal root diameter of Sorghum plants in hydroponic culture, even with decreased osmotic potential of roots. Also of interest is that Bakhat et al. (2009) found that corn supplied with 0.8 mM Na_2SiO_3 in solution culture under no stress conditions accumulated more leaf area and biomass than corn supplied with no silicon under the same conditions.

Shubhodaya Mycorrhiza ameliorates soil conditions as it interacts with the rhizosphere hence solubilizing fixed nutrients thus increasing their efficiency use. Finally, Shubhodaya Mycorrhiza harnesses and enhances acquisition of micro-nutrients such as Zn, Cu and Fe from low fertility soils, improves water use efficiency at low matric potential and generally improves the wellbeing of plant through proper photosynthetic activity (Karuku *et al.*, 2014c).

Specific objectives of the study were (i) Conduct trials to evaluate effectiveness of Privi-Silixol foliar fertilizer in comparison and interactively with conventional fertilizers and Shubhodaya mycorrhiza on economic yield, biomass and chlorophyll production of Cowpea, Common beans, Cabbage, Maize, and rice(ii)Evaluate effectiveness of Privi-Silixol fertilizer on improving plants immunity/ resistance to attack by pest and diseases and (iii) Determine N, P, K uptake of selected crops treated with Privi-Silixol.

MATERIALS AND METHODS

Study Site

The study was conducted at the University of Nairobi, Upper Kabete Field Station farm located at 1°15′S and 36° 44′ E at an altitude of 1940 m asl. The area is representative, in terms of soils and climate, of large areas of the Central Kenya highlands. According to the Kenya Soil Survey agro climatic zonation methodology (Sombroek *et al.*, 1982), the climate of the study area can be characterized as semi-humid. The area experiences a bimodal rainfall distribution with long rains in March–May and the short rains in October – December with mean annual rainfall of 1006 mm (Karuku *et al.,* 2012, Karuku *et al.*, 2014a,b). The ratio of annual average rainfall to annual potential evaporation, r/Eo is 58% (Karuku *et al.,* 2012, Karuku *et al.,* 2014a&b). The geology of the area is composed of the Nairobi Trachyte of the Tertiary age while the soils are well-drained, very deep (> 180 cm), dark red to dark reddish brown, friable clay (Gachene, 1989) and are classified as humic Nitisol (FAO, 1990; WRB, 2015). The land is cultivated for horticultural crops such as kales (*Brassica oleracea*), tomatoes (*Lycopersicon esculentum*), cabbage (*Brassica oleracea*), carrots, (*Daucus carota*), onions (*Allium fistulosum*), fruit trees such as avocadoes (*Persea americana*) and coffee (*Coffea arabica*).

Privi-Silixol as a nutrient

Privi-silixol is manufactured by Privi-Pharma in Mumbai-India. It is an Orthosilicic acid measured as silica (Si w/v) which is 0.8%v/v with 48% (v/v) stabilizers. The chemistry of formulation is well guarded by the manufacturers with only specifications being 500ml Privi-Silixol per hectare per application

and the crop should be given 3 applications at 15 days interval.

Characterization Privi-silixol product

Laboratory tests were conducted on Privi-Silixol fertilizer which is a new product in Kenya to ascertain chemical constitution of the product and whether it contains the various chemical nutrients stated on the package by the manufacturers. Purity of the product was assessed to ascertain presence/absence of pollutants such as heavy metals in the Privi-Silixol fertilizer as follows: (a) Lead and Cadmium Graphite Furnace Atomic Absorption Spectrometry (Schlemmer and Radziuk, 1999) and Mercury Cold Vapour Atomic Absorption Spectrometry (Shrader and Hobbins, 2010).This is a major requirement by Kenya Plant Health Inspectorate Service (KEPHIS) for new products entering the Kenyan market.

The bioassay of Privi-Silixol showed the content of heavy metals was very low as to affect the soil or bio-accumulate in plant tissues (Lead = 0.10, Cadmiun = 0.004, Mercury = 0.15 and Zinc = 0.85 ppm). NIOSH at CDC has set a Recommended Exposure Limit (REL) of $50\mu g/m^3$ to be maintained so that worker blood lead remains $<60\mu g/dL$ of whole blood (*http://www.cdc.gov/niosh/npg/npgd0368.html, 2017*). Uncontaminated soil contains lead concentrations less than 50 ppm but soil lead levels in many urban areas exceed 200 ppm (*http://www.cdc.gov/niosh/npg/npgd0368.html, 2017*). The allowable levels of lead in fertilizers are 5ppm according to Washington State Standards for Metals. (*http://www.ecy.wa.gov/programs/hwtr/dangermat/fert_standards.html, 2017*). Plant uptake factors are low ranging from 0.01-0.1 hence amount in these bio-fertilizers are negligible.

Privi-Silixol fertilizer trials

Trials evaluated the response of Rice, Maize, Cowpea, Common bean and Cabbages to Privi-Silixol alone; di ammonium phosphate (DAP) fertilizer in combination Privi-Silixol at full and at half recommended rates of the study site and Shubhodaya Mycorrhiza in combination with Privi-silixol comprising seven treatments replicated four times under a neutral media of acid washed sand. Parameters assessed included N, P and K uptake, chlorophyll content in leaves, above and below ground biomass and economic yield at harvest. In addition, disease and pest resistance, lodging incidences (falling on the side as in wheat, rice and maize especially with a slight wind) and draught tolerance were also evaluated.

Experimental layout and design for greenhouse trial

The experimental design was a Completely Randomized Design (CRD). The test crops used in the trials were Maize (*Zea mays*), Common beans (*Phaseolus vulgaris*), Cabbages *(Brassica oleracea var. capitata)*, Cowpeas *(Vigna unguiculata)* and Rice *(Oryza sativa)*. The experiment consisted of 7 treatments (T) replicated (R) four times in pots containing 5kg soil and the design was as in Table 1.

Table 1: Experimental design and layout

RI	RII	RIII	RIV
T7	T3	T5	T2
T2	T4	T1	T3
T5	T6	T7	T4
T1	T2	T6	T5
T6	T7	T4	T1
T4	T1	T3	T6
T3	T5	T2	T7

Legend: T1: Plant alone (0F 0PS)-Control; T2: Plant + Privi-Silixol only (0F PS); T3: Plant + ½ recommended DAP dose + Privi-Silixol (1/2F PS); T4: Plant + Full recommended DAP rate + Privi-Silixol (F PS); T5: Plant + Full recommended DAP dose (200 DAP kg/ha);T6: Plant + ½ recommended DAP dose (100 DAP kg/ha); T7: Plant + Shubhodaya Mycorrhizae (SM) alone; R: Replicates I-IV

River sand was obtained and then acid washed, rinsed with deionized water to remove any nutrients before being used as the neutral medium for the trials. This allowed for the proper administration of nutrients in the treatments.

Chlorophyll content

Chlorophyll content of the fully expanded leaf on the plant was measured with a Minolta SPAD-502 chlorophyll meter. SPAD chlorophyll readings were taken on each leaf, avoiding the midrib, and then averaged per plant. SPAD reading was done when the plant had four fully developed leaves.

Application rates

Each acre requires an application rate of 1 liter of Privi-Silixol dissolved in 1000 liters of water per application. The whole growing cycle required 3 applications hence 3 liters per hectare per season. And taking a furrow slice at 15cm depth, an acre is assumed to have $1x10^6$ kg of soils. Using this relationship and the bulk density of the sand/soil, the amount of Privi-Silixol or DAP fertilizer to add was calculated using the weight of the 10 kg sand in the pot. The parameters assessed were Stover residue biomass and Economic Yield at harvest; NPK uptake; Chlorophyll content and Disease and pest tolerance for plants.

Efficacy of privi-silixol in disease management

Maize. Leaves with *Exserohilum turcicum* which causes northern leave blight were collected from the university farm and inoculum was prepared using Muiru *et al.* (2010) procedure. Disease scoring was done on a scale of 1-5 where 1: Light infection, moderate number of lesions on lower leaves only, 2: Moderate infection, abundant lesions are on lower leaves, few on middle leaves. 3: Heavy infection, lesions are abundant on lower and middle leaves, extending to upper leaves. 4: Very heavy infection, lesions abundant on almost all leaves plants prematurely dry or killed by the disease.

Common beans. Isolation and inoculation of common beans with *Colletotrichum lindemuthianum* which causes bean anthracnose was using a method described by Wahome *et al.* (2011). The bean plants were inoculated by spraying at the sixth week using an atomizer. The plants were covered with plastic bags, and incubated under cool conditions overnight. A 1 to 5 severity scale, based on visual observation of the percentage of the organ presenting symptoms, was adopted. A score of 1 represented no observed symptoms while 5 corresponded to 100% of the organ covered by brown typical lesions of anthracnose

Cowpea. *Colletotrichum destructivum* was obtained from anthracnose lesions on the leaves of cowpea plants in the University of Nairobi farm. Inoculum preparation followed the procedure outlined by Sun and Zhang (2009). The cowpea plants were inoculated by spraying at the sixth week using an atomizer. The plants were covered with plastic bags; they were incubated under cool conditions overnight. A 1 to 5 severity scale, based on visual observation of the percentage of the organ presenting symptoms, was adopted. A score of 1 represented no observed symptoms while 5 corresponded to 100% of the organ covered by brown typical lesions of anthracnose.

Cabbage. *Xanthomonas campestris* pv campestris which causes black rot in brassica family was isolated, purified and then multiplied on nutrient agar. After 24 hours in the incubation chamber at 25 °C, the colonies were harvested using a spatula into a one liter beaker containing 200 ml of sterile distilled water. The bacterial cell suspension was adjusted using hemacytometer to $1x10^4$ colony forming units. The suspension was sprayed on two months old cabbage plants in the green house using a hand atomizer. The plants were monitored for the rate of establishment and progress of the necrotic lesions typical to black rot for two weeks. Disease reactions were rated from 10 days after inoculation with a 0 to 5 scale: 0 = disease free, 1 = trace infection, 2 = 0.5 to 0.9, 3 = 1.4, 4 = 1.5 to 1.9, and 5 = more than 2 cm 2 of leaf diseased.

Early blight inoculum preparation and inoculation. Tomato leaves showing *Alternaria solani* lesions were used to prepare *Alternaria inoculums* using Reni et al. (2007) procedure. The same procedure was followed for inoculation.

Rice. Blast infected rice tissues were used to prepare the inoculum in the laboratory. The tissues were surface sterilized and cultured in 9mm petri dish containing Potato dextrose agar (PDA). The spores were harvested and the suspension adjusted to 1×10^6 and then rice at three leave stage was inoculated using a hand atomizer. Inoculated seedlings were incubated overnight at 16°C in a humid chamber and subsequently transferred to the glasshouse for the disease development. The plants were monitored for three weeks. Based on the spots scoring were rated on the basis of severity as: 0-10% neck infection as resistant (1), 11- 25% infection as moderately resistant (2), 26–40% infection as mildly susceptible (3), 41-70% infection as moderately susceptible (4) and 71-100% as susceptible (5)

Privi-silixol in Pest management

Mite resistance experiment for maize, common bean, cabbages and rice crops. The trial to ascertain the efficacy of Privi-Silixol was carried out using the procedure described by Gatarayiha et al. (2010).

Pest resistance in cowpea. The experiment was laid out in a completely randomized design in green house. The seeds were sown in ten pots, 2 seeds per two liter plastic pot containing sterile (sand). At 5 weeks after germination, aphids were artificially introduced. An assessment for aphid population was carried out and on the 6 week and before treatments was applied to

establish the infestation and stabilization levels. Visual examination on the degree of infestation was done. The cow pea aphid, *Aphis fabae*, numbers were rated using a categorical index due to the fact that their numbers are usually exceptionally high, thus a challenge to enumerate, where 0 = None; 1 = few scattered individuals; 2 = few isolated colonies; 3 = Several isolated colonies; 4 = Large isolated colonies; and 5 = Large continuous colonies. The same key was used to assess the aphid population after folia spray. Aphid damage was classified as wilted. The control experiments both the negative and positive controls were carried out in nearby greenhouses independently. After application of potassium silicate, the Aphid population was monitored weekly for three weeks. Data on the aphid population was recorded using the above key.

RESULTS AND DISCUSSION

Table 2 shows chlorophyll content in bean, cabbage, maize and rice crops under different fertility treatments. Generally the plots treated with Privi-Silixol alone or in combination with diammonium phosphate showed higher chlorophyll content than those not treated with Privi-Silixol in all crop species. Chlorophyll is an indicator of photosynthetic activity and currently it's being linked to high N content in plant tissues.

Table 3 shows dry biomass of selected crops at harvest under different treatments. At harvest, rice biomass with full DAP recommended rate + Privi-Silixol treatment performed better compared to all other treatments. Plots treated with Privi-Silixol + ½DAP had 99 % more biomass than control and 46% more than ½DAP treated plots alone while Privi-Silixol + full DAP had 138% and 57% above control and full DAP alone, respectively in cowpeas. In rice crop, Privi-Silixol + ½DAP had 150% more biomass weight above control and 79.8% more than ½DAP recommended treated plots alone while Privi-Silixol + full DAP recommended rate had 199.2% and 84.1 % more biomass above control and full DAP recommended rate alone, respectively.

Table 2: Chlorophyll content on selected crops under different fertility level treatments and with Privi-Silixol

Treatment	1	2	3	4	5	6	7	mean	F-value	LSD	SED
Bean	12.18	26.38	20.75	22.25	19.48	25.63	12.40	19.86	P≤0.001	3.14	1.50
Cabbage	17.83	30.53	31.93	33.00	28.08	29.03	29.90	28.18	P≤0.001	2.81	1.34
Rice	27.92	23.70	32.02	33.35	28.30	27.72	25.02	28.29	P≤0.001	2.38	1.13
Maize	16.95	24.88	32.97	34.00	24.22	25.30	15.50	24.83	P≤0.001	3.62	1.73

Legend: Crop + the following treatments:1: Control, 2: Privi-Silixol only, 3: Silixol + 100 kg/ha di ammonium phosphate (DAP), 4: Privi-Silixol +200 kg/ha DAP recommended rate, 5: 200 kg/ha DAP recommended rate, 6: 100 kg/ha DAP half recommended rate, 7: Mycorrhiza only; GM: general mean, LSD: least square difference; SED: standard error

Table 3: Dry biomass at harvest of selected crops under different fertility treatments (g/Pot)

Treatment/crop	1	2	3	4	5	6	7	Mean	F-value	LSD	SED
Cowpea	2.05	2.63	4.08	4.88	3.10	2.80	2.40	3.13	P ≤0.004	1.36	0.65
Rice	1.28	1.70	3.20	3.83	2.08	1.78	1.53	2.20	P ≤0.001	1.05	0.50
Maize	5.30	5.47	12.35	13.95	9.40	8.35	5.22	8.58	P ≤0.004	3.65	1.74

Legend: Crop + the following treatments: - 1: Control 2: Privi-Silixol only 3: Privi-Silixol + 100 kg/ha DAP 4: Silixol +200 kg/ha DAP recommended rate: 5: 200 kg/ha DAP recommended rate 6: 100 kg/ha DAP half recommended rate 7: Mycorrhiza only; Lsd least square difference; Sed standard error

The same trend was observed in maize crop where, Privi-Silixol + ½DAP recommended had 94.5% more biomass above control and 30.3 % more than ½DAP recommended alone. Privi-Silixol + full DAP recommended had 100.6 % and 40.4 % more biomass above control and full DAP recommended rate alone, respectively and the data was significantly different at (p≤0.00)1 in all cases. These results are in agreement with those obtained by Kaerlek (2012) who, when investigating the effect of Silicon on plant growth and drought stress tolerance observed that the total corn plant mass increased by over 20% and the effect was statistically significant (p<0.05).Similar results were observed by Jawahar et al. (2015) at Annamalai University Experimental Farm, Annamalai Nagar, Tamil Nadu, India when studying the effect of silixol granules on growth and yield of rice. Among the different treatment imposed, 100 % recommended dose of fertilizers + silixol granules @ 37.5kgha[-1] recorded higher values for growth (plant height, number of tillers plant[-1], root length, root volume, leaf area index and dry matter production), yield attributing (number of panicles m[-2], number of grains panicle[-1] and test weight) characters and yield (grain and straw) of rice, respectively. Thus, it is concluded that soil application of silixol granules along with 100 % recommended dose of fertilizers holds immense potentiality to boost the productivity and profitability of rice. This was well articulated in this study where full recommended DAP rate plus Privi-silixol gave the highest performance in all parameters investigated.

Table 4 shows NPK content in maize tissues under different treatments. Highest NPK uptake was observed in plots treated with Privi-Silixol + ½DAP and Privi-Silixol + full DAP recommended rates in maize crop. Plots treated with Privi-Silixol + ½DAP had 183, 928.5 and 1364% more NPK uptake, respectively above control and 7.9, 27.4 and 4.0% above that of ½DAP alone treated plots, respectively while Privi-Silixol + full DAP recommended plots had 202, 1002 and 1561% and 16.9, 7.7 and 6.4% more NPK uptake above control and full DAP alone treated plots, respectively. Similar findings were reported by Datnoff and Rodrigues (2005) where silicon alone was associated with a gain in grain weight over the control, 37% (N at 50kgha[-1]) to 40% (N at 75kgha[-1]). The beneficial effects of Si to plants under biotic and/or

abiotic stresses have been reported to occur in a wide variety of crops such as rice, oat, barley, wheat, cucumber, and sugarcane. Leaves, stems, and culms of plants, especially rice grown in the presence of Si, show an erect growth thereby the distribution of light within the canopy is greatly improved (Alvarez and Datnoff, 2001; Elawad and Green, 1979; Epstein, 1991; Ma and Takahashi, 1991 and Savant et al., 1997). Silicon increases rice resistance to lodging and drought and dry matter accumulation in cucumber and rice (Adatia and Besford, 1986, Epstein, 1991; Lee et al., 1981). Silicon can positively affect the activity of some enzymes involved in the photosynthesis in rice and turf grass (Savant, et al., 1997; Schmidt et al., 1999) as well as reduce the senescence of rice leaves (Kang, 1980) hence the increased biomass. Silicon also lowers the electrolyte leakage from rice leaves and, therefore promotes greater photosynthetic activity in plants grown under water deficit or heat stress (Agarie et al., 1998) thus increasing biomass and yields of the crop.

Generally, Si-fertilization was shown to improve drought and salt stress tolerance, but the effects were inconsistent among crops. Silicon increased total maize plant biomass by up to 20% and the effect was statistically significant (p<0.05) (Kaerlek, 2012). Silicon increased water use efficiency (plant biomass accumulated divided by mass of water used) in corn by up to 36% and the effect was statistically significant (p<0.05) in one out of four trials. Collectively, these results indicate an effect of silicon in drought and salinity stress tolerance, but additional studies on the rate and onset of drought are needed to determine interacting factors and better understand the inconsistent results (Kaerlek, 2012).

Eneji et al. (2008) found that 1000 mg kg[-1] potassium silicate (K_2SiO_3) application to the soil of four grass species under deficit irrigation (half of field capacity) "produced the greatest biomass yield responses across species," as compared to calcium silicate ($CaSiO_3$) or silica gel. According to Gunes et al. (2008), sodium silicate applied to the soil mitigated the adverse dry mass reduction effects of drought in 6 of 12 sunflower cultivars. Pulz et al. (2008) found that using calcium and magnesium silicates in the place of dolomitic limestone (in areas with acidic soil) increased potato plant height, decreased stem lodging (weak lower stems), and increased the yield of marketable tubers in drought conditions (soil $\Psi = -0.05$ MPa). In addition to observations of reduced occurrence of stalk lodging and an increase of mean tuber mass in potatoes, Crusciol et al. (2009) found that the application of 284.4 mg dm^{-3}Ca and Mg silicate to the soil increased proline concentrations under drought conditions. Sonobe et al. (2009) found no effect of silicon on unstressed hydroponic sorghum, but found that 50 ppm (approximately 0.8 mM) silicon ameliorated dry mass reduction in hydroponic sorghum exposed to polyethylene glycol water stress. Chen et al. (2011) found that applying 1.5 mM silicon to drought stressed rice significantly ($P<0.05$) increased total root length, surface area, volume, and root activity, even to the extent that these parameters were equivalent to those observed in non-stressed plants in many cases.

Table 5 shows NPK content in bean tissues under different treatments. Highest uptake was observed in treatments of Silixol +200 kg/ha DAP recommended rates in bean crop just like in maize. Silixol + 100 kg/ha DAP also had high N and P uptake though K was slightly lower than 100 kg/ha DAP, 200 kg/ha DAP

recommended rate and Mycorrhiza only. The same trend observed with maize above is replicated with common beans.

Table 6 shows NPK content in rice tissues under different treatments. Highest uptake observed in treatments of Privi-Silixol +200 kg/ha DAP recommended rates in bean crop just like in maize. Privi-Silixol + 100 kg/ha DAP also had high N and P uptake though K was slightly lower than 200 kg/ha DAP recommended rate and Mycorrhiza only. The same trend observed with maize and beans above is observed with Rice crop. Similar reports were recorded by Sajal et al. (2016) in a field experiment conducted to study the effect of silicon (diatomaceous earth, DE) fertilization on growth, yield, and nutrient uptake of rice during the kharif season of 2012 and 2013 in the new alluvial zone of West Bengal, India.

Results showed that application of silicon significantly increased grain and straw yield as well as yield attributing parameters such as plant height (cm), number of tillers m-2, number of panicle m-2, and 1000-grain weight (g) of rice.

Table 4: NPK content in maize tissue under different fertility treatments six weeks after emergence

Treatment/crop	1	2	3	4	5	6	7	Mean	F-value	LSD	SED
% N	0.600	0.600	1.700	1.812	1.550	1.575	1.000	3.13	P ≤0.001	0.19	0.09
K ppm	508	708	5225	5600	5200	4100	4525	2.20	P ≤0.001	1032	496
P ppm	29.2	46.0	427.5	485.0	455.8	411.2	184.5	8.58	P ≤0.004	73.8	35.5

Legend: Crop + the following treatments:-1: Control 2: Privi-Silixol only 3: Privi-Silixol + 100 kg/ha DAP 4: Silixol +200 kg/ha DAP recommended rate; 5:200 kg/ha DAP recommended rate 6:100 kg/ha DAP half recommended rate 7: Mycorrhiza only; Lsd least square difference; Sed standard error

Table 5: NPK content in common bean tissues under different fertility six weeks after emergence

Treatment/crop	1	2	3	4	5	6	7	Mean	F-value	LSD	SED
% N	0.500	0.718	2.150	3.025	2.550	1.700	1.475	1.731	P ≤0.001	0.60	0.27
K ppm	396	531	2775	4300	3775	3050	4275	2729	P ≤0.001	1168	562
P ppm	62	52	448	562	505	418	304	336	P ≤0.004	91.3	43.9

Legend: Crop + the following treatments:-1: Control 2: Privi-Silixol only 3: Privi-Silixol + 100 kg/ha DAP 4: Silixol +200 kg/ha DAP recommended rate; 5: 200 kg/ha DAP recommended rate 6: 100 kg/ha DAP half recommended rate 7: Mycorrhiza only, Lsd least square difference; Sed standard error

Table 6: NPK content in rice tissues under different fertility treatments six weeks after emergence

Treatment/crop	1	2	3	4	5	6	7	Mean	F-value	LSD	SED
% N	0.625	0.525	2.425	3.350	2.375	1.675	1.200	1.739	P ≤ 0.001	0.69	0.33
K ppm	452	598	3625	5100	3925	2775	4025	2929	P ≤ 0.001	1038	499
P ppm	62	49	550	630	490	405	322	358	P ≤ 0.001	123.4	59.3

Legend: Crop + the following treatments:-1: Control 2: Privi-Silixol only 3: Privi-Silixol+ 100 kg/ha DAP 4: Silixol +200 kg/ha DAP recommended rate; 5: 200 kg/ha DAP recommended rate 6: 100 kg/ha DAP half recommended rate 7: Mycorrhiza only, Lsd least square difference; Sed standard error

The greatest grain and straw yields were observed in the treatment T6 (DE at 600kgha−1 in combination with standard fertilizer practice (SFP). The concentration and uptake of silicon, nitrogen (N), phosphorus (P), and potassium (K) in grain and straw were also greater under this treatment compared to others. It was concluded that application of DE at 600kgha−1 along with SFP resulted increased grain, straw, and uptake of NPK.

Table 7 shows pest and disease resistance in maize crop under various treatments. Privi-Silixol alone, Privi-Silixol + 100kg and 200 kgha⁻¹ DAP had highest pest and disease resistance ratings of slightly infested or diseased.

Table 7: Pest and disease resistance ratings in maize under different fertility treatments.

Treatment/ Disease or Pest	Average ratings	
	Mites	Blight
Control	4	5
Privi-Silixol only	2	2
Privi-Silixol + 100kgha⁻¹ DAP	2	2
Privi-Silixol + 200kgha⁻¹ DAP	3	2
200kgha⁻¹ DAP	3	2
100kgha⁻¹ DAP	3	3
Mycorrhiza only	4	4

Legend 1: Pest and disease free- not infested/diseased; 2: Slightly infested or diseased; 3: Moderately infested /diseased; 4: Fairly infested /diseased and 5: Fully infested/diseased

Table 8 shows pest and disease resistance in rice crop under various treatments. Privi-Silixol alone, Privi-silixol + 100kg and 200 kgha⁻¹ DAP had higher pest and disease resistance ratings of moderately and slightly infested or diseased, respectively (3 & 2 ratings). Onodera (1917) compared the chemical composition of the rice plants infected with blast with that of healthy ones grown in the same paddy field. He observed that diseased plants always contained less Si in comparison to healthy ones obtained from the same field, and that the natural Si content found in rice tissue depended on the paddy field in which the plants had been grown. His finding did not necessarily mean that blast infection was reduced by the Si content of the rice plants or that plants with less Si content were more susceptible. His results did show that there was a relationship between Si content and blast susceptibility.

In 1995 and 1996 experiments, Si was incorporated prior to seeding at 0 and 1000 kgha⁻¹ (80). Plots that were treated in 1995 (residual Si) were compared to plots receiving a fresh or current year application of Si

in 1996 to study the residual effect. Two foliar applications of edifenfos, sprayed at 20 and 35 days after planting, were made and followed by three applications of tricyclazole. Leaf blast was evaluated as percent area of individual leaves and neck blast was rated as percent incidence of 100 panicles. In both 1995 and 1996, ratings of leaf blast for Si alone (residual and fresh applications) and Si plus edifenfos (residual and fresh applications) were 50 to 68% lower than those in the non-treated control plots (Datnoff and Rodrigues, 2005).

Table 9 shows resistance to mites and anthracnose by common beans and cowpeas treated with privy-silixol or in combination with fertilizers. Privi-Silixol alone, Privi-silixol + 100 kg and 200 kg/ha DAP had higher pest and disease resistance ratings of moderately and slightly infested or diseased, respectively (3 & 2 ratings).

Table 8: Pest and disease resistance ratings in Rice crop under different fertility treatments (materials and methods for rice).

Treatment/ Disease or Pest	Average ratings	
	Mites	Blast
Control	4	4
Privi-Silixol only	3	2
Privi-Silixol + 100 kgha⁻¹ DAP	3	2
Privi-Silixol + 200 kgha⁻¹ DAP	3	3
200 kgha⁻¹ DAP	4	3
100 kgha⁻¹ DAP	5	4
Mycorrhiza only	5	4

Legend: 1: Pest and disease free- not infested/diseased; 2: Slightly infested or diseased; 3: Moderately infested /diseased; 4: Fairly infested /diseased and 5: Fully infested/diseased

Table 9: Pest and disease resistance ratings in common bean crop under different fertility treatments.

Treatment/ Disease or Pest	Average ratings	
	Mites	Anthracnose
Control	3	4
Privi-Silixol only	2	2
Privi-Silixol + 100 kgha⁻¹ DAP	2	2
Privi-Silixol + 200 kgha⁻¹ DAP	2	2
200 kgha⁻¹ DAP	3	3
100 kgha⁻¹ DAP	3	3
Mycorrhiza only	4	4

Legend: 1: Pest and disease free- not infested/diseased; 2: Slightly infested or diseased; 3: Moderately infested /diseased; 4: Fairly infested /diseased and 5: Fully infested/diseased

Table 10 shows resistance to aphids and anthracnose by cowpeas treated with Privi-silixol or in combination with fertilizers. Privi-Silixol alone, Privi-silixol + 100 kg and 200 kg/ha DAP had higher pest and disease resistance ratings of moderately and slightly infested or diseased, respectively (3 and 2 ratings).

Under cabbages (Table 11) mites fairly infested the crop and black rot was moderate under privy-silixol treatments.

Table 10: Pest and disease resistance ratings in cowpea crop under different fertility treatments

Treatment/ Disease or Pest	Average ratings	
	Aphids	anthracnose
Control	3	4
Privi-Silixol only	2	2
Privi-Silixol + 100 kgha^{-1} DAP	2	3
Privi-Silixol + 200 kgha^{-1} DAP	2	2
200 kgha^{-1} DAP	4	4
100 kgha^{-1} DAP	3	4
Mycorrhiza only	4	4

Legend: 1: Pest and disease free- not infested/diseased; 2: Slightly infested or diseased; 3: Moderately infested /diseased; 4: Fairly infested /diseased and 5: Fully infested/diseased

Table 11: Mites and black rot resistance ratings in cabbages under different fertility treatments

Treatment/ Disease or Pest	Average ratings	
	Mites	Black rot
Control	5	5
Privi-Silixol only	3	3
Privi-Silixol + 100 kgha^{-1} DAP	2	3
Privi-Silixol + 200 kgha^{-1} DAP	3	3
200 kgha^{-1} DAP	5	4
100 kgha^{-1} DAP	5	5
Mycorrhiza only	5	5

Control Chart Ratings: 1: Pest and disease free- not infested/diseased; 2: Slightly infested or diseased; 3: Moderately infested /diseased; 4: Fairly infested /diseased and 5: Fully infested/diseased

CONCLUSIONS

Privi-Silixol increased efficiency utilization of conventional fertilizers. Where used with Diammonium Phosphate (DAP) at full or half recommended rates, there was increased biomass in all crops tested, increased NPK uptake and increased pest and disease resistance. There were few disease incidences and pest infestation in the crops treated with Privi-silixol alone or in combination with DAP fertilizer. The chlorophyll content was higher indicating better photosynthetic activity. It can therefore be concluded that there is potential in the use of Privi-Silixol in combination with other sources of nutrients for maximum efficiency utilization of nutrients. This will cut down on use of excessive

fertilizers that has had a devastating effect on the health of our soils.

Acknowledgements
The authors acknowledge Novixa International Ltd for funding the research and to both the University of Nairobi and KEPHIS for the technical support during implementation.

REFERENCES

Adatia, M.H., Besford, A.T. 1986.The effects of silicon on cucumber plants grown in recirculating nutrient solution. Annals of Botany 58:343-351.

Agarie, S., Agata, W., Kaufman, P.B. 1998. Involvement of silicon in the senescence of rice leaves. Plant Production Science. 1:104-105.

Alvarez, J., Datnoff, L.E. 2001.The economic potential of silicon for integrated management and sustainable rice production. Crop Protection. 20:43-48

Alvarez, J., Snyder, G.H., Anderson, D.L., Jones, D.B. 1988. Economics of calcium silicate slag application in a rice-sugarcane rotation in the Everglades. Agricultural Systems 28:179–188.

Bakhat, H.F., Hanstein, S. and Schubert, S. 2009. Optimal level of silicon for maize (*Zea mays* L. c.v. Amadeo) growth in nutrient solution under controlled conditions. The Proceedings of the International Plant Nutrition Colloquium XVI, Davis, CA.

Bélanger, R.R., Benhamou, N., Menzies, J.G. 2003. Cytological evidence of an active role of silicon in wheat resistance to powdery mildew (*Blumeria graminis* f. sptritici) *Phytopathology*. 93:402–412.

Chen, W., Yao, X., Cai, K., Chen, J. 2011. Silicon alleviates drought stress of rice plants by improving plant water status, photosynthesis and mineral nutrient absorption. Biological Trace Element Research. 142: 67-76. doi: 10.1007/s12011-010-8742-x

Cornelis, J.T., Delvaux, B., and Titeux, H. 2010. Contrasting silicon uptakes by coniferous trees: a hydroponic experiment on young seedlings. Plant Soil 336:99–106.

Crusciol, C.A.C., Pulz, A.L., Lemos, L.B., Soratto, R.P., Lima, G.P.P. 2009. Effects of silicon and drought stress on tuber yield and leaf biochemical characteristics in potato. Crop Science, 49:949–954.

Datnoff, L.E. and Rodrigues, F.A. 2005. The Role of Silicon in Suppressing Rice Diseases. The American Phytopathological Society. APSnet Features. doi 10.1094/APSnetFeature-2005-0205.

Elawad, S.H., and Green, V.E. 1979. Silicon and the rice plant environment: A review of recent research. Riso 28:235-253.

Eneji, A.E., Inanaga, S., Muranaka, S., Li, J., Hattori, T., An, P., and Tsuji, W. 2008. Growth and nutrient use in four grasses under drought stress as mediated by silicon fertilizers. Journal Plant Nutrition. 31:355–365.

Epstein E. 1999. Silicon. Annual Review of Plant Physiology and Plant Molecular Biology. 50:641–664.

Epstein, E. 1991.The anomaly of silicon in plant biology. Proceedings National Academy Science. 91:11-17.

FAO. 1990. Soil map of the world, revised legend. Food and Agricultural Organization of United Nations, Rome.

Fauteux, F., Chain, F., Belzile, F., Menzies, J.G., Bélanger, R.R. 2006.The protective role of silicon in the *Arabidopsis*–powdery mildew pathosystem. Proceedings of the National Academy of Science. 103:17554–17559.

Gachene, C.K.K., Wanjogu, S.N. and Gicheru, P.T. 1989. The distribution, characteristics and some management aspects of sandy soils in Kenya. Paper presented at the International Symposium on Managing Sandy Soils, Jodhpur, India.

Gatarayiha, M.C, Laing, M.D., and Miller, R.M. 2010. Combining applications of potassium silicate and Beauveri abassiana to four crops to control two spotted spider mite, Tetrany chusurticae Koch. International Journal of Pest Management 56, 291–297

Gunes, A., Pilbeam, D.J., Inal, A. and Coban, S. 2008. Influence of silicon on sunflower cultivars under drought stress, I: growth, antioxidant mechanisms, and lipid peroxidation. Communications Soil Science Plant Analysis. 39:1885–1903.

Hemmi, T. 1933. Experimental studies on the relation of environmental factors to the occurrence and severity of blast disease in rice plants. Phytopathology Zeit. 6:305–324.

Jawahar, S., Vijayakumar, D., Bommera, R., Jain, N., and Jeevanandham. 2015. Effect of Silixol granules on growth and yield of Rice.

International Journal Current Research Academic Review. 3(12): 168-174.

Jones, L.H.P. and Handreck, K.A. 1967. Silica in soils, plants, and animals. In A.G. Norman (ed.) Advances in Agronomy. 19: 107–149.

Kaerlek, J.W. 2012. Effect of Silicon on Plant Growth and Draught Stress Tolerance. MSc Thesis. Utah State University. Logan, Utah.

Kang, Y.K. 1980. Silicon influence on physiological activities in rice. Ph.D. diss., University of Arkansas, Fayetteville, Arkansas.

Karuku, G.N., Gachene, C.K.K., Karanja, N., Cornelis, W., Verplancke, H. 2014b. Use of CROPWAT Model to Predict Water Use in Irrigated Tomato (*Lycopersicon esculentum*) Production at Kabete, Kenya. East African Agriculture Forestry Journal. 80(3): 175-183.

Karuku, G.N., Gachene, C.K.K., Karanja, N., Cornelis W., and Verplancke, H. 2014a. Effect of different cover crop residue management practices on soil moisture content under a tomato crop (*Lycopersicon esculentum*). Tropical and Subtropical Agroecosystems, 17: 509 -523.

Karuku, G.N., Gachene, C.K.K., Karanja, N., Cornelis, W., Verplancke, H., and Kironchi, G. 2012. Soil hydraulic properties of a Nitisol in Kabete, Kenya. Tropical and Subtropical Agroecosystems, 15: 595-609.

Karuku, G.N., Diaz, W., Kimotho, J., Anyika, F., Kalele, D., Munguti, J, Mongare, P. 2014c. Evaluating the effectiveness of Shubhodaya Mycorrhiza-VAM (Glomus spp) as a Bio-fertilizer for increased yields and P-uptake for selected crops in five different Agro-ecological zones and soil types, in Kenya. For Kenya Plant Health Inspection Service (KEPHIS) under the Kenya Standing Committee for imports and Exports (KSCIE). 2012-2014.

Kauss, H., Seehaus, K., Franke, R., Gilbert, S., Dietrich, R.A., Kröger, N. 2003.Silica deposition by a strongly cationic proline-rich protein from systemically resistant cucumber plants. The Plant Journal. 33:87–95.

Korndörfer, G.H. and Lepsch, I., 2001.Effect of silicon on plant growth and crop yield. In L.E. Datnoff, G.H. Snyder, and G.H. Korndörfer (ed.) Silicon in agriculture. Studies in Plant Science. 8 p.133–147.

Lee, S.K., Sohn, E.Y., Hamayun, M., Yoon, J.Y., and Lee, I.J. 2010. Effect of silicon on growth and salinity stress of soybean plant grown under hydroponic system. Agroforestry Systems 80:333–340.

Lee, T.S., Hsu, L.S., Wang, C.C., and Jeng, Y.H. 1981. Amelioration of soil fertility for reducing brown spot incidence in the patty field of Taiwan. Journal Agricultural Research China 30:35-49.

Liang, Y., Sun, W., Zhu, Y., Christe, P. 2007. Mechanisms of silicon-mediated alleviation of abiotic stress in higher plants: a review. Environmental Pollution. 147:422–428.

Ma, J., and Takahashi, E. 1991. Effect of silicate on phosphate availability for rice in a P-deficient soil. Plant Soil 133:151-155.

Ma, J.F. and Yamaji, N. 2006. Silicon uptake and accumulation in higher plants. Trends Plant Science 11:392–397.

Ma, J.F., Miyaki, Y., and Takahashi, E. 2001. Silicon as a beneficial element for crop plants p. 17 – 39. In L.E. Datnoff, G.H. Snyder, and G.H. Korndörfer (ed.) Silicon in agriculture. Elsevier, New York.

Moussa, H.R. 2006. Influence of exogenous application of silicon on physiological response of salt stressed maize (Zea mays L.) International Journal of Agriculture and Biology. 2006; 8:293–297.

Muiru, W.M., Koopman, B., Tiedemann, A.V., Mutitu, E.W. and Kimenju, J.W. 2010. Race typing evaluation and aggressiveness of Exserohilum turcum isolates of Kenyan, German and Australian origin, World Journal of Agricultural Sciences, 2010.

Narayan, K.S., Gasper, H.K., Lawrence, E.D. and George, H.S. 1999. Silicon nutrition and sugarcane production: A review. Journal Plant Nutrition. 22: 1853-1903.

Onodera, I. 1917. Chemical studies on rice blast. Journal Scientific Agricultural Society. 180:606-617.

Pulz, A.L., Crusciol, C.A.C., Lemos, L.B., Soratto, R.P. 2008. Influência de silicato e calcárionanutrição, produtividade e qualidade da batata sob deficiênciahídrica. Revista Brasileira de Ciência do Solo, Viçosa, MG, 32: 1651-1659.

Reni, C., Groenwold, R., Stam, P., Voorrips, R.E. 2007. Assessment of early blight (Alternaria solani) resistance in tomato using a droplet inoculation method. Journal General Plant Pathology. 73:96–103.

Richmond, K.E. and Sussman, M. 2003. Got silicon?

The non-essential beneficial plant nutrient. Current Opinion Plant Biology. 6:268–272.

Rodriguez, F.A, Benhamou, N, Datnoff, L.E, Jones, J.B, Bélanger, R.R. 2003. Ultra-structural and

Sajal Pati, Biplab Pal, Shrikant Badole, Gora Chand Hazra and Biswapati Manda. 2016. Effect of Silicon Fertilization on Growth, Yield, and Nutrient Uptake of Rice, Communications in Soil Science and Plant analysis, 47: 3, 2847-290

Savant, N.K., Snyder, G.H., and Datnoff, L.E. 1997. Silicon management and sustainable rice production. In: Advances in Agronomy, D. L. Sparks (ed). 58: 151-199

Schlemmer, G. and Radziuk, B. 1999. Analytical Graphite Furnace Atomic Absorption Spectrometry. A Laboratory Guide Book Biomethods, 1999.

Schmidt, R.E., Zhang, X., and Chalmers, D.R. 1999. Response of photosynthesis and superoxide dismutase to silica applied to creeping bent grass grown under two fertility levels. Journal Plant Nutrition. 22:1763-1773.

Shen, X., Zhou, Y., Duan, L., Li, Z., Eneji, A.E., Li, J. 2010. Silicon effects on photosynthesis and antioxidant parameters of soybean seedlings under drought and ultraviolet-B radiation. Journal Plant Physiology. 167(15):1248-1252. doi: 10.1016/j.jplph.2010.04.011.

Shrader, D.E. and Hobbins, W.B. 2010. The Determination of Mercury by Cold Vapor Atomic Absorption. Agilent Technologies: www.agilent.com/chem.

Sombroek, W.G., Braun, H.M.H. and Van Der Pouw, B.J.A. 1980. The Exploratory Soil Map and Agro climatic Zone Map of Kenya. Report No. E1. Kenya Soil Survey, Nairobi, Kenya.

Song, Z.L, Wang, H.L., Strong, P.J., Li, Z.M. and Jiang, P.K. 2012. Plant impact on the coupled terrestrial biogeochemical cycles of Silicon and Carbon: Implications for biogeochemical carbon sequestration. Earth Sci. Rev. 155: 319-331.

Sonobe, K., Hattori, T., An, P., Tsuji, W., Eneji, E., Tanaka, K., and Inanaga, S. 2009. Diurnal variations in photosynthesis, stomatal conductance and leaf water relation in sorghum grown with or without silicon under water stress. Journal Plant Nutrition. 32:433–442

Sonobe, K., Taiichiro, H., Ping, A., Wataru, T., Anthony, E.E., Sohei, K., Yukio, K., Kiyoshi, T., Shinobu, I. 2010. Effect of silicon application on sorghum root responses.to water stress. Plant Nutrition. 34 71–82. 10.1080/01904167.2011.531360

Sun, H., Zhang, J.Z. 2009. *Colletotrichum destructivum* from cowpea infecting Arabidopsis thaliana and its identity to *C. Higginsianum*.European Journal Plant Pathology, 2009.

Tubaña, B.S. and Heckman, J.R. 2015. Silicon in Soils and Plants. Springer International Publishing Switzerland 2015 F.A. Rodrigues, L.E. Datnoff (eds.), Silicon and Plant Diseases, DOI 10.1007/978-3-319-22930-0_2

Van Soest, P.J. 2006. Rice straw, the role of silica and treatments to improve quality. Animal Feed Science Technology. 130:137–171.

Wahome, S.W., Kimani, P.M., Muthomi, J.W., Narla, R.D. and Buruchara, R. 2011. Multiple Disease Resistance in Snap Bean Genotypes in Kenya. African Crop Science Journal, 19:289-302.

Wang, L., Nie, Q., Li, M., Zhang, F., Zhuang, J., Yang, W. 2005. Biosilicified structures for cooling plant leaves: a mechanism of highly efficient mid infra-red thermal emission. Applied Physics Letters. 87:194105.

Williams, R.F., and R.E. Shapter. 1955. A comparative study of growth and nutrition in barley and rye as affected by low-water treatment. Australian Journal Biological Science. 8:435–466.

WRB 2015. World Reference Base for Soil Resources 2014 (update 2015). International soil classification system for naming soils and creating legends for soil maps. World Soil Resources Reports 106. FAO, Rome, Italy.

Yoshida, S., Ohnishi, Y., Kitagishi, K. 1962. Histochemistry of silicon in rice plant III. The presence of cuticle-silica double layer in the epidermal tissue. Soil Science Plant Nutrition. 8:1–5

Zhu, Z., Wei, G., Li, J., Qian, Q., Yu, J. 2004. Silicon alleviates salt stress and increases antioxidant enzymes activity in leaves of salt-stressed cucumber (*Cucumis sativus* L.) Plant Science 167:527–533.

15

Arbuscular Mycorrhizal Fungi Alleviate Salt Stress on Sweet (*Ocimum Basilicum* L.) Seedlings

Yuneisy Milagro Agüero-Fernández[1]; Luis Guillermo Hernández-Montiel[1];
Bernardo Murillo-Amador[1*]; Alejandra Nieto-Garibay[1]; Enrique Troyo-
Diéguez[1]; Ramón Zulueta-Rodríguez[2], Carlos Michel Ojeda-Silvera[1]

[1] *Centro de Investigaciones Biológicas del Noroeste S.C., La Paz, Baja
California Sur, México. Email: yaguero@pg.cibnor.mx,
lhernandez@cibnor.mx, anieto04@cibnor.mx, etroyo04@cibnor.mx,
cojeda@cibnor.mx, bmurillo04@cibnor.mx.*
[2] *Facultad de Ciencias Agrícolas, Universidad Veracruzana, Campus Xalapa,
Xalapa, Veracruz, México Ramón. rzulueta36@hotmail.com*
Corresponding author

SUMMARY

Abiotic stress due to salinity is considered one of the main problems facing agriculture that has the detrimental effect of restricting the growth and development of plants. The objective of this study was to evaluate the effect of arbuscular mycorrhizal fungus species (*Rhizophagus fasciculatum*) as reliever of NaCl-stress in sweet basil seedlings in emergence stage. A completely randomized design with factorial arrangement with four replications was used; first factor was sweet basil varieties (Nufar, Genovese, and Napoletano), factor two were NaCl concentrations (0, 50, and 100 mM) and factor three was inoculation of seeds with *Rhizophagus fasciculatum* (+AMF) an arbuscular mycorrhizal fungi and non-inoculated seeds (−AMF). Chemical composition of the substrate, rate and emergence percentage, plant height, root length, fresh and dry biomass of aerial part and root, spore count, and colonization percentage were assessed. The results showed differences in all variables with Napoletano showing higher values of all variables at 0, 50 or 100 mM NaCl with AMF. Majority of variables decreased with NaCl concentrations. The substrate was suitable for growth of *Rhizophagus fasciculatum* and for sweet basil. No root colonization was found in any seedlings inoculated with AMF; however, the micrograph of the inoculated plants showed vegetative mycelium, a mycorrhizal structure that shows the initiation of the colonization process.
Keywords: *Ocimum basilicum*; emergence rate; emergence percentage; salt tolerance; NaCl; colonization.

INTRODUCTION

The intensive and disproportionate exploitation of resources has been the main tendency in productive agroecosystems for the past decades worldwide. Depleted and saturated soils have been caused by the excessive use of fertilizers, and salinity conditions are examples of the current situation (Aguado et al., 2012). Soil salinity is one of the main adverse factors crops have faced around the world (Das et al., 2015). This problem is a big concern in arid and semiarid areas where low rainfall, high evapotranspiration rates, and soil salinity are some of the limiting factors (Akram et al., 2010). Several environmental factors as salinity have an unfavorable impact on plant growth (Hashem et al., 2015a). Stress tolerance in plants is a complex phenomenon involving many changes at biochemical and physiological level. However, the mechanisms behind stress tolerance seem to be affected by the colonization of arbuscular mycorrhizal fungi (AMF) (Al-Karaki, 2000). Some studies have shown that inoculation with AMF greatly enhances growth, water uptake and essential nutrients for plants even in saline conditions (Zhu et al., 2012). The arbuscular mycorrhizal symbiosis is the evident result of the interaction between plant roots and a fungus, which is an example of the morphological alterations that roots have with the purpose of adapting themselves to the presence of a symbiont. In recent years, publications about the benefits of AMF in different crops have increased (Pérez et al., 2011).

There are mycorrhizal inoculants that are applied in small quantities, so they integrate to cultural practices (González et al., 2011; Harikumar, 2017) allowing them to become a promissory practice of biological basis for agricultural production. Sweet basil (Ocimum basilicum L.) is a crop with high demand in the international market. Among the crop limitations, its establishment is in areas where soil salinity is high (Batista-Sánchez et al., 2015). In the Peninsula of Baja California, Mexico, salinity in soil and water is one of the main problems agriculture confronts influencing in the decrease of basil yield, so studies with ecological alternatives should be performed to determine the tolerance of basil varieties able to establish in these conditions. It has been shown that arbuscular mycorrhizal fungi (AMF) are able to promote the growth of some plants and their tolerance to salinity The AMF use various mechanisms that favor the acquisition of nutrients, the production of plant growth hormones, the development of the rhizosphere, and even soil conditions (Qiang-Sheng et al., 2010; Harris-Valle et al., 2011; Mujica-Pérez and Fuentes-Martínez, 2012; Khalil, 2013; Sinclair et al., 2014; Hashem et al., 2015b; Elhindi et al., 2016; Khalid, et al., 2017; Wang et al., 2018). In studies carried out by Reyes-Pérez et al. (2014) and Batista-

Sanchez et al. (2017) to determine the effect of NaCl on the growth and development of basil plants observed that Napoletano showed tolerance to NaCl-stress. The aim of this study was to evaluate the effect of the Rhizophagus fasciculatum strain, characterized as AMF as possible stress reliever by NaCl in the emergence rate, emergence percentage and morphometric variables of seedlings of different basil varieties.

MATERIALS AND METHODS

Ethics statement

The research conducted herein did not involve measurements with humans or animals. The study site is not considered a protected area. For location/activities, no specific permissions were required and the studies did not involve endangered or protected species. However, to carry out research activities on the lands administered by Centro de Investigaciones Biológicas del Noroeste, S.C. (CIBNOR®), permission was granted by Ing. Saul Edel Briseño-Ruiz, manager of the experimental open-field and shade-enclosure areas at CIBNOR®. The seeds of the basil commercial varieties used were obtained from a store of agricultural products at La Paz, Baja California Sur, México. The varieties of Ocimum basilicum L. used in the present study are not pondered an endangered species and their use therefore had insignificant effects on broader ecosystem functioning.

Study area and characteristics of the place

The experiment was developed in a shadow mesh structure (40% of shading, black color, model 20 mesh) in the experimental field of CIBNOR located northwest of La Paz, BCS, Mexico, 24° 08' 10.03 N and 110° 25' 35.31 W, 7 m.a.s.l. The average temperature, maximum and minimum inside the shadow mesh during the experimental period were 25.59 ± 4.14, 34.7 ± 4.12 and 16.9 ± 4.11 °C, respectively with $69.29 \pm 13.31\%$ of relative humidity. Data of the climatological variables recorded during the study period were obtained from a portable weather station (Vantage Pro2® Davis Instruments, USA), which was placed inside the shadow mesh structure. The experimental site has a weather of Bw (h') hw (e) considered as semiarid with xeric vegetation (García, 2004).

Genetic material

Basil varieties used were Napoletano, Genovese, and Nufar, which were reproduced by seeds (Vis Seed Company®, Arcadia, Cal., USA). Prior to the experiment, the quality of seed varieties was evaluated through a germination test according to the suggested methodology by ISTA (2010).

Experimental conditions

An arbuscular mycorrhizal fungi from a monosporic crop was used, containing the *Rhizophagus fasciculatum* species for inoculating the seed lot at seedtime using the method of coating seeds proposed by Fernández *et al.* (1999) and another seed lot was uninoculated as the control. The inoculum used in the experiment had a spore content count of 25-50 spores g^{-1} in the substrate. The seeds were sown in polystyrene trays with 200 cavities, which contained sterilized vermiculite as substrate. Saline treatments based on NaCl (0, 50, and 100 mM NaCl) were applied daily (08:30 h). The quantity applied to each treatment was 500 mL per tray in order to make sure the solution drained through the holes and avoided the accumulation of salts in the substrate, which could increase salt concentration. The emergence of each seedling was recorded daily, and the final emergence percentage was determined at 14 d. The emergence rate (M) is the speed that a seed germinated over a period of time (in days) was calculated using Maguire (1962) equation: $M = n_1/n_2 + t_1/t_2 + \ldots n_{30}/t_{14}$; where $n_1, n_{2\ldots}n_{30}$ is the number of germinated seeds in the times $t_1, t_2 \ldots t_{14}$ (in days).

Chemical analysis of the substrate

Samples of substrate were taken and sieved through a 10 mesh (2 mm). The pH and electrical conductivity (EC) were measured with a soil relationship solution of 1:5. The pH was measured using a potentiometer (Hanna®, model 211, USA). The EC (dS cm^{-1}) was measured using a conductimeter (Hach®, Model Sension5, Loveland, Colorado, USA). Extractable phosphorus (P, mg kg^{-1}) was determined from the aqueous extract with a soil relationship solution of 1:5 and using Multiskan Acent® (Model Labsystems, No. 354, Finland). Extractable potassium (K, mg kg^{-1}) was determined using a flame atomic absorption spectrophotometer (GBC®, model Avanta, Australia). Extractable calcium and magnesium were measured by complexometry by the volumetric titration method (EDTA titration with 0.01 N). Organic matter content (OM, %) was determined by the Walkley and Black method using a 35 mesh (0.5 mm). Total nitrogen (N) was determined by the Dumas method (Leco®, Model FP-528, USA) using a 100 mesh (0.150 mm).

Morphometric values evaluated

Emerged seedlings were maintained during 21 days and after this period, 10 seedlings per replication were randomly selected; then, plant height (cm), root length (cm), fresh and dry biomass of aerial part (g), fresh and dry biomass of root (g) were determined by the destructive method, dividing each seedling in root and aerial part, and weighing each one separately using an analytical balance (Mettler Toledo®, AG204, USA). Once the fresh weight of roots and aerial part were obtained, they were placed in paper bags and then in a drying oven (Shel-Lab®, FX-5, series-1000203, USA) at 70 °C until constant weight (about 72 h). Subsequently, they were weighed on analytical balance (Mettler Toledo®, AG204, USA) expressing the dry matter weight in grams.

Extraction and spores counting of the inoculum and percentage of colonization

The AMF spore extraction was developed by the wet sieving and decanting technique described by Daniels and Skipper (1982) and modified by Utobo *et al.* (2011) and the number of spores in the inoculum were quantified with a dissecting microscope. The percentage of mycorrhizal colonization was determined by the methodology described by Hashem *et al.* (2014). Colonization was assessed taking into account the presence of vesicles, arbuscules and/or coenocytic hyphaes (hyphaes) typical of the AMF.

Experimental design

The experiment was developed using a completely randomized design with factorial arrangement with four replications of 30 seeds each, using three factors in study, first factor was the varieties of sweet basil (Nufar, Genovese, and Napoletano), factor two was the NaCl concentrations (0, 50, and 100 mM) and factor three was mycorrhizal inoculation treatments [*Rhizophagus fasciculatum* (+ AMF) and non-inoculated (− AMF)].

Statistical analysis

Analysis of variance and differences among means of each factor and variable was performed (Tukey HSD $p<0.05$). To determine the linear or curvilinear relationship among NaCl concentrations and the response of all variables, a polynomial regression analysis was performed. The emergency percentage data were transformed by arcsine (Little and Hills, 1989; Statsoft, 2011). The statistical analyzes were carried out with Statistica® v. 10.0 for Windows® (Steel and Torrie, 1995).

RESULTS AND DISCUSSION

Napoletano showed greatest emergence rate (ER) followed by Nufar and Genovese; ER showed greatest values in 0 and 50 mM and lowest in 100 mM but inoculated seeds with AMF showed greatest ER (Table 1). Varieties × NaCl interaction showed greatest ER in Napoletano and Nufar with 0 and 50 mM while Genovese showed lowest ER in 50 mM

(Table 2). The interaction NaCl × AMF showed greatest ER in 0 mM with AMF while the lowest ER was with 100 mM without AMF (Table 2). Varieties × NaCl × AMF interaction showed that ER was higher in Napoletano with 0 and 50 mM with AMF, Nufar with 0 mM with AMF, while Genovese showed lowest ER in 100 mM without AMF (Table 3). One of the primary effects of salt stress is to delay seedling emergence (Martínez-Villavicencio et al., 2011). A significant decrease of ER in the varieties was showed as NaCl concentrations increased, that is attributed to the salinity detrimental effect on crops in early stages (Tavakkoli et al., 2011). In this study, ER was greatest in seedlings where seeds were inoculated with *Rhizophagus fasciculatum* (+ AMF) but no was due to inoculation, since in emergence, no arbuscules were found, which is a mycorrhizal structure that must be present for the plant to get the benefits of this endophyte (Colla et al., 2008). Studies carried out by Mata-Fernández et al. (2014) reported that the most common effect of salt stress is the reduction in the ability to absorb water. Napoletano showed greatest emergence percentage (EP) followed by Nufar and Genovese; EP was greatest in 0 and 50 mM and lowest in 100 mM but increased in those seeds treated with AMF (Table 1). Varieties × NaCl interaction showed greatest EP in Napoletano with 0 and 50 mM while Genovese showed lowest EP in 50 mM (Table 2). Varieties × AMF showed greatest EP in Napoletano with AMF but EP showed lowest values in Genovese without AMF (Table 2). The interaction NaCl × AMF showed greatest EP in 0 mM with AMF and lowest EP in 100 mM without AMF (Table 2).

Table 1. Average values by each factor (varieties, NaCl and AMF). The values correspond to the rate and emergence percentage and morphometric variables of sweet basil seedlings under NaCl stress from seeds inoculated with *Rhizophagus fasciculatum* (+ AMF) and non-inoculated (− AMF).

Varieties	ER	EP (%)	PH (cm)	RL (cm)	FBAP (g)	DBAP (g)	FRB (g)	DRB (g)
Genovese	2.59±0.73 c	61.66±12.93 c	1.18±0.15 c	2.02±0.51 c	0.148±0.02 c	0.018±0.007 b	0.041±0.02 c	0.010±0.007 b
Napoletano	4.01±0.84 a	81.94±11.87 a	2.12±1.24 a	4.32±0.97 a	0.334±0.13 a	0.121±0.014 a	0.163±0.11 a	0.079±0.049 a
Nufar	3.63±0.78 b	78.05±6.24 b	1.60±0.26 b	3.97±1.0 b	0.223±0.05 b	0.059±0.011 b	0.096±0.06 b	0.014±0.012 b
Significance	***	***	***	***	***	***	***	***
NaCl (mM)	ER	EP (%)	PH (cm)	RL (cm)	FBAP (g)	DBAP (g)	FRB (g)	DRB (g)
0	3.64±1.14 a	75.69±15.96 a	2.18±1.22 a	4.07±1.54 a	0.278±0.15 a	0.071±0.013 ab	0.166±0.116 a	0.043±0.079 a
50	3.48±0.99 a	75.13±14.24 a	1.40±0.29 b	3.28±1.12 b	0.219±0.07 b	0.028±0.011 b	0.081±0.051 b	0.017±0.009 b
100	3.10±0.73 b	70.83±10.82 b	1.32±0.20 b	2.96±1.06 b	0.208±0.05 b	0.099±0.006 a	0.054±0.024 c	0.012±0.008 b
Significance	***	***	***	***	***	***	***	***
AMF	ER	EP (%)	PH (cm)	RL (cm)	FBAP (g)	DBAP (g)	FRB (g)	DRB (g)
+ AMF	4.07±0.78 a	81.38±10.03 a	1.86±1.08 a	3.86±1.39 a	0.264±0.125 a	0.088±0.013 a	0.119±0.107 a	0.034±0.066 a
− AMF	2.75±0.68 b	66.38±13.08 b	1.40±0.31 b	3.02±1.13 b	0.206±0.045 b	0.050±0.011 b	0.082±0.051 b	0.014±0.007 b
Significance	***	***	***	***	***	*	***	***

ER = emergence rate; EP = emergence percentage; PH = plant height; RL= root length; FBAP = fresh biomass of aerial part; BSPA = dry biomass of aerial part; FRB = fresh root biomass; DRB = dry root biomass; NaCl = sodium chloride; AMF = Arbuscular mycorrhizal fungi; −AMF= without AMF (control), + AMF= with AMF (1 g of the AMF inoculum). Values in each column followed by the same letter(s) are not significantly different at $p \leq 0.05$ (Tukey HSD). Significance level: *= $p \leq 0.05$; ***= $p \leq 0.001$ (means ± SD).

Table 2. Average values of rate and percentage of emergence and morphometric variables regarding the interactions varieties × NaCl, varieties × AMF and NaCl × AMF in sweet basil seedlings under NaCl stress from seeds inoculated with *Rhizophagus fasciculatum* (+ AMF) and non-inoculated (− AMF).

Varieties	NaCl	ER	EP (%)	PH (cm)	RL (cm)	FBAP (g)	DBAP (g)	FRB (g)	DRB (g)
Genovese	0	2.62±0.76d	62.08±14.02c	1.34±0.05cd	2.26±0.46e	0.180±0.013c	0.023±0.005c	0.052±0.018d	0.010±0.007b
	50	2.51±0.65d	59.16±11.65c	1.08±0.08d	1.98±0.15e	0.167±0.14c	0.021±0.004c	0.036±0.018d	0.016±0.008b
	100	2.63±0.87d	63.75±14.30c	1.12±0.14cd	1.81±0.70e	0.098±0.024d	0.010±0.005c	0.033±0.010d	0.012±0.003b
Napoletano	0	4.30±0.95a	87.50±13.18a	3.40±1.46a	4.83±1.22ab	0.414±0.18a	0.049±0.01bc	0.292±0.10a	0.101±0.12a
	50	4.24±0.70a	86.66±7.35a	1.53±0.20c	4.22±0.80bc	0.325±0.06b	0.040±0.007bc	0.127±0.04bc	0.025±0.008b
	100	3.47±0.66bc	71.66±7.35b	1.42±0.09c	3.91±0.68c	0.263±0.02c	0.274±0.005a	0.072±0.02d	0.020±0.007b
Nufar	0	3.99±0.99ab	77.50±9.55b	1.80±0.23b	5.14±0.61a	0.242±0.08c	0.141±0.01b	0.153±0.06b	0.017±0.009b
	50	3.70±0.73ab	79.58±3.30b	1.59±0.22bc	3.63±0.57cd	0.236±0.02c	0.024±0.008c	0.081±0.02cd	0.009±0.005b
	100	3.21±0.35c	77.08±4.86b	1.42±0.17c	3.15±0.37d	0.192±0.01c	0.012±0.005c	0.056±0.01d	0.007±0.007b
Significance		***	***	***	***	***	***	***	**

Varieties	AMF	EP (%)	PH (cm)	FBAP (g)	DBAP (g)	FRB (g)	DRB (g)
Genovese	+ AMF	73.05±7.17c	1.22±0.146d	0.167±0.012c	0.020±0.006b	0.043±0.019c	0.015±0.008b
	− AMF	50.27±3.88d	1.13±0.145d	0.130±0.011d	0.016±0.007b	0.038±0.016c	0.010±0.002b
Napoletano	+ AMF	89.44±9.83a	2.64±1.581a	0.415±0.133a	0.124±0.014a	0.189±0.134a	0.077±0.005a
	− AMF	74.44±8.68c	1.60±0.342bc	0.253±0.028b	0.118±0.013a	0.138±0.057b	0.021±0.005b
Nufar	+ AMF	81.66±5.03b	1.73±0.241b	0.249±0.054b	0.101±0.012a	0.124±0.065b	0.013±0.001b
	− AMF	74.44±5.38c	1.47±0.223c	0.198±0.042b	0.017±0.007b	0.069±0.030c	0.010±0.008b
Significance		***	***	***	*	**	**

NaCl (mM)	AMF	ER	EP (%)	PH (cm)	FBAP (g)	DBAP (g)	FRB (g)
0	+ AMF	4.44±0.89a	86.38±11.59a	2.66±1.58a	0.360±0.173a	0.113±0.013a	0.212±0.138a
	− AMF	2.83±0.71de	65.00±12.19cd	1.70±0.32b	0.197±0.043cd	0.029±0.009bc	0.119±0.065b
50	+ AMF	4.10±0.79b	80.00±10.25bc	1.52±0.32bc	0.209±0.083bcd	0.031±0.01bc	0.083±0.062bc
	−AMF	2.87±0.78d	70.27±16.36c	1.27±0.18d	0.230±0.047b	0.025±0.011c	0.079±0.033bc
100	+ AMF	3.65±0.40c	77.77±6.09b	1.41±0.19cd	0.224±0.045bc	0.101±0.005ab	0.061±0.029c
	− AMF	2.55±0.52e	63.88±10.13d	1.23±0.16d	0.191±0.042d	0.097±0.005abc	0.047±0.017c
Significance		***	***	***	***	*	***

NaCl =sodium chloride (mM); + AMF = with AMF (1 g of the AMF inoculum); −AMF = without AMF (control); ER = emergence rate; EP = emergence percentage; PH = plant height; RL = root length; FBAP = fresh biomass of aerial part; DBAP = dry biomass of aerial part; FRB = fresh root biomass; DRB = dry root biomass. Values in each column followed by the same letter(s) are not significantly different at $p \leq 0,05$ (Tukey HSD). Significance level: *= $p \leq 0.05$. **= $p \leq 0.01$. ***= $p \leq 0.001$ (means ± SD).

The EP decrease in concentrations of 100 mM, attributed to the harmful effects caused by NaCl in early development stages such as emergence. Batista-Sánchez *et al.* (2015) suggested NaCl limits germination, emergence, and growth because most seedlings are sensitive to this condition. Seeds were inoculated (+ AMF) showed greatest EP; however, this result is mostly attributed to tolerance of the varieties to NaCl and not just to inoculation, as no root colonization was found in this stage. Emergence percentage decreased as NaCl concentrations increased, this result coincides with those reported by Parés *et al.* (2008). Napoletano showed greatest EP because it is a NaCl tolerant variety (Reyes-Pérez *et al.*, 2013).

Napoletano had greatest plant height (PH) followed by Nufar and Genovese; pH was greatest in 0 mM and decreased as NaCl concentrations increased but was greatest in seedlings whose seeds were inoculated with AMF (Table 1). Varieties × NaCl interaction showed that PH was greatest in Napoletano in 0 mM, and in all varieties decreased as NaCl concentrations increased (Table 2). In the interaction varieties × AMF, Napoletano with AMF showed greatest PH while Genovese without AMF showed the lowest PH (Table 2). As far as the interaction NaCl × AMF, the PH was higher in 0 mM with AMF and decreased in 100 mM without AMF (Table 2). Concerning the interaction varieties × NaCl × AMF, Napoletano in 0 mM with AMF showed greatest PH whereas the PH decreased significantly in Genovese in 100 mM without AMF (Table 3). Lower plant height as NaCl increased interferes with mineral nutrition and cell metabolism (Amini et al., 2007), also NaCl stress modify the water uptake through the roots, phenomenon known as osmotic component (Fatma et al. 2014). Seedlings of inoculated (+ AMF) seeds showed greatest PH, but was not due to the inoculation.

Root length (RL) was longest in Napoletano followed by Nufar and Genovese and was longest in those plants treated with 0 mM and decreased as the concentrations of NaCl increased but the inoculation with AMF stimulated the root length (Table 1). Varieties × NaCl interaction showed that root was longest in 0 mM in both Nufar and Napoletano, decreasing as the NaCl increased (Table 2). The interaction varieties × NaCl × AMF showed that root was longest in Napoletano in 0 mM with AMF while lowest LR was showed by Genovese in 100 mM and without AMF (Table 3). NaCl stress causing reduced causes structural changes in meristematic cells and inducing structural changes in root and ion exhaust, and alterations of cell membranes (Abd-Allah et al., 2015). The results of this research are consistent with the assertions of Fatma et al. (2014) that NaCl causes osmotic stress caused by salinity, plants reacts with a wide range of physiological responses at molecular and cellular level, including changes in plant development and morphology as an increase or decrease in root growth and changes in life cycle.

Fresh biomass of aerial part (FBAP) was greatest in Napoletano followed by Nufar and Genovese; was greatest in 0 mM and decreased as NaCl increased but was greatest in seedlings with seeds inoculated with AMF (Table 1). Varieties × NaCl interaction showed that FBAP was greatest in Napoletano at 0 mM but decreased in all varieties as NaCl increased (Table 2). In the interaction varieties × AMF, FBAP was greatest in Napoletano with AMF and lowest in Genovese without AMF (Table 2). The interaction

NaCl × AMF showed greatest FBAP in 0 mM with AMF and lowest in 100 mM without AMF (Table 2). Varieties × NaCl × AMF indicated that FBAP increased their values in Napoletano with 0 mM with AMF and was lowest in Genovese with 50 mM with AMF (Table 3). Studies related to FBAP indicate NaCl causes biochemical changes such as increase of abscisic acid synthesis and osmoprotectants solute, physiological such as the change of membrane permeability, ions and water, reduction of transpiration and photosynthesis (Chávez and González, 2009), as well as water availability for plants (Elhindi et al., 2016). Other studies have attributed this effect to the impact of salt stress on growth and the effect of osmotic stress in the root zone of seedlings which entails a reduction in stem weight coinciding with a reduction in fresh biomass (Reyes-Pérez et al., 2013).

Dry biomass of aerial part (DBAP) was greatest in Napoletano while Nufar and Genovese showed lowest values; DBAP showed greatest values at 100 mM and lowest at 50 mM and increased in those seedlings whose seeds were inoculated with AMF (Table 1). Regarding the analysis of the interactions, varieties × NaCl revealed that DBAP was greatest in Napoletano in 100 mM and decreased in Nufar and Genovese decreasing slightly as NaCl increased (Table 2). Varieties × AMF showed that DBAP was greatest in Napoletano with and without AMF and Nufar with AMF (Table 2). The interaction NaCl × HMA showed that DBAP was greatest in 0 mM with AMF, but DBAP exhibited similar values in 100 mM with AMF (Table 2). In the interaction varieties × NaCl × AMF, the DBAP exhibited highest values in Napoletano with 100 mM with and without AMF and in Nufar with 0 mM with AMF (Table 3). The DBAP was greatest in Napoletano at 100 mM with AMF, which is attributed to the tolerance of this variety to salinity (Reyes-Pérez et al., 2013) while DBAP decreased in Genovese and Nufar as NaCl increased. This result agrees with those reported by Parés et al. (2008). Another result was obtained (Baracaldo et al., 2014) has registered a decrease in the growth of leaves, in terms of dry weight, when increasing the concentration of salinity. In the same way, these results coincide with those indicated for other species, such as coquia (García et al., 2011), wheat (Argentel et al., 2006) and guava (Casierra-Posada, 2006), in which there was a lower accumulation of dry biomass as the concentration of salts in the irrigation water increased.

Table 3. Average values of rate and emergence percentage and morphometric variables regarding the interaction varieties × NaCl × AMF of sweet basil seedlings under NaCl stress from seeds inoculated with *Rhizophagus fasciculatum* (+ AMF) and non-inoculated (− AMF).

Varieties	NaCl (mM)	AMF	ER	PH (cm)	RL (cm)	FBAP (g)	DBAP (g)	FRB (g)	DRB (g)
Genovese	0	+ AMF	3.29±0.56 f	1.36±0.04 defgh	2.52±0.57 fgh	0.185±0.01 fgh	0.026±0.006 b	0.059±0.020 def	0.009±0.001 b
	50	+ AMF	3.10±0.54 ef	1.10±0.09 gh	2.04±0.20 hi	0.019±0.01 i	0.022±0.001 b	0.034±0.018 ef	0.021±0.010 b
	100	+ AMF	3.41±0.73 ef	1.21±0.15 fgh	2.40±0.41 ghi	0.185±0.01 fgh	0.013±0.001 b	0.036±0.008 ef	0.016±0.001 b
	0	− AMF	1.94±0.56 g	1.32±0.05 fgh	2.00±0.01 hi	0.174±0.01 gh	0.021±0.005 b	0.045±0.015 ef	0.011±0.001 b
	50	− AMF	1.92±0.98 g	1.06±0.07 gh	1.92±0.06 hi	0.176±0.01 gh	0.019±0.004 b	0.039±0.021 ef	0.011±0.001 b
	100	− AMF	1.85±1.78 g	1.02±0.01 h	1.23±0.2 j	0.150±0.01 h	0.008±0.007 b	0.031±0.013 f	0.008±0.003 b
Napoletano	0	+ AMF	5.15±0.81 a	4.75±0.42 a	5.88±0.60 a	0.581±0.03 a	0.056±0.008 b	0.373±0.048 a	0.181±0.111 a
	50	+ AMF	4.84±0.67 b	1.72±0.03 bcde	4.70±0.56 abc	0.377±0.02 b	0.042±0.010 b	0.118±0.044 cde	0.027±0.005 b
	100	+ AMF	4.06±0.64 cd	1.45±0.13 defg	4.17±0.25 cd	0.286±0.01 cd	0.275±0.006 a	0.077±0.038 def	0.022±0.005 b
	0	− AMF	3.46±0.87 def	2.06±0.05 b	3.78±0.44 cde	0.247±0.02 de	0.042±0.006 b	0.211±0.022 b	0.022±0.005 b
	50	− AMF	3.65±0.31 de	1.34±0.02 efgh	3.75±0.76 cde	0.273±0.03 cd	0.038±0.001 b	0.135±0.022 bcd	0.022±0.006 b
	100	− AMF	2.88±0.11 f	1.40±0.02 defgh	3.65±0.92 cdef	0.241±0.02 def	0.273±0.005 a	0.067±0.055 def	0.018±0.005 b
Nufar	0	+ AMF	4.89±0.95 ab	1.87±0.32 bc	5.62±0.48 ab	0.313±0.03 c	0.259±0.006 a	0.204±0.049 bc	0.013±0.006 b
	50	+ AMF	4.36±0.81 bc	1.76±0.14 bcd	3.98±0.64 cde	0.232±0.01 defg	0.031±0.005 b	0.098±0.005 def	0.013±0.001 b
	100	+ AMF	3.49±0.27 def	1.56±0.14 cdef	3.42±0.11 defg	0.202±0.008 efgh	0.014±0.005 b	0.069±0.013 def	0.005±0.008 b
	0	− AMF	3.08±0.59 ef	1.73±0.11 bcd	4.66±0.19 bc	0.171±0.04 gh	0.022±0.002 b	0.102±0.005 def	0.021±0.001 b
	50	− AMF	3.03±0.33 ef	1.42±0.15 defgh	3.28±0.18 defg	0.240±0.02 def	0.017±0.006 b	0.064±0.023 def	0.006±0.003 b
	100	− AMF	2.92±0.39 f	1.28±0.04 fgh	2.88±0.34efgh	0.183±0.01 fgh	0.011±0.006 b	0.042±0.006 ef	0.003±0.001 b
Significance level			***	***	***	***	*	**	**

NaCl = sodium chloride (mM); + AMF = with AMF (1 g of the AMF inoculum); − AMF = without AMF (control); ER = emergence rate; EP = emergence percentage; PH = plant height; RL= root length; FBAP = fresh biomass of aerial part; DBAP = dry biomass of aerial part, FRB = fresh root biomass; DRB = dry root biomass. Values in each column followed by the same letter(s) are not significantly different at $p \le 0.05$ (Tukey HSD). Significance level: *= $p \le 0.05$; **= $p \le 0.01$; ***= $p \le 0.001$ (means ± SD).

Napoletano had greater fresh root biomass (FRB) and was greater in 0 mM decreasing as the concentrations of NaCl increased; FRB was greatest in seedlings whose seeds were inoculated with AMF (Table 1). In regard to the analysis of the interactions, varieties × NaCl showed that FRB was greatest in Napoletano in 0 mM and all varieties showed a trend to decrease as NaCl increased (Table 2). The interaction varieties × AMF revealed that all varieties increased FRB with AMF, with greatest values in Napoletano (Table 2). In the interaction NaCl × AMF, FRB was always greatest in seedlings whose seeds were inoculated with AMF even though the greatest value was in 0 mM (Table 2). Varieties × NaCl × AMF showed that FRB was greatest in Napoletano at 0 mM with AMF while lowest FRB was showed by Genovese at 100 mM without AMF (Table 3).

NaCl stress causes a decrease in germination and dry weight of stem and root and an increase in stem/root ratio (Ceccoli et al., 2011). High levels of Na^+ and Cl^- in the plant cause premature death of young tissues and produce marginal chlorosis on the leaves, which modifies the photosynthetically active area of seedlings, growth of stem and root and ion content in the plant (Batista-Sánchez et al., 2015).

Dry root biomass (DRB) was greatest in Napoletano regarding to Nufar and Genovese and showed greatest values in 0 mM, decreased as NaCl increased; nevertheless seedlings from seeds inoculated with AMF showed greatest DRB (Table 1). Varieties × NaCl interaction showed that Napoletano in 0 mM had greatest DRB (Table 2). In the interaction varieties × AMF, DRB exhibited greatest values in Napoletano with AMF while lowest DRB was showed in Nufar and Genovese without AMF (Table

2). Varieties × NaCl × AMF interaction showed that DRB was greatest in Napoletano in 0 mM with AMF while Genovese in 100 mM without AMF displayed lowest values (Table 3). The accumulation of dry biomass is widely used as a measure of plant growth because it expresses a balance between total productions of photoassimilates and breathing. Studies related to DRB of Chloris gayana under NaCl concentrations (0, 100 and 200 mM) dropped by 43% at concentrations of 200 mM compared to the control (Ceccoli et al., 2011), NaCl is known to reduce dry root biomass (Chávez and González, 2009).

The polynomial regression analysis shows that the relationship of NaCl concentrations and rate and percentage of emergence was nonlinear or quadratic. The relationship of NaCl concentrations with height of seedlings, length of root, fresh biomass of aerial part, fresh and dry biomass of root was linear while dry biomass of aerial part showed a quadratic relationship. These results are consistent with what was reported by Miranda et al. (2012) who observed a differentiated effect on growth variables under different saline solutions (Table 4).

Non-colonized roots were observed in the emergence stage of sweet basil seedlings. At 21 days a short period of AMF symbiosis was established; although at the beginning of the process was observed the site where the mycelium of Rhizophagus fasciculatum (+ AMF) makes contact with seedlings roots (Fig. 1). In seedlings roots from non-inoculated seeds (− AMF) no signs of colonization were detected (Fig. 2). The beneficial effects of this symbiosis happen because of a complex molecular interchange between the two symbionts (Camarena-Gutiérrez, 2012).

Table 4. Linear or curvilinear relationship among NaCl and all variables measured in sweet basil seedlings inoculated with AMF. Relationship determined by polynomial regression analysis.

Variables	Adjustment model	Correlation coefficient (R)	Coefficient of determination (R^2)	Number of data (N)	F-value (ANOVA)	P value	t value
PH (cm)	$2.066 - 0.008 \times NaCl$	-0.43	0.185	72	15.949	0.000) ***	-3.993
RL (cm)	$3.999 - 0.0112 \times NaCl$	-0.35	0.119	72	9.534	0.002 **	-3.087
FBAP (g)	$0.271 - 7E^{-3} \times NaCl$	-0.26	0.065	72	4.893	0.030 *	-2.212
DBAP (g)	$0.2285 - 0.214 \times NaCl + 0.057 \times NaCl\ ^{\wedge}2$	0.075	0.274	72	4.754	0.032 *	2.180
FRB (g)	$0.156 - 0.001 \times NaCl$	-0.51	0.263	72	25.002	0.0004 ***	-5.000
DRB (g)	$0.039 - 0.3E^{-3} \times NaCl$	-0.25	0.064	72	4.806	0.0316 *	-2.192

PH = plant height; RL= root length; FBAP = fresh biomass of aerial part; DBAP = dry biomass of aerial part, FRB = fresh root biomass; DRB = dry root biomass.

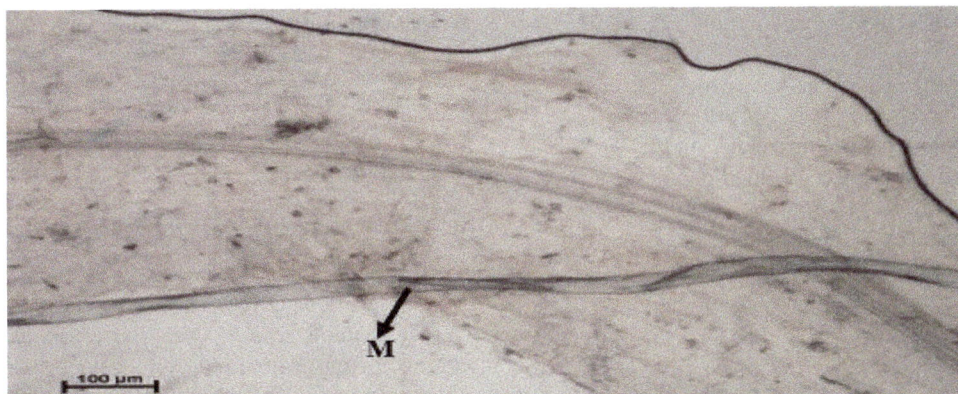

Figure 1. Micrograph of inoculated basil roots with *Rhizophagus fasciculatum* (+AMF). M= Vegetative mycelium.

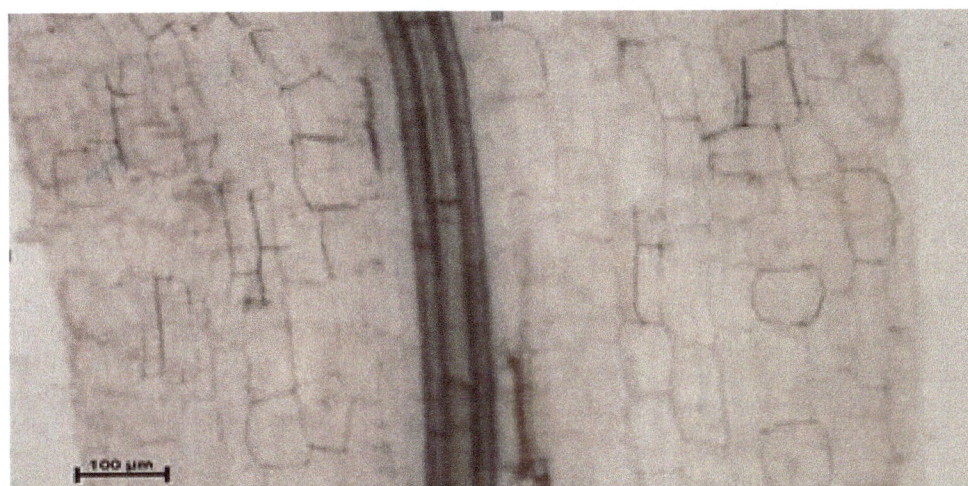

Figure 2. Micrograph of non-inoculated basil seedlings roots (−AMF).

CONCLUSIONS

No root colonization was found in any seedlings whose seeds were inoculated with AMF. Vegetative mycelium structure was observed in inoculated basil roots with *Rhizophagus fasciculatum* (+AMF). This mycelium is able to interconnect the roots of plants with the soil, which allows the flow of water and nutrients between the roots. The number of spores found in the inoculum is suitable to use in studies with AMF. The substrate used is suitable for the development of *Rhizophagus fasciculatum* and for sweet basil. Napoletano showed the highest values in most of the variables evaluated. Plant height, root length, fresh biomass of aerial part, fresh and dry root biomass showed linear relationship with NaCl concentrations while dry root biomass showed quadratic relationship. Rate and percentage of emergence do not shows linear or curvilinear relationship with NaCl concentrations.

Acknowledgements

Thanks to Lidia Hirales-Lucero and Manuel Salvador Trasviña-Castro for technical support. Diana Dorantes for English edition is gratefully acknowledged. Dr. Luis Guillermo Hernandez-Montiel and Dr. Bernardo Murillo-Amador are main thesis advisors of Yuneisy Milagro Agüero-Fernández, a Ph.D. student. This work was supported by CIBNOR, SEP-CONACYT (grants No. 236240 and No. 143), JICA-SATREPS-JST-CONACYT and CONACYT (national problems 2017 grant 4631). We are thankfully with Consejo Nacional de Ciencia y Tecnología-Mexico that supported Dr. Murillo-Amador by modality of sabbatical stays in foreign within the framework of the national call "support for sabbatical stays related to the consolidation of groups of research and/or strengthening of the national postgraduate program".

REFERENCES

Abd-Allah, E.F., Hashem, A., Alqarawi, A.A., Bahkali, A.H., Alwhibi, M.S. 2015. Enhancing growth performance and systemic acquired resistance of medicinal plant *Sesbania sesban* (L.) Merr using arbuscular mycorrhizal fungi under salt stress. Saudi Journal of Biological Science. 22(3):274-283. doi: 10.1016/j.sjbs.2015.03.004.

Aguado, S.G., Rascón, C.Q., Luna, B.A. 2012. Impacto económico y ambiental del empleo de fertilizantes químicos', ed. Aguado, S.G.A., Introducción al uso y manejo de los biofertilizantes en la agricultura, edit. INIFAP/SAGARPA, México, pp. 1-22.

Akram, M., Ashraf, M.Y., Ahmad, R., Waraich, E.A., Iqbal, J., Mohsan, M. 2010. Screening for salt tolerance in maize (*Zea mays* L.) hybrids at an early stage. Pakistan Journal of Botany. 42:141-151.

Al-karaki, G.N. 2000. Growth of mycorrhizal tomato and mineral acquisition under salt stress. Mycorrhiza. 10:51-54.

Amini, F., Ehsanpour, A.A., Hoang, Q.T., Shin, J.S. 2007. Protein pattern changes in tomato under in vitro salt stress. Russian Journal of Plant Phisiology. 54(4):464-472.

Argentel, L., González, L. M., Plana, R. 2006. Efecto de altas concentraciones salinas sobre la germinación y el crecimiento del trigo variedad Cuba-C-204 (*Triticum aestivum*). Cultivos Tropicales. 27:45-48.

Baracaldo, A. R. Carvajal, A. Romero, A. Prieto, F. García, G. Fischer y D. Miranda. 2014. Waterlogging affects the growth and biomass production of chonto tomatoes (*Solanum lycopersicum* L.), cultivated under shading. Rev. Colomb. Cienc. Hortic. 8(1), 92-102. Doi: 10.17584/rcch.2014v8i1.2803

Batista Sánchez, D., B. Murillo Amador, A. Nieto Garibay, L. Alcaraz Meléndez, E. Troyo Diéguez, L. Hernández Montiel y C. M. Ojeda Silvera. 2017. Mitigación de NaCl por efecto de un bioestimulante en la germinación de *Ocimum basilicum* L. Terra Latinoamericana 35:309-320.

Batista-Sánchez, D., Nieto-Garibay, A., Alcaraz-Meléndez, Troyo-Dieguéz, E., Hernández-Montiel L.G., Ojeda Silvera, C.M., Murillo-Amador, B. 2015. Uso del FitoMas-E® como atenuante del estrés salino (NaCl) durante la emergencia y crecimiento inicial de *Ocimum basilicum* L. Nova Scientia. 7(3):265-284.

Camarena-Gutiérrez, G. 2012. Interacción planta-hongos micorrízicos arbusculares. Revista Chapingo. Ciencias Forestales y del Ambiente. doi: 10.5154/r.rchscfa.2011.11.093 http://www.chapingo.mx/revistas.

Casierra-Posada, F. 2006. Distribución y producción total de materia seca en guayabo (*Psidium guajava* L. cv. Palmira ICA-1) bajo estrés salino Orinoquia. Universidad de Los Llanos Meta, Colombia. 10(2):59-66.

Ceccoli, G., Ramos, J.C., Ortega, I.L., Acosta, M.J., Perreta, M.G. 2011. Salinity induced anatomical and morphological changes in *Chloris gayana* Kunth roots. Biocell. 35(1):9-17.

Chávez, L., González, L.M., 2009. Mecanismos moleculares involucrados en la tolerancia de las plantas a la salinidad. Instituto de Investigaciones Agropecuarias. Granma. Cuba. pp. 231-256.

Colla, G., Rouphael, Y., Cardarelli, M., Tullido, M., Rivera, A.M., Rea, A. 2008. Alleviation of salt stress by arbuscular mycorrhizal in zucchini plants grown at low and high phosphorus concentration. Biology and Fertility of Soils. 44:501-509.

Das, P., Nutan, K.K., Pareek, S.L.S., Pareek, A. 2015. Understanding salinity responses and adopting omics based approaches to generate salinity tolerant cultivars of rice. Frontiers in Plant Science. 6:712. http://dx.doi.org/10.3389/fpls.00712.

Elhindi, K.M., El-Din, A.S. and Elgorban, A.M. (2016). The impact of arbuscular mycorrhizal fungi in mitigating salt-induced adverse effects in sweet basil (*Ocimum basilicum* L.). Saudi Journal of Biological Sciences. 24:170-179.

Fatma, M., Masood, M.A.A., Khan, N.A. 2014. Excess sulfur supplementation improves photosynthesis and growth in mustard under salt stress through increased production of glutathione. Environmental and Experimental Botany. 107:55-63.

Fernández, F., Gómez, R., Vanegas, L.F., Martínez, M.A., Blanca de la Noval., Rivera, R. 1999. Metodología de recubrimiento de semillas con inóculo micorrizógeno. Patente Cubana No. 22641.

García, E. 2004. Modificaciones al sistema de

clasificación climática de Köppen. Instituto de Geografía. Universidad Nacional Autónoma de México. 98 p.

García, J.P.S., Valdés, A., Facio, F., Arce, L., Burciaga, Hilda, C. 2011. Calidad fi siológica de semillas de coquia (Kochia scoparia (L.) Roth) a diferentes niveles de salinidad con NaCl. Agraria. Nueva Época. 8(3):12-17.

González, P.J., Rivera, R., Arzola, J., Morgan, O., Ramírez, J.F. 2011. Efecto de la inoculación de la cepa de hongo micorrízico arbuscular *Glomus hoi-like* en la respuesta de *Brachiaria híbrido* cv. Mulato *II* (CIAT 36087) a la fertilización orgánica y nitrogenada. Cultivos Tropicales. 32(4):05-12.

Harikumar, V.S. 2017. Biometric parameters of field grown sesame influenced by arbuscular mycorrhizal inoculation, rock phosphate fertilization and irrigation. Tropical and Subtropical Agroecosystems. 20:195-202

Harris-Valle, C., Esqueda, M., Valenzuela- Soto, E., Castellanos, A. (2011). Tolerancia a sequía y salinidad en *Cucurbita pepo*. var. *pepo* asociado con hongos micorrízicos arbusculares del desierto sonorense. Agrociencia. 45(8):959-970.

Hashem, A., Abd-Allah, E.F., Alqarawi, A.A., Al-Whibi, Mona, S., Alenazi, M.M., Egamberdieva, D., Ahmad, P. 2015a. Arbuscular mycorrhizal fungi mitigates NaCl adverse effects on *Solanum lycopersicum* L. Pakistan Journal of Botany. 47(1):327-340.

Hashem, A., Abd-Allah, E.F., Alqarawi, A.A., El-Didamony, G., Alwhibi-Mona, S., Egamberdieva, D., Ahmad, P. 2014. Alleviation of adverse impact of salinity on faba bean (*Vicia faba* L.) by arbuscular mycorrhizal fungi. Pakistan Journal of Botany. 46(6):2003-2013.

Hashem, A., Abd-Allah, E.F., Alqarawi, A.A., Dilfuza, E. 2015b. Induction of salt stress tolerance in cowpea (*Vigna unguiculata* L.) Walp by arbuscular mycorrhizal fungi. Legume Research. 38(5): 579-58.

ISTA (International Rules for Seed Testing). 2010. Rules proposals for the International Rules for Seed Testing 2010 Edition, OM Rules Proposals for the International Rules for Seed Testing 2010 Edition.doc. Approved by ECOM Decision. No. 498. 51 p.

Khalid, M.E., S.E.D. Ahmed, M. E. Abdallah.

(2017). Elgorban. The impact of arbuscular mycorrhizal fungi in mitigating salt-induced adverse effects in sweet basil (*Ocimum basilicum* L). Saudi Journal of Biological Sciences. 24:170-179.

Khalil, H.A. (2013). Influence of vesicular-arbuscular mycorrhizal fungi (*Glomus spp.*) on the response of grapevines rootstocks to salt stress. Asian Journal of Crop Science. 5(4):393-404.

Little, T. M.; Hills, F. J.1989. 'Statistical methods in agricultural research'. Versión en español. 'Métodos estadísticos para la investigación en la agricultura'. Ed. Trillas. México. 128 p.

Maguire, J.D. 1962. Speed of germination - Aid in selection and evaluation for seedling emergence and vigor. Crop Science. 2:176-177.

Martínez-Villavicencio, N., López-Alonso, C.V., Basurto-Sotel, M., Pérez-Leal, R. 2011. Efecto de la salinidad en el desarrollo vegetativo. Tecnociencia Chihuahua. 5(3):156-161.

Mata-Fernández, I., Rodríguez-Gamiño, M.L., López-Blanco, J., Vela-Correa, G. 2014. Dinámica de la salinidad en los suelos. Revista Digital del Departamento. El Hombre y su Ambiente. 1(5):26-35.

Miranda, D., Ulrichs, C., Fischer, G. 2012. Efecto del cloruro de sodio (NaCl) sobre el crecimiento y colonización micorrízica en uchuva (*Physalis peruviana* L.). Avances de la investigación agronómica II. Facultad de Agronomía, Universidad Nacional de Colombia, Bogotá, Colombia. pp. 15-25.

Mujica-Pérez, Y. y Fuentes-Martínez, A.G. (2012). Efecto a la biofertilización con hongos micorrízicos arbusculares (HMA) en el cultivo del tomate en condiciones de estrés abiótico. Cultivos Tropicales. 33(4):40-46.

Parés, J., Arizaleta, M., Sanabria, M.E., García, G. 2008. Efecto de los niveles de salinidad sobre densidad estomática, índice estomático y el grosor foliar en plantas de (*Carica papaya* L). Acta Botánica de Venezuela. 31(1):27-34.

Pérez, A.C., Sierra, J.R., Montes, V.D. 2011. Hongos formadores de micorrízas arbusculares: una alternativa biológica para la sostenibilidad de los agroecosistemas de praderas en el Caribe colombiano. Revista Colombiana de Ciencia Animal. 3(2):366-385.

Qiang-Sheng, W., Ying-Ning, Z. and Xin-Hua, H.

(2010). Contributions of arbuscular mycorrhizal fungi to growth, photosynthesis, root morphology and ionic balance of citrus seedlings under salt stress. Acta Physiol Plant. 32:297-304.

Reyes-Pérez, J.J., Murillo-Amador, B., Nieto-Garibay, A., Troyo-Diéguez. E., Reynaldo-Escobar, M.I., Rueda-Puente, E.O., García-Hernández, J.L. 2013. Tolerancia a la salinidad en variedades de albahaca (*Ocimum basilicum* L.) en las etapas de germinación, emergencia y crecimiento inicial. Universidad y Ciencia. 2:101-112.

Reyes-Pérez, J.J., Murillo-Amador, B., Nieto-Garibay, A., Troyo-Diéguez, E., Reynaldo-Escobar, M.I., Rueda-Puente, E.O. 2014. Crecimiento y desarrollo de variedades de albahaca (*Ocimum basilicum* L.) en condiciones de salinidad. Terra Latinoamericana. 1:35-45.

Sinclair, G., Charest, C., Dalpe, Y. and Khanizadeh, S. (2014). Influence of colonization by arbuscular mycorrhizal fungi on three strawberry cultivars under salty conditions. Agricultural and Food Science. 23:146-158.

StatSoft, Inc. 2011. Statistica. System reference. StatSoft, Inc., Tulsa, Oklahoma, USA.

Steel, G. D. R.; Torrie, J. H. 1995. 'Bioestadística. Principios y procedimientos'. Ed. McGraw Hill. México. 92 p.

Tavakkoli, E., Fatehi, F., Coventry, S., Rengasamy, P., Mcdonal, G. 2011. Additive effects of Na^+ and Cl^- ions on barley growth under salinity stress. Journal of Experimental Botany. 62:2189-2203.

Utobo, E.B., Ogbodo, E.N., Nwobaga, A.C. 2011. Techniques for extraction and quantification of arbuscular mycorrhizal fungi. Libyan Agriculture Research Center Journal International. 2:68-78.

Wang, Y., Wang, M., Li Y, Wu A, Huang J (2018) Effects of arbuscular mycorrhizal fungi on growth and nitrogen uptake of *Chrysanthemum morifolium* under salt stress. PLoS ONE 13(4): e0196408. https://doi.org/10.1371/journal.pone.0196408.

Zhu, X.C., Song, F.B., Liu, S.Q., Zhou, T.D.X. 2012. Arbuscular mycorrhizae improves photosynthesis and water status of *Zea mays* L. under drought stress. Plant Soil and Environment. 58:186-191.

16

Morphology, Phenology, Nutrients and Yield of Six Accessions of *Tropaeolum Tuberosum* Ruiz y Pav (Mashua)[1]

M. Valle-Parra[1], P. Pomboza-Tamaquiza[1]*, M. Buenaño-Sánchez[1], D. Guevara-Freire[1], P. Chasi-Vizuete[2], C. Vásquez[1] and M. Pérez-Salinas[1]

[1]Facultad de Ciencias Agropecuarias, Universidad Técnica de Ambato, Cantón Cevallos vía a Quero, sector el Tambo- La Universidad, 1801334, Tungurahua, Ecuador. Email: pp.pomboza@uta.edu.ec
[2]Universidad Técnica de Cotopaxi. Faculty of Agricultural Sciences and Natural Resources. Av. San Felipe s/n Latacunga, Ecuador
*Corresponding author

SUMMARY

The aim of this work was to characterize the morphology, phenology, main nutrients and yield, of six accessions of mashua (*Tropaeolum tuberosum* Ruiz y Pav) from the central region of Ecuador. The trial was carried out in Cevallos-Ecuador, at 2865 m.a.s.l, single parcels were installed at each accession. The morphological characteristics observed were related to foliage, leaf, stem, flower and tuber. Also, the duration of the four phenological phases were registered; the yield; and the macronutrients of the tuber were analyzed. Analysis were carried out regarding the main components, the variance and conglomerates. The results reveal that the variables associated with the flower and tuber were the most useful for identifying the accessions. The accession Poza Rondador registered the longest duration of the cultivation cycle (282 days), the highest usage of water (Kc 1.1) and the highest content of nutrients (protein 18.25%, phosphorus 0.73% and potassium 2.3%), whilst the Amarilla registered the shortest cultivation cycle (169 days) and the lowest amount of nutrients (protein 11.19%, phosphorus 0.42% and potassium 0.99%). The rest of the accessions varied between these ranges. The results suggest the need to promote the cultivation of accessions with higher content of nutrients. On the other hand, they also reveal the need to study secondary metabolites, and to identify accessions with potential to create nutraceutical foods.

Keywords: mashua, phenology, morphology, Andean tubers, nutrients, crop coefficient

INTRODUCTION

Mashua is an ancestral Andean crop, with high genetic diversity, used with increasingly less frequency in food and known amongst farmers for its effectiveness in treating illnesses of the urinary tract and prostate. This crop is found throughout several Andean countries (Bolivia, Peru, Ecuador and Columbia) and has many local names like Isaño, Añu, Mashua and Cubio, amongst others. The tubers can present many colors such as Amarilla, Blanca, and Morada, and they are cooked after being left out in the sun to improve their flavor.

Preliminary investigations report that mashua is an edible tuber originating in the Andes, and was domesticated by native peoples since pre-Incan times (Barrera *et al.*, 2004). Mashua is an annual species cultivated in elevated zones of the Andean region, in dry lands associated with typical crops like ocas (*Oxalis tuberosa* Mol) and potatoes, forming part of agricultural farming systems (Grau et al., 2003). In Bolivia mashua is cultivated between April and June, the cultivation cycle lasts around 7 to 8 months (Gonzales et al., 2003).

In relation to genetic diversity, mashua presents an ample genetic variability that responds to the ecologic and cultural characteristics of the region (Malice y Baudoin, 2009). However, there are reports of genetic erosion, on average a loss of 46.5% of genetic variability is estimated. In Ecuador, the province of Cañar has the highest values of 61.1% (Tapia y Estrella, 2001). In Chimborazo, in 2002 70 ecotypes of mashua where found in local fairs in the community of Huanconas (Tapia et al., 2003). On the other hand, INIAP has characterized 100 samples by morphology and 80 by agronomical characteristics (Tapia y Estrella, 2001). However, cultivations in some provinces of Ecuador diminish, and among the causes is the change in food habits, the presence of monoculture, deficient support and technical assistance to the autochthonous crops and the ignorance regarding ancestral practices (Tapia et al., 2003). Also, the low demand of mashua in markets demotivates the producers who are seen force into planting more cost effective crops (Espinosa, 2004). Regarding nutrition the crop played in important role in the communities of the region (Roca y Manrique, 2005). Currently, in rural areas it is still used as food (Quispe et al., 2015). On the other hand, in mashua accessions in the bank of germoplasm of INIAP the energetic content is reported to be between 4.19 and 4.64 Kcal/g, protein between 7.22 and 13.99, starch between 20.01 y 79.46 and total sugars between 6.77 y 55.23 (Espín et al., 2003). These values coincide with another report: humidity 88.7%, ash 4.81%, proteins 9.17%, fiber 5.86%, phosphorous 0.32%, potassium 1.99% and energy 440 Kcl/100g (Villacrés

et al., 2016). These properties are attributed also to the content of carbohydrates, fiber, ascorbic acid, vitamin A and C, and calories (Manrique et al., 2014; Roca y Manrique, 2005). These nutritional characteristics reflect the importance in rural nourishment.

Regarding medicinal use, mashua tubers possess bioactive substances that, when ingested, influence cellular activity and physiological mechanisms. Likewise, it contains phytochemicals (Glucosinolates, anthocyanins and carotenoids) that protect the crops from plagues and illnesses (Grau et al., 2003), among these the main phenolic compounds identified are the anthocyanins (Chirinos et al,. 2008). On the other hand, glucosinolates have medicinal properties, which are used by farmers in traditional medicine (Flores, 2011). When these are hydrolyzed they become isothiocyanates, sulfuranes, nitriles, and thiocyanates by the action of the enzyme myrosinase (Rincón, 2014), these have antibiotic, insecticide, nematicide, anticarcinogenic, and diuretic properties (Manrique et al., 2014). Furthermore, they effectively control prostate hyperplasia (Aire et al., 2013), and reduce sperm mobility without causing toxicity (Vásquez et al., 2012). Thus, it can be seen that mashua is a species with many studies. However, in the accessions of the central zone of Ecuador the existing ecotypes and their characteristics are unknown. In this context the investigation had the objective of characterizing six mashua accessions in their morphology, phenology, nutrients and yield.

MATERIALS AND METHODS

The investigation was carried out in the Experimental Farm Querochaca (Cevallos - Ecuador) at an altitude of 2865 m.a.s.l. The average annual temperature registered over the last five years was 13.6°C; the average yearly rainfall 465mm; environmental humidity 75.15% and wind velocity 1.7 m/s (INAMHI, 2016).

The mashua accessions were collected: In the province of Tungurahua, Amarilla (A) and Morada (M); In Cotopaxi, Blanca (B): and in Chimborazo Milicia roja (MR), Verde amarilla (VA) and Poza Rondador (PR) (vulgar name as peasant know the accessions of Mashua in the Andean region of Ecuador). These materials where planted between May 2016 and February 2017, in single parcels (15 x 3.2 m), two tubers per point, with 0.6 m between plants and 0.8 m between grooves. The fertilization was done using guinea pig manure.

The morphological characterization was completed with qualitative parameters related to the color of vegetative and reproductive structures, using the atlas of vegetable colors as recommended by Kuppers

(1979). These parameters where referenced in the leaf (color of the midrib, underside and veins); in the stalk; in the flower (pedicel, sepals, underside of the sepals, petals and spur); in the predominant color of the surface of the tuber (Cpst); in the secondary color of the surface of the tuber (Csst); in the predominant color of the pulp of the tuber (Cppt); and in the secondary color of the pulp of the tuber (Cspt) (Manrique et al., 2014) .

Quantitative variables were also registered, such as: plant height (Ap) length (Ls) and width of the sepals (As), length (Lpe) and width of the petals (Ape), length (Le) and width of the spur (Ae), length (Lf) and width of the flower (Af), length (Lh) and width of the leaf (Ah), length (Lfr) and width of the fruit (Afr), length of the axils (La), length of the petiole (Lp), diameter of the stalk (Dt), diameter of the tuber (Dtu), length of the tuber (Ltu), weight of 10 tubers (P 10 tu) and indent of the eye of the tuber (Hotu) according to the descriptors proposed by Manrique et al. (2014). The quantitative and qualitative data were registered in the flowering phase with the exception of the data related to the tubers which were collected at harvest. On the other hand, the yield (R) was calculated based on production.

The phenological phases were determined according to the duration of the four stages established by the FAO which are: initial, development, intermediate and final. The crop coefficient (Kc) was calculated, with data from the Abacus and formulas for initial, intermediate and final Kc, as propose by Allen et al. (2006). The nutritional analysis of the tubers, was carried out after harvest. For nitrogen (N) the Dumas method was used, as was the elemental analyzer CHN 628 of Leco brand; for phosphorus (P_2O_5), the

colorimetric vanadate molybdate method and the UV spectrophotometer GENESYS 20 brand were used; for K, Ca, Mg, Zn the method of atomic absorption spectroscopy and atomic absorption spectrophotometer Perkin Elmer 100 brand were used, the gravimetric method and an OHAUS brand analytic balance, and for energy the thermochemical method with an IKA C 6000 brand isoperibolic calorimeter.

RESULTS

Morphological characterization

In the qualitative variables that were evaluated, differences were found between the accessions, predominant and secondary color of the surface of the tuber and in the primary and secondary color of the pulp of the tuber (Table 1). The Blanca and Verde amarilla accessions showed similarity in the predominant color of the surface of the tuber. According to the atlas of colors by Kuppers (1979) the corresponding nomenclature is N00A50M10 and N10A99M00, but they differ in the coloration of the point of the tuber in the Verde amarilla variety. On the other hand, the Amarilla and Milicia Roja accessions were similar in the predominant coloration of the surface of the tuber (A99M50C00 and A99M50C10). However, the Milicia Roja presented scores around the indent of the eye, and the surface of the tuber. The tone of the predominant color of the surface of the Morada and Poza Rondador accessions were similar (N00C00A60 and N00A90M00), they differed in the indent of the eye. Regarding the color of the flower (Cf), variation in the intensity was observed between the analyzed accessions.

Table 1. Nomenclature of the color of the tubers of the accessions

Ac	B	A	M	MR	PR	VA
Cpst	N00A50M10	A99M50C00	N00C00A60	A99M50C10	N00A90M00	N10A99M00
Csst	N00A50M00	A99M60C10	N60A60M90	N10A50M80	N00A50M60	N00C10A99
Cppt	A10M00C00	A90M30C00	N00A10M00	N00A60M00	N00A50M00	N00A40M00
Cspt	A60M00C00	A50M00C00	N40A50M50	N10A60M00	N00A40M00	N00A60M00
Cf	A60M90C00	A60M90C10	A99M90C20	A70M99C20	A50M99C80	A99M90C30
Photo						

Caption: Ac= Accession; B= Blanca; A= Amarilla; M= Morada; MR= Milicia Roja; PR=Poza Rondador; VA= Verde amarilla

In regard to the quantitative variables, only 10 out of 21 presented significant difference between the accessions. The tallest registered plant was the Verde amarilla accession, followed by the Poza Rondador, while the smallest were Blanca and Amarilla. Regarding the size of the spur of the flower, the Verde amarilla variety had the longest length, while the Morada and Poza Rondador varieties presented the lowest value. On the other hand, the Poza Rondador accession exhibited higher values for the length and width of the leaf, diameter of the stalk and yield. However. The lowest values were observed in the Amarilla (length of the leaf), Morada (width of the leaf and diameter of the stalk) and Verde amarilla (yield). Also, the Blanca accession present the highest value of eye indentation and weight of 10 tubers. Meanwhile, the lowest value was seen in the Morada and Verde amarilla accessions (eye indentation of the tuber) and Verde amarilla (weight per 10 tubers) (Table 2).

Table 2. Morphological descriptors of the accusations of Mashua

	B	A	M	MR	PR	VA
Ap (cm)	41.3a	57.91b	64.23b	63.72b	80.66c	87.44c
Ls (cm)	1.33bc	1.46c	1.24ab	1.45c	1.15a	1.32abc
As (cm)	0.48a	0.49a	0.44a	0.58ab	0.68b	0.48a
Lpe (cm)	1.50b	1.6b	1.2a	1.62b	1.23a	1.18a
Ape (cm)	0.78ab	0.83ab	0.69a	0.81ab	0.89b	0.93b
Le (cm)	1.84ab	1.97bc	1.72a	1.96bc	1.72a	2.11c
Ae (cm)	0.31a	0.34ab	0.35ab	0.45c	0.37b	0.38b
Lf (cm)	1.58ab	1.72b	1.44a	1.75b	1.47a	1.7b
Af (cm)	0.9b	0.97bc	0.74a	1.11c	1.01bc	0.95b
Lh (cm)	4.36abc	3.88a	4.03ab	4.79c	5.39d	4.45bc
Ah (cm)	5.13b	4.49a	4.67ab	5.92c	6.61d	5.93c
Lfr (cm)	0.87b	0.9bc	0.9bc	0.95bc	0.73a	1.02c
Afr (cm)	1.52b	1.45b	1.36b	1,.5b	1.05b	1.59a
La (cm)	0.30a	0.28a	0.31a	0.32a	0.30a	0.26a
Lp (cm)	14.14ab	13.93ab	12.76a	15.43ab	16.66ab	17.38b
Dt (cm)	0.54bc	0.44ab	0.40a	0.5abc	0.57c	0.53bc
Dtu (cm)	3.8b	3.24a	3.05a	3.92b	2.84a	2.88a
Ltu (cm)	11.38a	12.84a	12.09a	12.03a	11.74a	12.01a
Hotu (cm)	0.37d	0.3bc	0.20a	0.33cd	0.24ab	0.2a
P 10 tu (Kg)	0.85d	0.50a	0.6b	0.7c	0.5a	0.49a
R (Tn/Ha)	42.70bc	33.33ab	35.35b	23.85a	47.92c	24.79a

Hight of the plant (Ap), length (Ls) and width of the sepals (As), length (Lpe) and width of the petals (Ape), length (Le) and width of the spur (Ae), length (Lf) and width of the flower (Af), length (Lh) and width of the leaf (Ah), length (Lfr) and width of the fruit (Afr), length of the axils (La), length of the petiole (Lp), diameter of the stem (Dt), diameter of the tuber (Dtu), length of the tuber (Ltu), weight per 10 tubers (P 10 tu), indent of the eye of the tuber (Hotu) yield (R). The letters a, b, c, and d indicate significant difference according to the probability $p < 0.05$

According to the analysis of main components the variables that best explain the differences are: flower width (Af), indent of the eye of the tuber (Hotu), spur width (Ae) and weight per 10 tubers (P 10 tu), which allowed the identification of four varieties. The Amarilla and Milicia Roja accessions apparently are the same variety, and also the Blanca and Morada accessions seem to be another variety (Figure 1).

PCA case scores

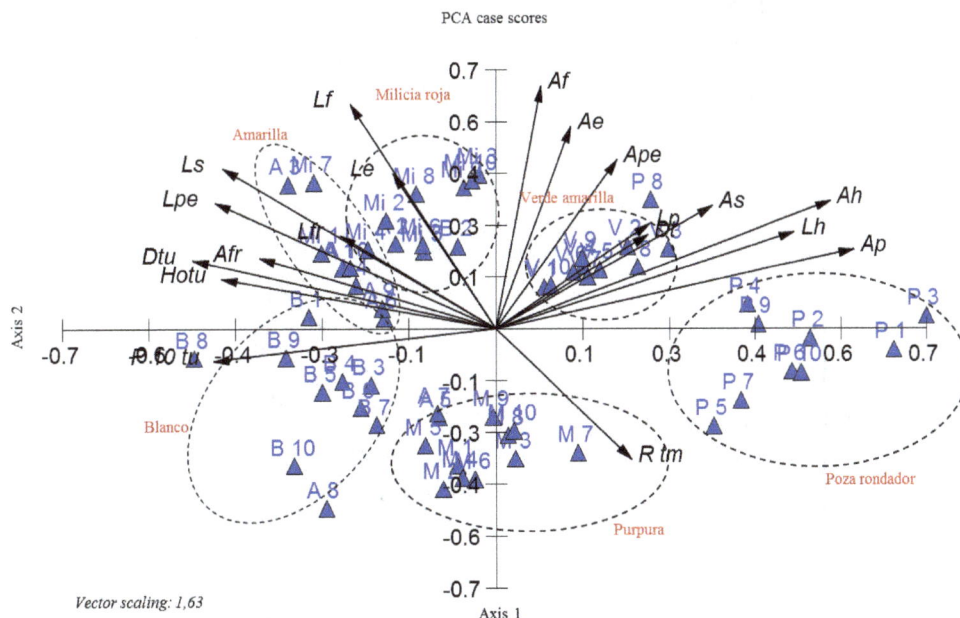

Figure 1. Analysis of main components based on quantitative variables of mashua accessions. (Blanca, Amarilla, Milícia Roja, Verde Amarrilla, Poza Rondador and Morada, vulgar name as peasant know the accessions of Mashua in the Andean region of Ecuador)

Dendogram

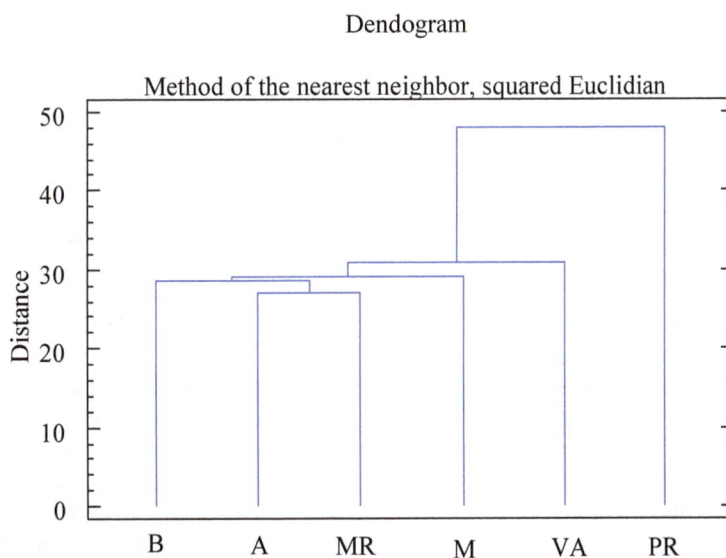

Figure 2. Analysis of the conglomerates based on the quantitative variables of mashua accessions. B=Blanca; A= Amarilla; M= Morada; MR= Milicia roja; PR=Poza Rondador; VA= Verde amarilla

In the analysis of conglomerates two groups were observed, the first formed by the Blanca, Amarilla, Morada, Milicia Roja and Verde amarilla accessions. Meanwhile the second contains the Poza Rondador accession (Figure 2).

Phenology

In the initial phase, the Morada accession registered the higher number of days between the sowing and

germination (28), while the Milicia Roja presented the lowest number of days (16). In the development phase, the Poza Rondador accession showed the higher number of days from germination to tuberization (143), conversely the Blanca accession showed the least number of days (63). In the intermediate phase, the Poza Rondador accession took longer between the tuberization and flowering (18 days). Meanwhile the Morada, Milicia Roja and Verde amarilla accessions, reached this phase within

10 days. And in the final phase, between flowering and harvest, the Milicia Roja accession took longer (188 days), conversely the Amarilla accession took 70 days. In this manner, the Poza Rondador accession possessed the longest cultivation cycle (282 days) while the Amarilla accession was the shortest with 169 days (Figure 3)

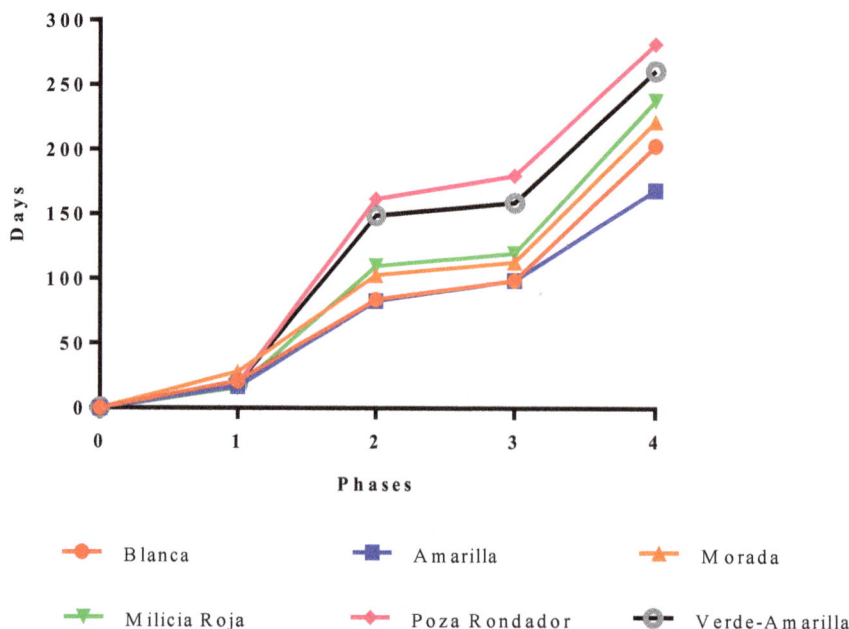

Figure 3. Duration (in days) of the phenological phases in each mashua accession.
Blanco, Amarillo, Milícia Roja, Verde Amarrilla, Poza Rondador y Purpura, is the vulgar name as peasant know the accessions of Mashua in the Andean region of Ecuador.

Relating to the value of the crop coefficient (Kc), no differences were detected in the initial Kc (0.51) between the six accessions. However, differences were detected in the middle and fj al Kc (Figure 4). The mean Kc value was 1.02 in White accession, 1,0 for Amarilla, Morada, and Milicia Roja accessions, 1.1 in Poza Rondador and 0.96 in Verde amarilla. The final Kc reported values of 0.73 in Blanca; 0.58 in Amarilla; 0.71 in Morada, Milicia Roja and Poza Rondador and 0.7 in Verde amarilla.

Chemical analysis of nutrients

The chemical compositions of the accessions, showed variable results in the evaluated elements. The higher values of energy were exhibited in the Milicia Roja accession and Blanca the lowest value. Regarding the content of phosphorus and calcium, the Verde amarilla accession showed high values, while the Amarilla accession had the lowest value of phosphorous and the Morada had the lowest amount of calcium. Finally, the lowest values of magnesium were detected in the Verde amarilla accession and the higher in the Morada and Milicia Roja accessions. In the latter, high values of zinc were found, and low values were found in the Verde amarilla accessions (Table 3).

DISCUSSION

The morphological characterization of the phytogenic resources, consist in determining a set of characteristics (qualitative and quantitative) that function as descriptors, that enable the taxonomical differentiation of the species (Hernández, 2013). In this, study (Province of Tungurahua), the qualitative descriptors that most helped in the identification of the differences between the mashua accessions, were related with the flower and the tuber. Also, Quispe et al. (2015) reported that the qualitative morphological characteristics related to the flower and the tuber, were of most use in the characterization of mashua accessions.

The analysis of the variance of the quantitative characteristics, revealed significant statistical differences in plant height, sepal length, spur length, flower width, length and width of the leaf, diameter of the stalk, indent of the eye of the tuber, weight per 10 tubers, and yield. In contrast, with the reports of

Malice y Baudoin (2009), in Cochabamba-Bolivia, who reported significant differences only in plant height and leaf width. The accessions studied in

Tungurahua, presented similar values to those obtained in Bolivia.

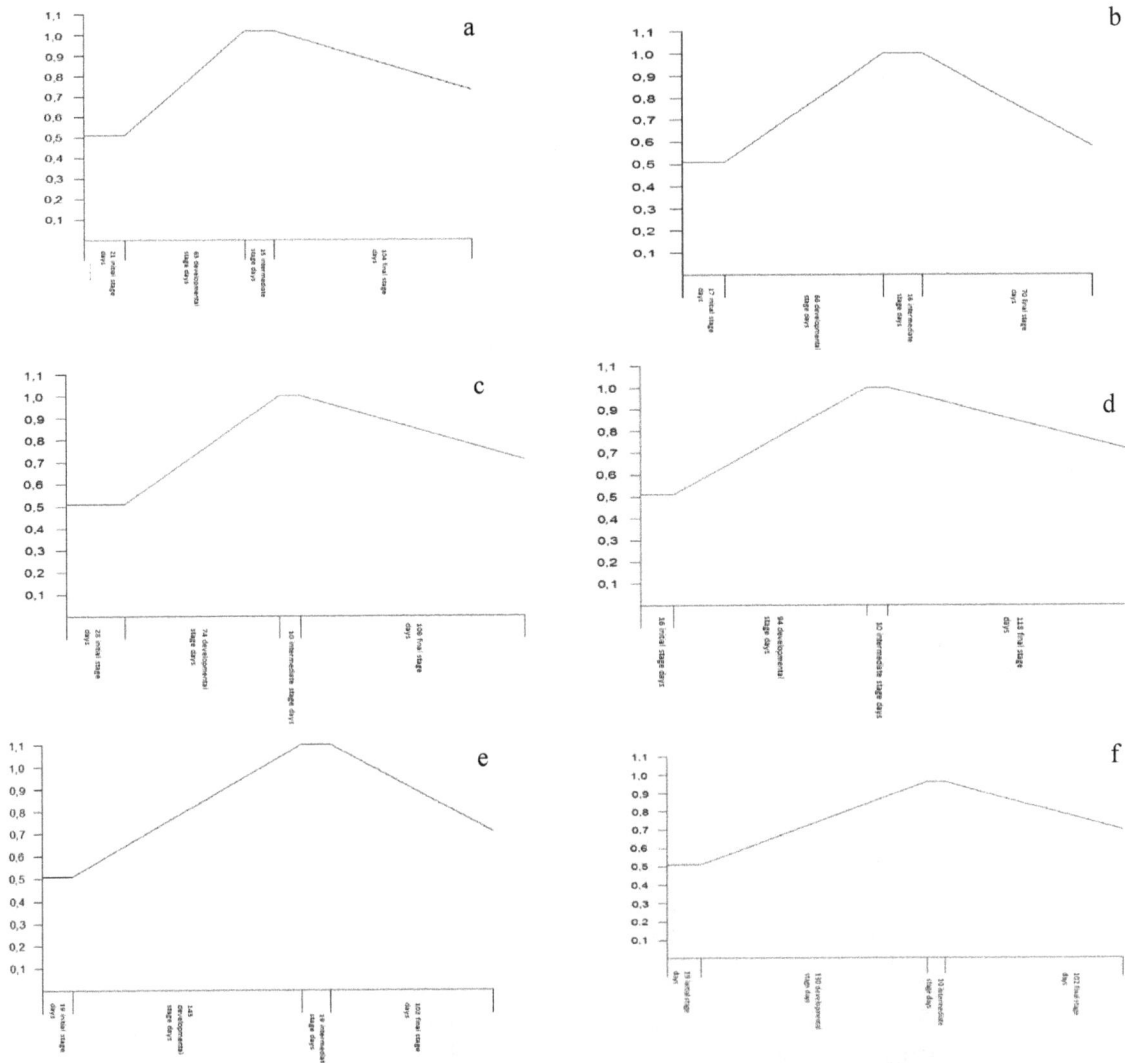

Figure 4. Values of Kc in the accessions: Blanca (a), Amarilla (b), Morada (c), Milicia roja (d), Poza Rondador (e), Verde amarilla (f) of mashua cultivated in Querochaca, Province of Tungurahua.

In so far as foliar morphology, one group of accessions (Blanca, Amarilla, Milicia Roja and Poza Rondador), were found to present both tribulated and pentalob leaves in an estimated 50% in each accession. Conversely, in the Bolivia accessions tribulate and pentalob plants were observed in 100% of the leaves (Malice and Baudoin, 2009). The analysis of main components indicated that of the 21 qualitative characteristics analyzed four contributed significantly, two related to the flower and two with the tuber. However, Quispe et al. (2015), found that out of the 44 characteristics studied, 25 of them were of greater contribution to the total variance. These were related to the flower, the tuber, the leaf and the

stem. Regarding the phenology, four phases were identified: emergence, tuberization, flowering and harvest, The Amarilla accession registered the lowest number of days in the cultivation cycle. Meanwhile, the Poza Rondador turned out to be the most belated, with an estimated difference of 113 days. These results suggest that the cultivation cycle is related to the genetic characteristics of the accession. Also, in the behavior of the accessions regarding the phenological phases, differences between the accessions in the second and fourth phase were found (Figure 3).

However, in previous works seven phenological phases were cited in this species: emergence, formation of stolons, tuberization, budding, flowering, fructification, ripening (Yzarra and López, 2011; Fries and Tapia, 2007). In the case study, four phenological phases are considered in accordance with the methodology proposed by Allen (FAO, 2006) who also facilitated the calculation of the crop coefficient (Kc). Regarding the crop coefficient, it was observed that in the first phase all the accessions had similar values, variance was found in the second,

third and fourth stages of the cultivation. Regarding this, Valverde (1998) mentions that the crop coefficient (Kc) depends on anatomic, morphologic and physiologic characteristics of the plant. This determines the volume of water that the plant absorbs, in relation to the vegetative state. These values can be useful for the planning of irrigation in each accession to achieve a more efficient irrigation.

Table 3. Content of nutrients of the Mashua accessions

Nutrients	B	A	M	MR	PR	VA
H (%)	82.73ab	78.04a	87.76cd	90.25cd	92.18d	86.61bc
E (Kcal/g)	3.66b	3.67b	3.94a	3.94a	3.9a	3.91a
P (%)	10.06b	11.19a	12.56b	16.25d	18.25e	14.13c
P2O5 (%)	0.55ab	0.42a	0.56abc	0.69bc	0.73bc	0.77c
K (%)	0.62a	0.99a	0.82a	0.66a	2.33b	1.26a
Ca (%)	0.08a	0.1a	0.025a	0.03a	0.65b	1.43c
Mg (%)	0.12b	0.11b	0.14b	0.14b	0.13b	0.03a
Zn (ppm)	9.6a	9.63a	17.0b	27.13c	8.33a	5.0a

Caption: humidity (H), energy (E), protein (P), phosphorous (P2O5), potassium (K), calcio (Ca), magnesium (Mg), zinc (Zn). The letters a, b, c, d, and e, indicate significant differences according to the probability $p<0.05$. B; Blanca, A; Amarilla, M; Morada, MR; Milicia roja, PR; Poza rondador, VA; Verde amarilla

Respecting the content of nutrients, it was observed that the Poza Rondador accession has the highest content of humidity, nitrogen and potassium. Meanwhile, the Blanca accession had the lowest content of nitrogen and potassium. This indicates a relationship between the accessions and the nutritional content. These values were similar to those reported by Espín et al (2003) and Villacrés et al. (2016). Furthermore, all the accession presented high calorie content. In relation to this, Ayala (2004) sustains that for each 100g of tuber contains 50 Kcal and 84.1% humidity. The characteristics of some accessions make mashua an alternative for the improvement of the nutrition of rural populations.

CONCLUSIONS

Regarding the morphology, the six accessions presented differences relating to the variables associated with the flower and tuber. In like manner, in the phenology (cultivation cycle) differences between all the accessions were observed, the Poza Rondador accession with 282 days and Amarilla accession with 169 days stood out. In relation to the Kc, differences were found in the second, third and fourth phase. In function of the cultivation cycle the Poza Rondador variety demanded the highest amount of water. Finally, the accessions that showcased the higher quantity of nutrients (N, humidity, P and K) were the Poza Rondador and Verde amarilla. Meanwhile, the accessions with the lowest quantity were the Amarilla and Blanca. The information that

was found reveals the presence of a few accessions that have good potential for commercial and food production.

REFERENCES

Aire G., Charaja R., De La Cruz H., Guillermo B., Gutarra M., Huamaní P., Pari R. 2013. Efecto de Tropaeolum tuberosum frente a la hiperplasia prostática benigna inducida en ratas Holtzman. CIMEL Ciencia E Investigación Médica Estudiantil Latinoamericana. 18: 1–13.

Allen G., Pereira L., Raes D., Smith M. 2006. Evapotranspiración del cultivo Guías para la determinación de los requerimientos de agua de los cultivos. FAO, 323. [Disponible en] ftp://ftp.fao.org/agl/aglw/docs/idp56s.pdf

Ayala G. 2004. Aporte de los Cultivos andinos a la Nutrición Humana. In Raíces Andinas: Contribuciones al conocimiento ya la capacitación. Universidad Nacional Mayor de San Marcos. Lima, Perú. (pp. 101–112).

Barrera V., Espinoza P., Tapia C., Monteros A., Valverde F. 2004. Raíces y tubérculos andinos : alternativas para la conservación y uso sostenible en el Ecuador. Centro Internacional de la Papa (CIP).

Chirinos R., Campos D., Costa N., Arbizu C., Pedreschi R., Larondelle Y. 2008. Phenolic profiles of andean mashua (*Tropaeolum tuberosum* Ruízy Pavón) tubers: Identification by HPLC-DAD and evaluation of their antioxidant activity. Food Chemistry: 106: 1285–1298.

Espín S., Villacrés E., Brito B. 2003. Raíces y Tubérculos Andinos : Alternativas. In V. Barrera, C. Tapia,y A. Monteros (Eds.), Raices y Tuberculos Andinos: Alternativas para la conservación y uso sostenible en el Ecuador (p. 187). Quito: INIAP.

Espinosa P. 2004. Consumo, Aceptabilidad y Oportunidad de Aumentar la Demanda Urbana de RTAs. In Conservación y uso de la biodiversidad de raíces y túberculos andinos (p. 187).

Flores D. 2011. Recopilación de saberes ancestrales sobre las especies andinas alimenticias: Mashua (*Tropaeolum tuberosum*), Melloco (*Ullucus tuberosus*), Oca (*Oxalis tuberosa*) y Quinua (*Chenopodium quinoa*) en la comunidad de Pesillo, Cayambe – Ecuador – 2010. Tesis de Pregrado. Universidad Politécnica Salesiana. [Disponible en] http://dspace.ups.edu.ec/bitstream/12345678 9/4596/6/UPS-YT00068.pdf

Grau A., Ortaga R., Nieto C., Hermann M. 2003. Mashua *Tropaeolum tuberosum* Ruiz and Pav. Promoting the Conservation and Use of Underutilized and Neglected Crops. Roma, Italia: International Plant Genetic Resources Institute.

Hernández A. 2013. Caracterización morfológica de recursos fitogenéticos. Revista Bio Ciencias. 2: 113–118.

INAMHI. 2016. Anuario metereológico. Quito.

Kuppers H. 1979. Atlas de los colores. (Blume, Ed.) (Primera). Barcelona. España.

Malice M., Baudoin J. 2009. Genetic diversity and germplasm conservation of three minor Andean tuber crop species. Biotechnology, Agronomy, Society and Environment. 13: 441–448.

Manrique I., Arbizu C., Vivanco F., Gonzales R., Ramírez C., Chávez O., Tay D., Ellis D. 2014. *Tropaeolum tuberosum* Ruízy Pav. (T. A. G. Educativa, Ed.). Lima - Perú.

Quispe C., Mansilla R., Chacón A., Blas R. 2015. Análisis de la variabilidad morfológica del " Añu " *Tropaeolum tuberosum* Ruizy Pavón procedente de nueve distritos de la región Cusco. Ecología Aplicada 14: 211-222.

Rincón A. 2014. Biosintesis De Los Glucosinolatos E Importancia Nutricional Humana Y Funciones De Protección a Las Plantas. Alimentos Hoy. 22: 64–80.

Roca W., Manrique I. 2005. Valorización de los recursos genéticos de raíces y tubérculos andinos para la nutrición y la salud. Agrociencia. 9: 195–201.

Roldan C., Carbajal A. 2012. Componentes bioactivos de los alimentos. Manual Práctico de Nutrición Y Salud. 31–36.

Tapia C., Estrella J., Monteros A., Valverde F., Nieto M., Córdova J. 2003. Caracterización fisico-quimica, nutricional y funcional de raices y tuberculos andinos. Manejo y conservación de RTAs in siyu en fincas de agricultores y ex situ en el banco de germosplasma de INIAP (Vol. 4).

Tapia C., Estrella J. 2001. Genetic erosion quantification in Ullucus (*Ullucus tuberosus* Caldas), Oca (*Oxalis tuberosa* Mol.) and Mashua (*Tropaeolum tuberosum* R.&P.) in agroecosystems of the provinces of Cañar, Chimborazo and Tungurahua – Ecuador. INIAP. Quito. [Disponible en] http://archive.unu.edu/env/plec/cbd/Montreal /abstracts/Tapia.pdf

Vásquez J., Gonzáles J., Pino, J. 2012. Disminución en los parámetros espermáticos de ratones tratados con el extracto hidroalcohólico de *Tropaeolum tuberosum* "mashua." Revista Peruana de Biología. 19: 89–93.

Villacrés, E., Quelal, M., Alvarez, J. 2016. Redescubriendo la oca y la mashua. Quito: INIAP.

Yzarra W., López F. 2011. Manual de observaciones fenológicas. (SENAMHI, Ed.). Perú.

17

Animal Welfare and Poultry Productivity

D.F. Avilés-Esquivel, **M.A. Montero, H. Zurita-Vásquez**
and M. Barros-Rodríguez

Facultad de Ciencias Agropecuarias, Universidad Técnica de Ambato, Carretera
Cevallos-Quero, 180350 Cevallos, Tungurahua, Ecuador.
Email: df.aviles@uta.edu.ec
*Corresponding author

SUMMARY

The aim of this review is to present production alternatives based on animal welfare standards and good management practices on broiler and posture breeding. The excessive intensive production of eggs and chicken meat comes from the demand for food from a growing world population. And it has forgotten good animal husbandry and management practices. Which causes an imbalance of the sympathetic nervous system and adrenal medullary tissue that control the response of the poultry to stress, raising catecholamine levels inducing the release of glucose causing liver, heart and neuronal failure. these effects become visible due to the increase of diseases since the immune system is depressed, food intake decreases, oxytocin inhibition, which results in reduced reproduction and even death of the poultry. The incorporation of animal welfare standards imposed in intensive breeding, as well as, the production of eggs and meat under production systems (grazing) and backyard production become a sustainable alternative that maximizes the welfare of the poultry

Keywords: *Gallus domesticus*; Backyard; Animal welfare; Production.

INTRODUCTION

The world poultry production has grown at an accelerated pace since 1961 (Nicol and Davies, 2013) driven by the high demand for white meat and eggs for the consumption of the world population, which has also grown by leaps and bounds since 1961 where there were approximately 3,036 thousand million human beings, which has allowed and influenced during the last 15 years its expansion, consolidation, development and globalization in most of the countries of the world (Glatz et al., 2013).

The Food and Agriculture Organization of the United Nations determined that the world production of poultry meat was 108.7 million tons (Giacomozzi, 2015), as well as the world production of eggs within the last two decades. from 64.2 million tons to 86.2 million tons during the period from 2010 to 2015, thus supplying the global demands (Giacomozzi, 2014), which are generated since the 1980s as a result of the development and population growth, the increase in per capita income and the demand for livestock products leading to the so-called "livestock revolution" (Steinfeld and Chilonda, 2006). Thanks to this livestock revolution, the council on animal welfare of farm animals that the English government created in 1979, prior to research carried out in intensive livestock farming by Professor Roger Branbell in 1960, was left aside. which has caused, that for several decades the animal welfare of the domestic species that the human being breeds has been neglected, among them the poultry, suffering thus, high levels of stress due to the over exploitation to which they are submitted in intensive systems managed worldwide.

Stress arises when there is an imbalance of the nervous system, in poultry there are three phases: alarm reaction, resistance or adaptation and exhaustion (Mohan, 2005; Medina, 2016) in general the mechanism influences the sympathetic nervous system and adrenal medullary tissue that control the response of the bird to stress, raising catecholamine levels (dopamine and nor-adrenaline) inducing the release of glucose causing liver, heart and neuronal failure (Selye, 1950; Siegel, 1980), these effects are visible in the increase of diseases since the immune system is depressed, food intake decreases, oxytoxine inhibition, reduction of reproduction and even death (Brake, 1985; Freeman, 1987; Maxwell, 1993).

Animal welfare

Although all species of domestic and wild animals are different to humans, they have the ability to experience and express different emotional states, which is why it is extremely important to implement ethical considerations within their production or exploitation guaranteed, thus a satisfactory level of animal welfare, the dictionary of the Royal Academy of the Spanish Language determines that welfare is the set of things and activities necessary for the development of situations that generate a satisfactory living condition (Nicol and Davies, 2013; Van Horne and Achterbosch, 2008), the World Organization for Animal Health (OIE) defines animal welfare as the way in which an animal confronts all the conditions of its environment (Meluzzi and Giordani, 1989; Rojas et al., 2005).

Animal welfare has been formally established as a discipline since 1965, and in 1979 the Farm Animal Welfare Council of the United Kingdom sets five freedoms and principles for the origin of animal welfare, taking into account the physical and behavioral needs of animals. they are: freedom from not suffering hunger or thirst, freedom from not suffering discomfort, freedom from not presenting pain, wounds or illnesses, freedom to express a natural and innate behavior and freedom from developing fear or anguish (Rojas et al., 2005; Dottavio and Di Masso, 2010; Nicol and Davies, 2013;).

In recent years the issue of animal welfare within poultry production has gained strength and significant importance in the European Union, the United States (Van Horne and Achterbosch, 2008), Canada and in some developing countries such as: Argentina, Bolivia, Brazil , Cuba, El Salvador, Mexico, Paraguay, Peru and Uruguay (Rojas et al., 2005) the same ones that propose viable production alternatives, improving especially the management practices, care and production of poultry (Nicol and Davies, 2013) increasing notoriously the productivity and the quality of the products as well as, that standards of animal welfare are fulfilled which allows the widening of the range in sales and exports (Rojas et al., 2005).

In 2012, the European Union prohibits the use of conventional cages for laying hen within commercial egg production, where it is approved that equipment and cages must present conditions that favor the manifestation of natural and innate behaviors of poultry (Nicol Davies, 2013), thus replacing traditional cages or lodgings with composite cages, which have an area of $750cm^2$ (Van Horne and Achterbosch, 2008) and are made up of a perch, a bed, a nest box, their respective feeders and drinking troughs, constituting an alternative system to the intensive productions in which the poultry had access to a minimum area of 550 cm^2 and lacked all comfort and well-being (Aseprhu and Prhucam, 2013) as can be seen in Figure 1. For commercial poultry production (meat) a maximum level of density was also determined in relation to the number of poultry per square meter, setting a density of 33kg to 39kg/m^2

if animal welfare conditions are present at a considerable level and range within the production system (Van Horne and Achterbosch, 2008; Vizcaíno and Betancourt, 2013), given that in intensive production systems there are large levels of poultry integration per m^2 which generate biological disturbances and behaviors related to inappropriate management in different stages during the development of the poultry, sometimes causing death (Dottavio and Di Masso, 2010).

Intensive poultry production

The obtaining of products of massive consumption (meat and eggs) is characterized by contributing with the minimum of investments to obtain the maximum of the possible yields (Dottavio and Di Masso, 2010), in countries of the Andean region it has been adopted with little creativity the intensive breeding of birds with the objective of increasing the economy, where maximizing the density of poultry per m^2 is a priority without taking animal welfare into account. What generates unnatural and precarious living conditions for birds (Meluzzi and Giordani, 1989; Dottavio and Di Masso, 2010) depriving them of all kinds of innate behavioral expression such as the agitation of their wings, snacking and bathing with sand (Van Horne and Achterbosch, 2008; Nicol and Davies, 2013); Other factors that affect the welfare of the birds are: the limited supply of the amount of food, inadequate

handling during transportation and slaughter. In general, all these bad conditions lead to metabolic disorders that mainly damage the bone system, cardiovascular functions and low immune system in poultry (Dottavio and Di Masso, 2010). As a result, high levels of stress are triggered, which are directly proportional to the high mortality of birds (Nicol and Davies, 2013).

Production alternatives

Within the production of eggs, various systems have been implemented as alternatives to improve the welfare of laying birds (Muñoz and Vellojin, 2002), these systems eradicate all or part of the use of cages (Fanatico, 2007), within which They have established 3 types called: Production in soil, the same one that foresees a density of 9 birds per m^2 (Vizcaíno and Betancourt, 2013), bed, feeder, drinker, perches and nesting, all these are on aviaries or a unique plant within of a warehouse or shed (Aseprhu and Prhucam, 2013), which has characteristics similar to that of intensive production (De Luca and Kuricic, 2015), within this production system can be obtained low rates of ectoparasites "lice" since the hens present higher levels of grooming and care of all the external parts of their body, thanks to the space that this system provides them (Castañeda and Gómez, 2010) (Figure 2).

Figure 1. Cages with excess laying hens by m^2

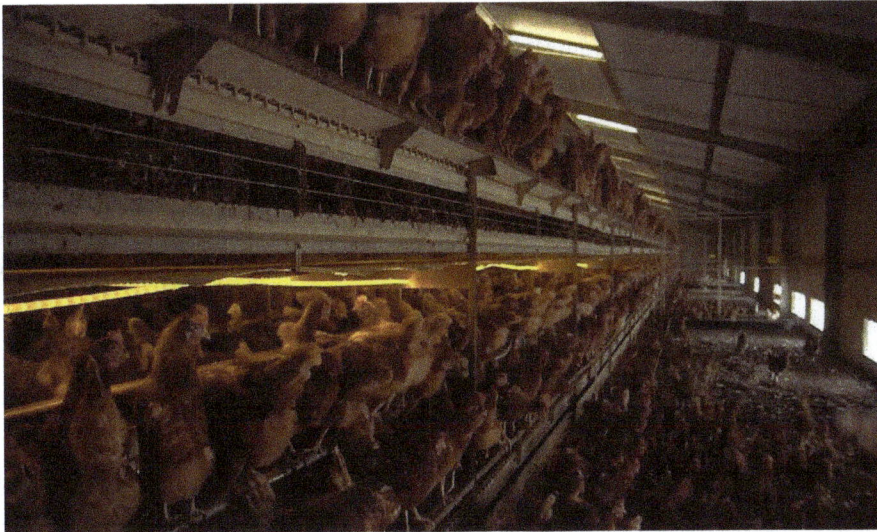

Figure 2. Hens in a modified barn coop with greater comfort

Backyard production; It is a complement to the production system in soil with the variation that poultry have access to parks (Castañeda and Gómez, 2010) or large areas of open land and covered with vegetation where they will proceed to graze (Fanatico, 2007; Dottavio and Di Masso, 2010), this activity is carried out by momentarily leaving the facilities by means of trapdoors or doors arranged longitudinally in the warehouse or production shed, the parks must be planned for a density of 1 hen per m² or a population of 2500 poultry per hectare (Aseprhu and Prhucam, 2013) (Figure 3).

Ecological production; is the combination of production systems in soil and jacket production in which a density of 6 poultry per m² is handled with the difference that the farm area within this production system must be declared in ecological production or that the area to be used For the pasture of the hen has not been subjected to treatments with any type of chemical products (Fanatico, 2007; Aseprhu and Prhucam, 2013), this system improves the poultry immune response, reduces stress levels and significantly reduces the percentage of Mortality and incidence of diseases (Muñoz and Vellojin, 2002; Castañeda and Gómez, 2010).

On the other hand, meat production has also adopted some characteristics of the production system of egg producing hen, allowing poultry to have free access to the farm (Fanatico, 2007) in such a way that they can express their own behavior. the animal species and thus conferring better quality meat (Dottavio and Di Masso, 2010). This production alternative promotes the use of probiotics or foods rich in beneficial microorganisms for the gastro-intestinal flora, natural

vitamins, fresh air, natural light (photoperiod), natural diet and balanced with fresh pastures (Castañeda and Gómez, 2010)

Influence of light

When poultry do not have access to natural light, it is recommended that they receive at least 8 hours of artificial light, according to the rules on animal welfare for poultry, during an experiment with chickens it was observed that, by offering them the possibility of choosing between light and dark, chickens chose to spend 80% of the time in the light (Savory and Duncan, 1982). Hughes and Black (1974) studied the negative effect of low intensities on the behavior of hens. Thus, chicks housed in ships with intensities between 17 and 22 lux were more timid and fearful of mobile forms than those housed in ships with intensities between 55 to 88 lux. In another study carried out by Martin (1989) it was evidenced that there was a higher percentage of picace at low intensities (50 lux) than high intensities (500 lux), since the hens at high intensities, had greater visual acuity, paid more attention to the soil (both in cages housed in cages as in soil) spending much more distracted pecking particles of soil and investigating their social environment, which evidently decreased social disputes and pecking between chickens.

The backyard: a viable alternative, ecological and of comfort

The poultry production, traditional, rustic or also called backyard is a livestock activity practiced since pre-Hispanic times that aims to use family labor to obtain products (meat and eggs) with high nutritional content in an ecological way and with high levels of well-being. animal (Soler, 2010; Molina, 2013; Hernandez et al., 2014), these products can be both for the consumption of the family nucleus and for commercialization, generating income capable of satisfying needs and improving the quality of life of producers (Hernández et al., 2014; Romero, 2013). In this sense, it constitutes an activity that is important within the rural regions of developing countries

(Molina, 2013) such as: Cuba, Chile, Costa Rica, Honduras, Mexico, Africa and developed countries such as: Asia and members of the European Union (Soler, 2010), because the birds are monogastric and small animals constitute an easy handling and great adaptation to the majority of existing climatic floors (Molina, 2013). Furthermore, this production system requires low investments in facilities, supplies and equipment for its implementation and operation (FAO and SAGARPA, 2007).

Figure 3. Breeding poultry with access to grazing.

The backyard consists of a small group of 4 to 15 poultry which live together and remain free in grazing during the day and at night crowded in a chicken coop during their rest, their food is varied naturally consisting of herbs, insects, pastures, forages , worms, worms, organic household waste (fruits, vegetables), grains, seeds and food supplements provided by the producer (FAO and SAGARPA, 2007; Soler, 2010; Molina, 2013; Hernández et al., 2014), poultry houses or facilities are built based on existing materials to the area or recycled materials (plastic bottles) and according to the budget of the farmers (Figure 4).

The size of the poultry house and the grazing area will depend directly on the number of poultry to produce these should have a density of 2-3 poultry/m^2 (FAO and SAGARPA, 2007), have feeders and drinkers that can be made by hand or acquire them of some commercial house. The nests can be built empirically as long as they offer comfort and the established dimensions (30cm wide x 40cm long and 35cm deep) perches for the rest of the poultry and finally a fence that surrounds these facilities to protect the integrity of the poultry in production, which avoids its dispersion and facilitate its management

(Hernández et al., 2014; Soler, 2010; FAO and SAGARPA, 2007).

The backyard viability is presented as a feasible production system since it can reach 2.26 kg of weight at sexual maturity (22 weeks) and an egg production of 2.3 eggs /hen/ week obtaining a peak of posture of 59.1% in red Rhode Island hens, considering that the feeding was not based on concentrate (Jerez-Salas and Carrillo Rodríguez, 2009). This alternative small-scale system proposed for the production of eggs fulfills a very important function, whose main purpose is self-consumption, the generation of complementary income and an element of savings in the units of peasant production. With respect, to the quality of the egg in the backyard Creole hens the results are similar to the eggs of hens of commercial stock (Segura-Correa et al., 2010). In this sense, Jerez-Salas et al. (2014) report higher (P <0.05) pigmentation of the yolk and omegas content in the eggs of hens with access to purslane (*Portulaca oleracea L.*) typical characteristics of the so-called "ranch eggs" themselves that are very desired by consumers.

Appropriate transportation

The transport of domestic animals to the farms or slaughterhouse influences the loss of weight due to mismanagement, triggering high levels of stress to which they are subjected during this activity (Kannan et al., 2000, Ali et al., 2006; Schwartzkopf- Genswein et al., 2012). The mobility of poultry through land transport and in conjunction with loading and unloading operations are frequent activities that constitute a new and unknown situation for them (Glatz et al., 2013), the simple fact of removing poultry from its controlled atmosphere of production, whether from one farm to another, from the farms to the place of slaughter or from the farms to the markets of markets and fairs, suppose conditions that generate stress, anxiety, fear, vertigo and alterations in activities or frequencies food (Ros, 2008; Arrebola et al., 2013).

During transport, the poultry are placed in loose boxes or modular trucks, which represent a large population of poultry in a small area, thus generating a very high density (Glatz et al., 2013; Arrebola et al., 2013). of this, they are exposed to noise and vibrations of which they are not accustomed, for example: the vibration of the vehicular engine and the noises of the outside of the farm, all this together causes the poultry to develop aggression behaviors among themselves giving as Result lesions, fractures or bone ruptures, muscle tears, bruises, bruises and sometimes extreme death (Aseprhu and Prhucam, 2013; Glatz et al., 2013; Arrebola et al., 2013).

Figure 4. Poultry reared in traditional or backyard system.

For all these problems of animal welfare the European Union determines that for long trips (more than eight hours) from the embarkation of the first poultry to the discharge of the last poultry, an adequate and abundant supply of water and food must be available, as well as as well as rest periods (Ros, 2008; Arrebola et al., 2013; Glatz et al., 2013) the vehicle should simulate a relatively stable environment and provide protection, comfort and safety to poultry, this regulation also integrates the maximum densities per poultry weight for transport, which are detailed in Table 1 (Aseprhu and Prhucam, 2013).

Heat stress

Several studies (Franco-Jiménez and Beck, 2007; Mertens et al., 2010; Havlicek and Salma, 2011; Gudev et al., 2011) have shown that thermal stress in hens, especially in poultry, makes their Heat dissipation system (conduction, convection and radiation) becomes less effective with the increase of room temperature, since they can not sweat due to the absence of sweat glands, therefore, the animal depends more and more on thermolysis by panting and metabolic changes to relieve heat stress, thus decreasing thermogenesis, limiting the provision of nutrients and proteins for egg formation, directly influencing the decrease in weight and egg size, quality of the skin (appearance of furrows), loss of the coloring of the yolk, loss of consistency of the albumin, making the yolk vulnerable to microbial contamination; the optimum temperatures in laying hens are from 21 to 25 ° C (Keener et al., 2006; Mashaly et al., 2010; Felver-Gant et al., 2012).

Interaction between animal welfare and poultry productivity

Numerous studies show that improving animal welfare can increase productivity and quality of meat, milk or eggs (Beattie et al., 1995; Boivin et al., 1998;

Briones et al., 2011; Schwartzkopf-Genswein et al., 2012). In this sense, when comparing intensive production systems versus production of hens reared in the open field, less parasitism (coccidia) and lower mortality from birth were observed in hens reared in the open field (evaluation for 13 weeks) (Briones et al., 2011). In another study conducted by Mohammed et al. (2013) reported higher ($P<0.05$) egg mass (50.66 vs 45.30 g egg/hen/day) higher feed intake (103.70 vs 97.67 g/day) and eggs with darker yellow buds in hens under one system of open field production vs cage production system, it should be noted that the concentrated feed supplied was the same for the two production systems.

Table 1. Density of poultry transported in cages or containers (Aseprhu and Prhucam, 2013; Ros, 2008; Arrebola et al., 2013).

Category	Space (cm^2)
One-day-old chicks	21-25 cm^2/chicken
chicken less than 1.6 kg	180-200 cm^2/kg
chicken of 1.6 kg to 3 kg	160cm^2/kg
chicken of 3 Kg to 5 kg	115cm^2/kg
chicken of more than 5 kg	105cm^2/kg

Perspectives

As a result of this review, the following questions arise:

Is consumer behaviour influenced by animal welfare awareness?

Could the animal welfare implement be promoted through NGOs in developing countries, given that their government seen to have little interest in this issue?

Could the backyard existing in rural areas of all countries, known as traditional breeding, be managed in a more technical way, conserving the empirical experiences of its owners?

CONCLUSION

The incorporation of animal welfare standards imposed on intensive farming, as well as the production of eggs and meat under production systems (grazing) and backyard production, become a sustainable alternative that maximizes the welfare of the poultry. providing better quality animal protein products and promoting the protection of animals and the ecosystem in future generations.

REFERENCES

Ali BH, Al-Qarawi, AA, Mousa, HM. 2006. Stress associated with road transportation in desert sheep and goats, and the effect of pretreatment with xylazine or sodium betaine. Research in Veterinary Science. 80 (3): 343-348.

Arrebola F, Ordoñez I, Morillo M. 2013. Bienestar Animal en el Transporte. Instituto de Investigación y Formación Agraria y Pesquera. Consejería de Agricultura, Pesca y Medio Ambiente de Andalucía. Sevilla, España. 150pp.

Aseprhu, Prhucam. 2013. Guía de buenas prácticas de manejo y bienestar animal en granjas avícolas de puesta. Gobierno de España. Ministerio de Agricultura, Alimentación y Medio Ambiente. 125pp.

Brake JT. 1985. Stress and flock health. Vineland Update. November. Vineland laboratories, Vineland NJ 08360.

Beattie VE, Walker N, Sneddon IA. 1995. Effects of environmental enrichment on behaviour and productivity of growing pigs. Animal Welfare. 4 (3): 207-220.

Boivin X, Garel JP, Durier CL, Neindre P. 1998. Is gentling by people rewarding for beef calves? Applied Animal Behaviour Science 61: 1-12.

Briones M, Avendaño L, Ulloa A, Arias, N, Alarcón N. 2011. Comparación del comportamiento de pollos de una línea de postura (Hy-Line) y de una línea Araucana, en condiciones de campo y de plantel comercial. Actas Iberoamericanas de Conservación Animal. 1 :397-400.

Castañeda C, Gómez J. 2010. Evaluación del bienestar animal y comparación de los parámetros productivos en gallinas ponedoras de la línea Hy-line Brown en tres modelos de producción: piso, jaula y pastoreo. Cien. Anim. 3: 14pp.

De Luca F, Kuricic E. 2015. Impacto sobre los indicadores productivos según el tipo de galpón utilizado en la producción de carne avícola. 16pp.

Dottavio A, Di Masso R. 2010. Mejoramiento avícola para sistemas productivos semi-intensivos que preservan el bienestar animal. Universidad Nacional de Rosario. Facultad de Ciencias Veterinarias. Casilda, Argentina. Journal of Basic & Applied Genetics. 21 (2). 10pp.

Fanatico A. 2007. Sistemas Avícolas Alternativos con Acceso a Pastura. ATTRA. National Sustainable Agriculture Information Service. 28pp.

Franco-Jimenez, D.J and Beck, M.M. 2007. Physiological Changes to Transient Exposure to Heat Stress Observed in Layding Hens. Poult Sci, 86 (3): 538-544.

FAO, SAGARPA. 2007. Programa Especial para la Seguridad Alimentaria: Producción y Manejo de aves de traspatio. México. 32pp.

Felver-Gant, J., Mack, L., Dennis, R., Eicher, S., Cheng, H. 2012. Geneticvariations alter physiological responses following heat stress in 2 strains of laying hens. Poult Sci, 91(7):1542-1541.

Freeman BM. 1987. The stress syndrome. World's Poult. Sci. J. 43 (1): 15-19.

Giacomozzi J. 2014. Situación actual de la industria del huevo: Aves-huevos-producción-precios. Ministerio de Agricultura de Chile. Oficina de Estudios y Políticas Agrarias. ODEPA. 8pp.

Giacomozzi J. 2015. Actualización del mercado avícola: Aves-pollo-pavo-producción-exportaciones-comercio. Ministerio de Agricultura de Chile. Oficina de Estudios y Políticas Agrarias. ODEPA. 9pp.

Glatz P, Pym R, Nicol C. 2013. Revisión del desarrollo Avícola: Alojamiento y manejo de las aves de corral en países en desarrollo, Transporte y sacrificio de las aves de corral. University of Bristol. School of Veterinary Science. Bristol, Reino Unido. FAO. 26-30, 128-130. 136pp.

Gudev, D., Popova-Ralcheva, S., Yanchev, I., Moneva, P., Petkov, E., Ignatova, M. 2011. Effect of betaina on egg performance some blood constituents in laying hens reared indoor under natural summer temperaturas and varying levels or air ammonia. Bulgarian Journal of Agricultural Science, 17(6): 859-866.

Havlicek, Z and Salma, P. 2011. Effect of heat stress on biochemical parameteres of hens. Proceedings of ECOpole, 5 (1): 57-60 .

Hernández R, Solís S, Martínez C. 2014. Proyecto Estratégico de Seguridad Alimentaria: Modulo de producción de huevo y carne de aves de traspatio para zonas frías. Secretaria de Agricultura, Ganadería, Desarrollo Rural, Pesca y Alimentación. Organización de las Naciones Unidas para la Alimentación y Agricultura. FAO. México. 30pp.

Hughes, B.O. and Black, A.J. 1974. The effect of environmental factors on activity, selected bahaviour patterns and fear of fowls in cages and pens. British Poultry Science, 15 (4): 375-380.

Jerez-Salas MP, Carrillo- Rodriguez JC. 2009. Producción de huevo de gallinas Rhode Island Rojas bajo un sistema alternativo de traspatio. Rev Bras. De Agroecologia. 4(2): 656-659.

Jerez-Salas MP, Camacho MA, Quijano-Vicente G, Lozano-Trejo S, Sosa-Montes E, Ruíz-Luna J. 2014. Características del huevo de gallinas de traspatio alimentadas con una formulación alternativa con o sin verdolaga (*Portulaca oleracea* L.). Actas Iberoamericanas de Conservación Animal. 4 :158-160.

Kannan G, Terrill TH, Kouakou B, Gazal OS, Gelaye S, Amoah EA, Samaké S. 2000. Transportation of goats: effects on physiological stress responses and live weight loss.. Journal of animal science 78:1450-1457.

Keener, K., McAvoy, K., Foegeding, J., Curtis, P., Anderson, K., Osborne, J. 2006. Effect of testing temperatura on internal egg quality measurements. Poult Sci, 85 (3): 550-555.

Nicol CJ, Davies A. 2013. Bienestar de las aves de corral en los países en desarrollo. University of Bristol. School of Veterinary Science. Bristol, Reino Unido. FAO. 1pp.

Martin, G. 1989. Federpickhaufigkeit in Abhangigkeit von Draht und Einstreuboden sowie von der Lichtintensitat. Kuratorium fur Technik und Bauwesen in der Landwirtschaft Schrift, 342: 108-133.

Mashaly, K., Hendricks, G., Kalama, M., Gehad, A., Abbas, A., Patterson, P. 2010. Effect of heat stress on production parameters and immune responses of comercial laying hens. Poult Sci, 83(6): 889-894.

Mertens, K., Vaesen, I., Loffel, J., Kemps, B., Kamers, B., Perianu, C., Zoons, J., Darius, P., Decuypere, E., De Baerdemaeker, J., De Ketelaere, B. 2010. The transmission color value: A novel egg quality measure for recording Shell color used for monitoring the stress and health status of a Brown layer flock. Poult Sci, 89 (3): 609-617.

Maxwell, MH. 1993. Avian blood leucocyte responses to stress. World's Poultry Sci. J. 49 (1):34-43.

Medina, B. 2016. Estrés en aves y un nuevo enfoque para su mitigación. Los avicultores y su entorno. México.18 (111) 15-22pp.

Meluzzi, A, Giordani G. 1989. El bienestar de las aves en la explotación intensiva de las ponedoras: Alojamientos. Universidad Autónoma de Barcelona. Rivista di Avicoltura. 58 (5): 13-19.

Mohan, J. 2005. Physiology of stress in poultry. Central Avian Research Intitute, Izatnagar UP, 11.

Mohammed, KAF., Sarmiento-Franco, LA., Santos-Ricalde, R., Solorio-Sanchez, JF. 2013. Egg production, egg quality and crop content of Rhode Island Red hens grazing on natural tropical vegetation. Tropical animal health and production. 45(2): 367-372.

Molina P. 2013. Comparación de dos sistemas de producción y de manejo sanitario de las aves criollas de traspatio en los municipios de Ignacio de la Llave y Teocelo. Universidad Veracruzana. Facultad de Medicina Veterinaria y Zootecnia. Veracruz, México. 47pp.

Muñoz J, Vellojin J. 2002. Diseño y Evaluación de un Sistema de Producción de Huevos con Gallinas bajo Pastoreo en el Trópico Húmedo. Universidad EARTTH. Costa Rica. 5pp.

Rojas H, Stuardo L, Benavides D. 2005. Políticas y prácticas de bienestar animal en los países de América: estudio preliminar. Sci. Tech. Off. Int. Epiz. 24 (2): 549-565.

Romero M. 2013. Producción Avícola en Pequeña Escala. Secretaria de Agricultura, Ganadería, Desarrollo Rural Pesca y Alimentación (SAGARPA). Subsecretaría de Desarrollo Rural Dirección General de Apoyos para el Desarrollo Rural. Xochimilco, México. 8pp.

Ros J. 2008. Bienestar Animal en el Transporte. Consejería de Agricultura y Agua de la Región de Murcia. Centro Integrado de Formación y Experiencias Agrarias de Lorca. Pictografía. Murcia, España. 48pp.

Savory, C.J. y Duncan, I.J.H. 1982. Voluntary regulation of lighting by domestic fowls in Skinner boxes. Applied Animal Ethology, 9(1): 73-81.

Schwartzkopf-Genswein KS, Faucitano L, Dadgar S, Shand P, González LA, Crowe TG. 2012. Road transport of cattle, swine and poultry in North America and its impacto in animal welfare, carcass and meat quality: A review. Meat Science. 92 (3): 227-243

Segura-Correa J, Gutierrez-Vázquez E, Juárez-Caratachea A, Santos-Ricalde R. 2010. Calidad del huevo de gallinas criollas criadas en traspatio en Michoacan, México. Tropical and Subtropical Agroecosystems. 12: 109-115.

Selye H. 1951. Stress. The Physiology and Pathology of exposure to stress. Hans Selye Montreal, Canada: Acta Endoerinologica, 113(2938): 462-463.

Siegel HS. 1980. Physiological stress in birds. BioSci. 30 (8): 529-534.

Soler D. 2010. Importancia de los Sistemas Avícolas Campesinos (Pollo De Engorde Y Gallina Ponedora) dentro de la Unidad Productiva y su aporte a la Seguridad Alimentaria. Pontificia Universidad Javeriana. Facultad De Estudios Ambientales Y Rurales. Duitama, Boyacá. 138pp.

Steinfeld, H., and Chilonda, P. 2006. Perspectiva Mundial: Producción de Carne. 16pp.

Van Horne, PLM, and Achterbosch, TJ. 2008. Animal welfare in poultry production systems: impact of EU standards on world trade. World's Poultry Science Journal. 64 (1): 40-52.

Vizcaíno D, Betancourt R. 2013. Guía de Buenas Practicas Avícolas Resolución Técnica N° 0017. Ministerio de Agricultura, Ganadería, Acuacultura y Pesca. Agencia Ecuatoriana de Aseguramiento de la Calidad del Agro. 1ed. Ideaz. Ecuador. 56pp.

18

Black Oat (*Avena Strigosa*) Silage for Small-Scale Dairy Systems in the Highlands of Central Mexico

Maria Mitsi Nallely Becerril-Gil[1], Felipe López-Gonzalez[1*], Julieta Gertrudis Estrada-Flores[1], Carlos Manuel Arriaga-Jordán[1]

[1]*Instituto de Ciencias Agropecuarias y Rurales (ICAR), Universidad Autónoma del Estado de México. Campus UAEM El Cerrillo, El Cerrillo Piedras Blancas, Toluca, Estado de México. C.P. 50090. Email:flopezg@uaemex.mx*
Corresponding author:

SUMMARY

Black oat (*Avena strigosa* cv. Saia) silage (BOS) as an alternative forage for the dry season in small-scale dairy systems was evaluated against maize silage (MSL) at 6.0 kg DM/cow/day. Treatments were evaluated through on farm participatory livestock research: T1=100 BOS, T2=66:34 BOS:MSL, T3=34:66 BOS:MSL, and T4=100 MSL fed to milking dairy cows that also received 4.5 kg DM/cow/day of a commercial compound dairy concentrate and 2.2 kg DM/cow/day of cut-and-carry pasture. Eight Holstein cows were allotted to a replicated 4X4 Latin Square design, with 14 day experimental periods. Daily milk yields and milk composition were measured during the last four days, and live weight and body condition score recorded on the last day of each period. Feeding costs were determined by partial budget analysis. There were no differences in milk yield (15.9±0.26 kg/cow/day), or milk composition with mean values for milkfat of 38.8±0.86 g/kg, milk protein 32.2±0.38 g/kg, and lactose 46.3±0.22 g/kg. There were also no differences in milk urea nitrogen (MUN) with a mean of 11.8±0.83 mg/dl, live weight 385.6±1.67 kg, or body condition score with a mean of 2.6±0.01. Feeding costs per kg milk were 33% higher in T1 and T2 than T4, with intermediate feeding costs in T3 (T1=0.88, T2= 0.85, T3= 0.74, T4= 0.66 R$/kg). Profit margins and income/feeding costs were all positive. Black oat silage may be an alternative forage in small-scale dairy systems in the dry season when maize silage cannot be cultivated or fails due to climate concerns.

Key words: alternative forages; cut-and-carry pasture; feeding costs; maize silage; participatory livestock research

INTRODUCTION

Small-scale dairy systems (SSDS) are a development option to alleviate poverty and enhance food production in developing countries (FAO, 2010). In Mexico, SSDS represent over 78% of specialised dairy farms, and produce 37% of the national milk supply (Hemme *et al.*, 2007). Dairy production is the primary economical activity of small farms in southwestern Paraná, Brazil (Pin *et al.*, 2011).

SSDS in Mexico are heterogeneous in both technological and agro-ecological terms, so that there is an ample variation on the productivity of each farm (Camacho-Vera *et al.*, 2017). In the central highlands of Mexico, small-scale dairy farms with access to some irrigation base the feeding strategies of their herds on small areas sown to temperate ryegrass/white clover cut-and-carry cultivated pastures (Fadul-Pacheco *et al.*, 2013), similar to SSDS in southeast Asia (Moran, 2005). This herbage is a high quality component of diets (Martínez-García *et al.*, 2015).

There is feed scarcity in the dry season since pastures reduce growth due to restricted irrigation. Limitations on the availability of water for irrigation may be worsened by possible effects of climate change due to alterations in the rainfall regimes (Victor *et al.*, 2014). Traditionally, small-scale dairy farmers complement their milking cows with straws (mainly maize stover), concentrates and maize grain which result in high feeding costs, so that conserved good quality forage improves performance and profitability of farms (Martínez-García *et al.*, 2015). Maize silage has been proven as a source of high quality forage for the dry season (Jaimez-García *et al.*, 2017).

However, in the face of climate change with erratic or less rainfall, there must be a diversification of forage crops, with short agricultural cycles and adaptable to adverse conditions (Thornton *et al.*, 2009). Black oat (*Avena strigosa*) is a short cycle small-grain cereal tolerant to drought conditions, poor soils, and has good quality for feeding cattle (Dial, 2014). Conserved as silage it may be an option for the dry season in small-scale dairy systems in temperate areas.
The objective was to evaluate the productive and economic effect of including black oat silage (BOS) in the feeding strategy of lactating dairy cows in SSDS, alone or mixed with maize silage, complemented with fresh cut-and-carry herbage and concentrate.

MATERIALS AND METHODS

The work took place in the municipality of Aculco, in the State of Mexico (that surrounds Mexico City) in Mexico, located between 20° 00' and 20° 17' North, and between 99° 40' y 100° 00' West, at an altitude of 2440 m. Climate is sub-humid temperate with mean temperatures between 10 and 18°C, and 700 to 1000 mm annual rainfall. The experiment took place from 13 March to 24 April 2016, during the dry season.

A hybrid dual purpose (grain and forage) maize variety was sown for silage, and managed according to the usual farmers' practice. Sowing date was 30 April 2015 with 25 kg/ha of seed (to achieve between 70,000 and 80,000 plants/ha), and harvested 151 days after sowing on 30 September 2015. The crop was fertilised with $130\ N - 90\ P_2O_5$ -60 K_2O kg/ha.

Black oat (*Avena* strigose) of the Saia variety was sown on 3 October 2015 with 120 kg seed/ha, fertilised with $82\ N - 46\ P_2O_5$ - 0 K_2O kg/ha, and harvested at 95 days after sowing on 8 January 2016.

The cut-and-carry pasture was five years old sown to annual and perennial ryegrass (*Lolium multiflorum* cv. Maximus and *L. perenne* cv. Bargala) at 35 kg grass seed/ha, and white clover (*Trifolium repens* cv. Ladino) at 3.0 kg seed/ha. The participating farmer utilises the pasture under cut-and-carry since it is far from the pen where he keeps his cows (next to the family house).

Eight multiparous Holstein cows with a mean initial live weight of 363 ± 19 kg and daily milk yield of 13.0 ± 1.2 kg cow/ day and 103 ± 60 days in lactation were selected for the experiment from the farmers' small herd. Cows were kept in confinement on an open pen half of which had a concrete floor. The rest was unpaved. Cows were milked twice daily (7:30 and 17:00 h) in a small milking shed within the same pen with a portable milking machine.

Milk yield was weighed with a spring balance, and composition determined with an ultrasound milk analyser, on the last four days of each experimental period, and a composite sample of the four days kept refrigerated to determine Milk Urea Nitrogen (MUN) following procedures described by Aguerre (2007).

Live weight and body condition score (1 – 5 scale) (Wildman *et al.*, 1982) were recorded on the last day of each period, using an electronic portable weighbridge for live weight.

Cows received 6.0 kg of silage, and the inclusion black oat silage (BOS) and maize silage (MSL) in the feeding strategy was evaluated in four treatments: T1=100 BOS, T2= 66:34 BOS: MSL, T3= 34:66 BOS:MSL, and T4= 100 MSL.

All cows also received 4.5 kg DM/cow/day of a commercial compound dairy concentrate and 2.2 kg DM/cow/day of cut-and-carry pasture.

The experiment took place on-farm following a participatory livestock research approach (Conroy, 2004).

Experimental design was a 4X4 replicated Latin Square. Cows were allotted to two groups of four (squares) taking into consideration days in milk and mean daily milk yield before the experiment. Treatment sequence in the first square was randomised and assigned as mirror image in the second square to minimise carry-over effects (Celis-Alvarez *et al.* 2016), and cows randomly allotted to treatment sequence.

Experimental periods were 14 days, with 10 days for adaptation to diets and four days for measurements and sampling following Pérez-Ramírez *et al.* (2012).

The analysis of variance model for the statistical analysis was (Kaps and Lamberson, 2004):
$Y_{ijkl} = \mu + S_i + C_{j(i)} + P_k + t_l + e_{ijkl}$,

Where: μ = General mean; S = effect due to squares. i = 1, 2; C = effect due to cows within squares j = 1, …, 4; P = effect due to experimental periods k = 1, …, 4; t = effect due to treatments. L = 1, … 4; and e = residual error term.

Tukey's test was applied if significant differences (P≤0.05) were found. Statistical procedures were performed using Minitab (version 14).

Herbage mass was estimated from four 0.64 m^2 quadrants cut to ground level by hand with shears every 14 days. Samples were taken on each experimental period for botanical composition: grasses, clover and other plants.

Particle size of silages (BOS and MSL) was determined with the Penn State Forage Particle Separator, following the methodology described by Heinrichs and Kononoff (2002), as ancillary for estimating silage conditions.

Chemical analyses of silages, herbage and concentrate samples followed established procedures (Anaya-Ortega et al. 2009) for dry matter (DM), organic matter (OM), crude protein (CP), neutral detergent fibre (NDF), and acid detergent fibre (ADF).

In vitro digestibility of dry matter (IVDDM), organic matter (IVDOM), and NDF were determined by the *in vitro* gas production technique (Theodorou *et al.* 1994). Estimated metabolizable energy (eME) was calculated from the equation (AFRC, 1993): eME (MJ/kg DM) = 0.0157 IVDOM (g/kg DM). Silages were analysed for pH both in fresh silage and extracted juice, as well as starch content in silages with a commercially available kit (Megazyme® product code K-TSTA-100A).

Digestibility of forages was estimated by the *in vitro* gas production technique (Menke and Steingass, 1988; Theodorou *et al.*, 1994). Four 160 ml glass bottles with 0.99 ± 0.01 g of each forage and concentrate were added with 90 ml of buffer solution and 10 ml of rumen liquor, and incubated at 39°C. Gas pressure recordings with a pressure transductor were for 120 h (at 1, 2, 3, 4, 5, 6, 7, 8, 12, 16, 20, 28, 36, 44, 52, 60, 72, 84, 96, and 120 h after incubation). The following variables were determined after 120 hours of incubation: *in vitro* digestibility of DM (DMIVD), *in vitro* digestibility of organic matter (OMIVD), and *in vitro* digestibility of neutral detergent fibre (NDFIVD) (Aragadvay-Yungán *et al.*, 2015).

The *in vitro* fermentation parameters were calculated using the Jessop and Herrero (1996) equation:
$GP = A \times (1-\exp(-c_A \times t)) + B \times (1-\exp(-c_B \times (t - lag))) \times (t > lag) \times t^{-1}$

Where:
A = Gas Production in 4.0 h (ml); B = Potential gas Production, c_A = rate of gas production of fraction A (hour); c_B = rate of gas production of fraction B (hour); lag = time before fermentation of the NDF fraction begins (h); t = time of incubation.

The economic analyses was by partial budget analysis for each treatment as has been done in other work (Celis-Alvarez *et al.* 2016). Only feeding costs and income from milk sales were included (Moran and Brower, 2014), to obtain margins over feed costs per cow and per kg of milk produced.

RESULTS

There were no significant differences (P>0.05) for any animal variable in the different treatments, but there were significant differences between periods (P<0.05). In the different treatments mean milk yield was 15.9 kg/cow/day, and mean values for milkfat was 38.8 g/kg, milk protein 32.2 g/kg, and lactose 46.3 g/kg, and MUN was 11.8 mg/dl. Mean live weight was 385.6 kg, and mean body condition score was 2.6. There are significant differences between periods (P<0.05) in milk yield being higher in period 2 and 3, MUN finding the lowest value in period 1 and the highest value in period 2, and live weight being higher in period 2 and 3 (Table 1).

Table 2 shows results for pasture variables. Mean herbage mass available per day was 96.99 kg DM/ha/day. In terms of botanical composition, the pasture had a mean composition of 46.6% grass and 51.7% white clover, with only 1.6% of other plants. Clover proportion diminished in Period 3, following a fall in Period 2 for pasture height, herbage mass and milk yield, mainly due to delays in the availability of irrigation. *In vitro* digestibilities (DMIVD, OMIVD, and NDFIVD) remained constant during Periods 1 and 2, falling in Period 3 and further down in Period 4. Table 3 shows results for the analysis of particle size. MSL had a particle size smaller than 19 mm but larger than 8 mm, representing the highest proportion in the mid. Most MSL (46.5 %) was retained in the mid sieve so that particle size was above 8 mm but under 19 mm. BOS had a smaller particle size since 31.3 % was retained in the mid sieve, and 38.8 % in the lower sieve, meaning a particle size smaller than 19 mm but larger than 1.7 mm.DM content was similar in both silages, but CP was 27 % higher in BOS than MSL;

and MSL had higher NDF and ADF than BOS. Starch content was similar in both silages. IVDDM and IVDOM were higher in BOS, although IVDNDF was higher in MSL. Fermentation was good in both silages, with pH slightly higher in BOS than in MSL. BOS had an eME similar to that of the CCP, although CP was 15 % higher in CCP (Table 4).

CDC had significantly higher A fraction, followed by CCP, BOS, and the lower A fraction was in MSL (P<0.05); although the highest rate of fermentation for the A fraction (C_A/h) was in MSL that was significantly different (P<0.05) from the other feeds that were not different among them (P>0.05).

Table 1. Animal performance of cow fed diets containing black oat silage or maize silage

	Treatments						
	T1	T2	T3	T4	Mean	SEM	P-value
Milk yield (kg cow⁻¹day⁻¹)	16.55	14.98	16.06	16.03	15.90	0.26	0.054
Milk fat (g kg⁻¹)	38.45	40.35	37.42	39.10	38.80	0.86	0.722
Milk protein (g kg⁻¹)	32.38	32.45	31.98	32.05	32.21	0.38	0.928
Lactose (g kg⁻¹)	46.60	46.14	46.06	46.51	46.30	0.22	0.258
MUN (g dL⁻¹)	11.87	11.70	12.09	11.28	11.80	0.83	0.844
LW (kg)	391.40	392.60	375.60	382.60	385.60	1.67	0.111
BCS (1-5)	2.6	2.6	2.6	2.6	2.6	0.00	0.897
	Periods						
	P1	P2	P3	P4	Mean	SEM	P-value
Milk yield (kg cow⁻¹day⁻¹)	15.03ᵃ	16.42ᵇ	16.45ᵇ	15.71ᵃ	15.90	0.26*	0.005
Milk fat (g kg⁻¹)	38.38	38.46	39.52	38.71	38.76	0.86NS	0.772
Milk protein (g kg⁻¹)	32.33	31.91	31.76	33.01	32.25	0.38NS	0.353
Lactose (g kg⁻¹)	46.88	46.20	46.01	46.20	46.32	0.22NS	0.052
MUN (g dL⁻¹)	7.31ᶜ	16.50ᵃ	11.74ᵇ	11.28ᵇ	11.70	0.83**	0.000
LW (kg)	381.10ᵇ	388.40ᵃ	389.00ᵃ	383.50ᵇ	385.50	1.67	0.049
BCS (1-5)	2.9	2.9	2.9	2.9	2.9	0.00	0.897

ᴺˢ P >0.05, * P <0.05, MUN= Milk urea nitrogen, LW= Live weight, BCS= Body condition score. T1=100 BOS (black oat silage), T2= 66:34 BOS: MSL (maize silage), T3= 34:66 BOS:MSL, and T4= 100 MSL

Table 2: Herbage availability, botanical composition and in vitro digestibility.

Period	Herbage Mass (kg DM/ha/day)	Botanical composition (%)				In vitro digestibility (g/kg DM)		
		Grass	Clover	Weeds		IVDM	IVDOM	IVDNDF
1	113.22	43.33	52.86	3.81		810.89ᵃ	805.44ᵃ	750.39ᵃ
2	78.56	43.09	56.91	0		818.87ᵃ	804.00ᵃ	748.39ᵃ
3	93.40	59.24	37.72	3.03		749.66ᵃᵇ	743.80ᵃᵇ	696.14ᵃᵇ
4	102.79	40.81	59.19	0		719.42ᵇ	713.96ᵇ	639.87ᵇ
Mean	96.99	46.61	51.67	1.61	SEM	2.03	2.06	2.14
					P-value	0.0078	0.0090	0.0588

ᵃᵇᶜ in columns (P<0.05)

Table 3: Particle size in black oat silage (BOS) and maize silage (MSL).

Particle size	Top sieve (19 mm)	Mid sieve (8 mm)	Lower sieve (1.67 mm)	Bottom tray
BOS	12.47 %	31.34 %	38.77 %	17.43 %
MSL	17.98 %	46.52 %	23.42 %	12.09 %

Table 4: Chemical composition, estimated Metabolizable Energy, and silage pH.

	BOS	MSL	CCP	CDC
DM	384.03	360.40	260.07	911.25
MO (g/kg DM)	989.65	994.62	987.77	992.65
CP (g/kg DM)	106.94	78.04	123.05	193.13
NDF (g/kg DM)	494.10	591.89	386.56	265.40
ADF (g/kg DM)	251.66	332.01	220.40	91.29
IVDDM (g/kg DM)	768.46	713.51	760.73	885.48
IVDOM (g/kg DM)	763.12	708.17	755.16	880.03
IVDNDF (g/kg DM)	651.34	711.72	695.46	710.56
eEM (MJ/kg DM)	11.98	11.12	11.85	13.81
Starch (g/kg DM)	291.15	297.89	-	-
pH	4.15	3.65	-	-

BOS= Black oat silage, MSL= Maize silage, CCP= Cut-and-carry pasture, CDC= Commercial dairy concentrate, DM = Dry Matter, OM= Organic Matter, CP= Crude Protein, NDF= Neutral Detergent Fibre, ADF= Acid Detergent Fibre, IVDDM= *In vitro* digestibility of DM, IVDOM= *In vitro* digestibility of OM, IVDNDF= *In vitro* digestibility of NDF, eEM= Estimated Metabolizable Energy.

There were no differences (P>0.05) between BOS and MSL in fraction B, which were significantly different (P<0.05) from CDC, which in turn was higher (P<0.05) than CCP which showed the lowest B fraction. The rate of B fermentation was slowest in the CCP (P<0.05) with no differences between silages and CDC (P>0.05).

BOS had the highest lag time to initiate fermentation of the B fraction significantly different (P<0.05) than MSL. The smallest lag time was for CDC, with an intermediate lag time in CCP between MSL and CDC (P>0.05).

Table 6 shows results for the partial budget analysis. There were lower feeding costs with T4 MSL (0.66 R$ kg^{-1}milk) compared to T1 BOS (0.88 R$/kg milk). Selling price was 1.05 R$/kg milk, so that all treatments had positive margins over feed costs, lower forage yields represented higher costs for BOS, so that T4 has margins over feed costs 33 % higher than T1. There are no differences between T1 and T2, and T3 is intermediate. Cost per kg DM for each feed per kg/DM were: BOS R$ 0.25, MSL R$ 0.20, CCP R$ 0.06, and CDC R$ 0.78.

Table 5: In vitro gas production parameters of Black oat silage, Maize silage, Cut-and-carry pasture and Commercial dairy concentrate.

Feed	A (ml gas g^{-1} DM)	C_A (h)	B (ml gas g^{-1} DM)	C_B (h)	Lag time (h)
BOS	49.71[c]	0.23[b]	202.55[a]	0.047[a]	3.48[a]
MSL	27.33[d]	0.59[a]	259.56[a]	0.042[b]	2.84[b]
CCP	69.75[b]	0.25[b]	182.87[b]	0.050[a]	2.64[bc]
CDC	96.34[a]	0.21[b]	195.09[c]	0.045	2.23[c]
SEM	0.87	0.09	0.91	0.02	0.14
P-value	<0.0001	0.0003	<0.0001	0.0212	0.0002

BOS= Black oat silage, MSL= Maize silage, CCP= Cut-and-carry pasture, CDC= Commercial dairy concentrate
NS P >0.05, * P <0.05, [a,b,c] in colums P<0.05.

Table 6: Partial budget analysis (R$).

	T1	T2	T3	T4
Cost of ration (R$/kg)	0.88	0.85	0.74	0.66
Feeding cost per cow (R$/cow)	10.32	10.32	8.96	7.70
Feeding cost / kg milk (R$/milk)	0.64	0.64	0.56	0.48
Selling price of milk (R$/kg)	1.05	1.05	1.05	1.05
Margin over feed costs for milk (R$/kg)	0.40	0.40	0.48	0.56
Margin over feed costs per cow (R$/cow)	6.93	5.36	7.87	9.03

R$= ???

DISCUSSION

Milk yields increased 22 % compared to milk yields before the commencement of the experiment, due to better feeding of the cows brought about by the evaluated treatments, with no differences among the four treatments (P>0.05). These results are lower than studies evaluating MSL as the only forage source for dairy cows in these systems (Jaimez-García et al., 2017), although it must be noted the small size of cows in the experiment herein reported. Milk yields are lower than reports for optimised feeding strategies based on quality forages for small-scale dairy farms, but higher that the yields obtained from traditional feeding strategies in those studies (Velarde-Guillén et al., 2017). Milk yields were also lower than reports from work in Vietnam with black oats Salgado et al. (2013), but in the experiment herein reported concentrates represented a lower proportion of the diet. As mentioned before, cows in this experiment are small, with live weight lower than the 435 kg reported by Celis-Alvarez et al. (2016) and 520 reported by Pincay-Figueroa et al. (2016). BCS was higher than those and other reports (Jaimez-García et al., (2017).

Milk composition was above minimum requirements established in Mexican standards for raw milk. Milk fat content was higher than reports by Garduño-Castro et al. (2009) and Celis-Alvarez et al. (2016), both evaluating common oat (Avena sativa) silages for grazing dairy cows in SSDS; but lower than reports by Jaimez-García et al. (2017) when MSL complemented grazing dairy cows in SSDS. Higher milk fat content in the experiment herein reported may have been due to the high forage and therefore component of the diet (Gabbi et al., 2013), that was 65:35 forage and concentrate ratio. Protein content was lower than reports by Jaimez-García et al. (2017) both when MSL was the only source of forage for milking dairy cows, as when MSL was a complement to grazing.

Milk urea nitrogen is an indicator of protein nutrition and the balance in energy and protein in the diet of milking dairy cows (Wattiaux et al., 2005). Mean MUN was 11.7 mg/dl, indicating adequate protein provision in the diet, within the range between 10 to 16 mg/dl reported as normal values by Wattiaux et al. (2005), but are lower to the 22.7 mg/dl reported by Stanislao-Atzori et al. (2009).

During the dry season, limited irrigation available to farmers limits the growth and productivity of pastures. Cut-and-carry pasture in this experiment represented 17 % of the diet, higher than the 7 % reported for these systems by Velarde-Guillén et al. (2017) in the dry season.

In terms of botanical composition, the cut-and-carry pasture had a high proportion of clover, nearly the same as the proportion of grass, favoured by the intermittent cutting that favours clover. The high clover proportion had a positive effect on the protein content of herbage and on the in vitro digestibility. As time progressed, herbage mass decreased with more mature herbage and therefore reduced IVDDM (Furusawa et al., 2013).

Heinrichs and Kononoff (2002) recommended that between 45 and 65 % of MSL remains in the mid sieve in the Penn State Box system to assess particle size in silage; and 40 % in the lower sieve, as an indirect indicator of good forage compaction. Larger particle sizes do not allow good compaction, and therefore hamper good fermentation patterns in silage. Both BOS and MSL in this experiment met the recommended proportions of particle size, so that both silages were adequately compacted, reflected in the good quality of the obtained silages.

DM content of BOS was that of grain in the milky stage with together with the CP content, which it influences the results of crude protein and are comparable to reports by (Sánchez-Gutiérrez et al., 2014, David et al., 2010).

NDF in BOS was lower than reports of black oat forage by Salgado et al. (2013) in Vietnam, David et al. (2010) in Brasil, and Sánchez-Gutiérrez et al. (2014) in north central Mexico, although ADF is comparable to reports by Salgado et al. (2013). IVDDM of BOS was higher than the digestibility reported by David et al. (2010). The nutritional quality of BOS was high.

In regards to MSL, it had higher DM content than repots by Khan et al., (2015); but lower in NDF and ADF but higher in CP than reports by Martínez-Fernández et al. (2014). Digestibility parameters (IVDDM, IVDOM, and IVDNDF) for MSL were higher than reported by Corral-Luna et al. (2011) and Aragadvay-Yungán et al. (2015); and starch content was similar to reports by Martínez-Fernández et al. (2014).

Interestingly, starch content was similar between BOS and MSL, and both had high digestibility values, such that eME was as high in BOS as in MSL and CCP. The content of starch in maize silage is becoming increasingly important, the starch of MSL is slowly degraded in the rumen, the non-degraded fraction of the starch is highly digested in the small intestine, the glucose and disaccharides are available for energy supply and can be converted into lactose and milk protein production (Martínez-Fernández et al., 2014).

Successful conservation of silages requires an ample supply of soluble carbohydrates for fermentation. Good fermentation is indicated by the pH of silages (Martínez-Fernández et al., 2013). BOS had a pH of

4.15 at a phenological stage between flowering and milky grain. David *et al.* (2010) reported pH values for black oat silage between 3.7 and 4.7. MSL had a pH of 3.69, within values of 3.5-4.4 reported by Khan *et al.* (2015) (pH) for well-preserved maize silage.

The *in vitro* gas production techniques enables the knowledge of ruminal kinetics, where the determination of the fermentation patterns of the carbohydrate fractions enable the correct estimation of the energy available in feeds (Calabrò *et al.,* 2003).

Fraction A of BOS had a high quantity of rapidly available carbohydrates that ferment into volatile fatty acids realising ATP as energy supply for microbial growth (Jessop and Herrero, 1998); although fermentation rate in MSL was much higher than in BOS, CCP or even CDC, indicating a high availability of rapidly fermented carbohydrates.

Contents of the B fraction in MSL was higher than reported by Aragadvay-Yungán *et al.* (2015) related to the higher NDF content. However, the fastest rate of fermentation of the B fraction was in CCP, followed by BOS. MSL and the concentrate had similar rates; indicating the high digestibility of NDF in CCP and BOS.
Lag time in BOS was lower than reported by David *et al.* (2010), and lag time for MSL was higher than reported by Aragadvay-Yungán *et al.* (2015).

The substrates of high degradability, low gas production, has the highest DM intake, higher efficiency in the synthesis of microbial protein. The voluntary intake is correlated to the characteristics of ruminal fermentation, especially in the NDF (Castro-Hernández *et al.* 2017).

Feeding costs are associated with the ratio of forage to concentrate, the quality of forages, and the dependency on external inputs that have an effect on the economic performance of farms (Casasnovas-Oliva and Aldanondo-Ochoa, 2014; Moran and Brower, 2014; Cortez-Arriola *et al.* 2016).

The forage to concentrate ratio in this work was 65:35, with a lower proportion of concentrate than reports by Salgado *et al.* (2013) in Vietnam with a 55:45 ratio; although these authors mentioned that farmers traditionally use more concentrates up to 40:60 ratios or up to 30:70 forage to concentrate ratios. In Malaysia Moran and Brower (2013) reported a 48:52 ratio of forage to concentrate in small-scale dairy farms.

Milk price paid to farmers in the experiment herein reported (1.05 R$/kg) was higher than reported by Garduño-Castro et al. (2009) at 0.83 R$/kg and Reiber et al. (2010) at 0.90 R$/kg in Honduras. Profit margins

in all treatments were positive, but BOS represented higher feeding costs, and lower margins.

CONCLUSIONS

Black oat silage was a quality forage that may be included in the feeding strategies of milking dairy cows alone or mixed with maize silage during the dry season. Due to its lower yields than maize, which represent 33% higher costs, it can be used when the maize crop cannot be cultivated or fails due to climate concerns; or as a complement since its frost resistance enables its growth in winter after the maize crop has been harvested, if irrigation is available.

REFERENCES

AFRC Animal and Food Research Council. 1993. Energy and Protein Requirements for Ruminants. An advisory manual prepared by the AFRC Technical Committee on response to nutrients. CAB International. Wallinford, UK, 159 p.

Aguerre, M. 2007. Determination of milk urea nitrogen. UW-Madison, Dairy Science Department.

Anaya-Ortega, J.P., Garduño-Castro, G., Espinoza-Ortega, A., Rojo-Rubio, R., Arriaga-Jordán, C. M. 2009. Silage from maize (*Zea mays*), annual ryegrass (*Lolium multiflorum*) or their mixture in the dry season feeding of grazing dairy cows in small-scale campesino dairy production systems in the Highlands of Mexico. Tropical Animal Health and Production. 41: 607–616. doi:10.1007/s11250-008-9231-5.

Aragadvay-Yungán, R.G., Rayas-Amor, A.A., Martínez-Castañeda, F.E. Arriaga-Jordán, C.M. 2015. In vitro evaluation of sunflower (*Helianthus annuus* L.) silage alone or combined with maize silage. Revista Mexicana de Ciencias Pecuarias. 6: 315-327. doi:10.22319/rmcp.v6i3.4094.

Calabrò, S., Zicarelli, F., Infascelli, F., Piccolo, V. 2003. Kinetics fermentation and gas production of the neutral detergent–soluble fraction of fresh forage, silage and hay of Avena sativa. Italian Journal of Animal Science. 2: 201-203. doi:10.1002/jsfa.2186.

Camacho-Vera, J.H., Cervantes-Escoto, F., Palacios-Rangel, M.I., Rosales-Noriega, F., Vargas-Canales, J.M. 2017. Factores determinantes del rendimiento en unidades de producción de lechería familiar. Revista Mexicana de Ciencias Pecuarias. 8: 23-29. doi: 10.22319/rmcp.v8i1.4313.

Casasnovas-Oliva, V., Aldanondo-Ochoa, A.M. 2014. Feed prices and production costs on Spanish dairy farms. Spanish Journal of Agricultural Research. 12: 291-30. doi:10.5424/sjar/2014122-4890.

Castro-Hernández, H., Domínguez-Vara, I.A., Morales-Almaráz, E., Huerta-Bravo, M. 2017. Composición química, contenido mineral y digestibilidad in vitro de raigrás (*Lolium perenne*) según intervalo de corte y época de crecimiento. Revista Mexicana de Ciencias Pecuarias. 8: 201-210. doi:10.22319/rmcp.v8i2.4445.

Celis-Álvarez, M.D., López-González, F., Martínez-García, C.G., Estrada-Flores, J.G., Arriaga-Jordán, C.M. 2016. Oat and ryegrass silage for small-scale dairy systems in the highlands of central Mexico. Tropical Animal Health and Production. 48: 1129-1134. doi:10.1007/s11250-016-1063-0.

Conroy, C. 2004. Participatory Livestock Research. ITDG Publishing, Bourtonon-Dunsmore, Warwickshire, U.K. doi:10.3362/9781780440316.000.

Corral-Luna, A., Domínguez-Díaz, D., Rodríguez-Almeida, F.A., Villalobos-Villalobos, G., Ortega-Gutiérrez, J.A., Muro-Reyes, A. 2011. Composición química y cinética de degradabilidad de ensilaje de maíz convencional y sorgo de nervadura café. Revista Brasileña de Ciencias Agrícolas. 6: 181-187. doi:10.5039/agraria.v6i1a973.

Cortez-Arriola, J.G., Jeroen, C.J., Rossing, W.A.H., Scholberg, J.M.S., Améndola-Massiotti, R.D., Tittonell, P. 2016. Alternative options for sustainable intensification of smallholder dairy farms in North-West Michoacán, Mexico. Agricultural Systems. 144: 22-32.

David, D.B.D., Nörnberg, J.L., Azevedo, E.B.D., Brüning, G., Kessler, J.D., Skonieski, F.R. 2010. Nutritional value of black and white oat cultivars ensiled in two phenological stages. Revista Brasileira de Zootecnia. 39: 1409-1417. doi:10.1016/j.agsy.2016.02.001.

Dial, H.L. 2014. Plant guide for black oat (*Avena strigosa* Schreb.) USDA-Natural Resources Conservation Service, Tucson Plant Materials Center, Tucson, AZ.

Fadul-Pacheco, L., Wattiaux, M.A., Espinoza-Ortega, A., Sánchez-Vera, E., Arriaga-Jordán, C.M. 2013. Evaluation of sustainability of smallholder dairy production systems in the highlands of Mexico during the rainy season. Agroecology and Sustainable Food Systems.

37: 882-901. doi:10.1080/21683565.2013.775990.

FAO. 2010. Status of and Prospects for Smallholder Milk Production – A Global Perspective; Hemme T y Otte J. Rome.

Flaten, O. 2002. Alternative rates of structural change in Norwegian dairy farming: impacts on costs of production and rural employment. Journal of Rural Studies. 18: 429-441. doi:10.1016/s0743-0167(02)00031-1.

Furusawa, S., Yoshihara, Y., Sato, S. 2013. Plant diversity, productivity and nutritive value change following abandonment of public pastures in Japan. Grassland Science. 59: 59-62. doi:10.1111/grs.12012.

Gabbi, A.M., McManus, C.M., Silva, A.V., Marques, L.T., Zanela, M.B., Stumpf, M.P., Fisher, V. 2013. Typology and physical–chemical characterization of bovine milk produced with different productions strategies. Agricultural Systems. 121: 130-134. doi:10.1016/j.agsy.2013.07.004.

Garduño-Castro, Y., Espinoza-Ortega, A., González-Esquivel, C.E., Mateo-Salazar, B., Arriaga-Jordán, C.M. 2009. Intercropped oats (*Avena sativa*) - common vetch (*Vicia sativa*) silage in the dry season for small-scale dairy systems in the Highland of Central Mexico. Tropical Animal Health and Production. 41: 827-834. doi:10.1007/s11250-008-9258-7.

Heinrichs J., Kononoff P. 2002. Evaluating particle size of forages and TMRs using the New Penn State Forage Particle Separator. Department of Animal and Dairy Science, Pennsylvania State University, Pennsylvania.

Hemme, T. 2007. IFCN Dairy Research Center. International Farm Comparison Network, Kiel, Germany.

INRA, 2010. Alimentación de bovinos, ovinos y caprinos: Necesidades de los animales y valores de los alimentos. Institut National de la Recherche Agronomique, Francia.

Jaimez-García, A.S., Heredia-Nava, D., Estrada-Flores, J.G., Vicente, F., Martínez-Fernández, A., López-González, F., Arriaga-Jordán, C. M. 2017. Maize silage as sole forage source for dairy cows in small-scale systems in the highlands of central Mexico. Indian Journal of Animal Science. 87: 752–756. doi:10.1007/s11250-016-1063-0.

Jessop, N. S., Herrero, M. 1996. Influence of soluble components on parameter estimation using the in vitro gas production technique. Journal of Animal Science. 62: 626-627.

Jessop, N. S., Herrero, M. 1998. Modelling fermentation in an in vitro gas production system: effects of microbial activity. In: In vitro techniques for measuring nutrient supply to ruminants. Deaville, E.R., Owen, E., Adesogan, A.T., Rymer, C., Huntington, J.A. and Lawrence, T.L.J. (eds). British Society of Animal Science, Edimburg. pp 81-84.

Kaps, M., Lamberson, W. 2004. Change-over designs with the effects of periods. Latin square. pp 301-305. In: Biostatistics for animal science. M. Kaps, W. Lamberson. CABI Publishing CAB International, Wallingford Oxfordshire UK. doi:10.1079/9780851998206.0294.

Khan, N.A., Yu, P., Ali, M., Cone, J.W., Hendriks, W.H. 2015. Nutritive value of maize silage in relation to dairy cow performance and milk quality. Journal of the Science of Food and Agriculture. 95: 238-252. doi:10.1002/jsfa.6703.

Martínez-Fernández, A., Soldado, A., De la Roza-Delgado, B., Vicente, F., González-Arrojo, M.A., Argamenteria, A. 2013. Modelling a quantitative ensilability index adapted to forages from wet temperate areas. Spanish Journal of Agricultural Research. 11: 455-462. doi:10.5424/sjar/2013112-3219.

Martínez-Fernández, A., Argamentería, G.A., De la Roza, D.B. 2014. Manejo de forrajes para ensilar. Servicio Regional de Investigación y Desarrollo Agroalimentaria, España. 280 pp.

Martínez-García, C.G., Rayas-Amor, A.A., Anaya-Ortega, J.P., Martínez-Castañeda F.E., Espinoza-Ortega, A., Prospero-Bernal, F., Arriaga-Jordán, C. M. 2015. Performance of small-scale dairy farms in the highlands of central Mexico during the dry season under traditional feeding strategies. Tropical Animal Health and Production. 47: 331-337. doi:10.1007/s11250-014-0724-0.

Menke, K.H., Steingass, H. 1988. Estimation of the energetic feed value obtained from chemical analyses and in vitro gas production using rumen fluid. Animal Research and Development, 28: 7-55.

Moran, J.B. 2005. Tropical dairy farming. Feeding management for small holder dairy farmers in the humid tropics. CSIRO Publishing. Australia.

Moran, J.B., Brouwer, J.W. 2013. Interrelationships between measures of cow and herd performance and farm profitability on Malaysian dairy farms. International Journal of Agriculture and Biosciences. 2: 221-233. www.ijagbio.com.

Moran, J.B., Brouwer, J.W. 2014. Quantifying the returns to investing in improved feeding management on dairy farms in Peninsular Malaysia. Animal Production Science. 54: 1354-1357. doi:10.1071/an14076.

Pérez-Ramírez, E., Peyraud, J.L., Delagarde, R. 2012. N-alkanes v. ytterbium/faecal index as two methods for estimating herbage intake of dairy cows fed on diets differing in the herbage: maize silage ratio and feeding level. Animal. 6: 232-244. doi:10.1017/s1751731111001480.

Pin, E.A., Soares, A.B., Possenti, J.C., Ferrazza, J.M. 2011. Forage production dynamics of winter annual grasses sown on different dates. Revista Brasileira de Zootecnia. 40: 509-517. doi:10.1590/S1516-35982011000300007.

Pincay-Figueroa, P.E., López-González, F., Velarde-Guillén, J., Heredia-Nava, D., Martínez-Castañeda, F.E., Vicente, F., Martínez-Fernández, A., Arriaga-Jordán, C.M. 2016. Cut and carry vs. grazing of cultivated pastures in smallscale dairy systems in the central highlands of Mexico. Journal of Agriculture and Enviromental for International Development. 110: 349-363. doi: 10.12895/jaeid.2016110.496.

Reiber, C., Schultze-Kraft, R., Peters, M., Lentes, P., Hoffmann, V. 2010. Promotion and adoption of silage technologies in drought constrained areas of Honduras. Tropical Grasslands. 44: 231–245. doi:10.17138/tgft(1)235-239.

Salgado, P., Thang, V.Q., Thu, T.V., Trach, N.X., Cuong, V.C., Lecomte, P., Richard, D. 2013. Oats (Avena strigosa) as winter forage for dairy cows in Vietnam: an on-farm study. Tropical Animal Health and Production. 45: 561-568. doi:10.1007/s11250-012-0260-8.

Sánchez-Gutiérrez, R.A., Gutiérrez-Bañuelos, H., Serna-Pérez, A., Gutiérrez-Luna, R., Espinoza-Canales, A.S. 2014. Producción y calidad de forraje de variedades de avena en condiciones de temporal en Zacatecas, México. Revista Mexicana de Ciencias Pecuarias. 5: 131-142. doi:10.22319/rmcp.v5i2.3220.

Stanislao-Atzori, A., Carta, P., Cannas, A. 2009. Monitoring CP usage in dairy cattle rations by using milk urea as indicator in a nitrate vulnerable area. Italian Journal of Animal

Science. 8: 346-361. doi:10.4081/ijas.2009.s2.253.

Theodorou, M. K., Williams, B.A., Dhanoa, D. M., McAllan, A. B. and France, J. 1994. A simple gas production method using a pressure transducer to determine the fermentation Kinetics of ruminant feeds. Animal Feed Science and Technology. 48: 185-197.

Thornton, P.K.J., Van de Steeg, A.N., Herrero, M. 2009. The impacts of climate change on livestock and livestock systems in developing countries: A review of what we know and what we need to know. Agricultural Systems. 101: 113-127. doi:10.1016/j.agsy.2009.05.002.

Velarde-Guillén, J., López-González, F., Estrada-Flores, J.G., Rayas-Amor, A.A., Heredia-Nava, D., Vicente, F., Martínez-Fernández, A., Arriaga-Jordán, C. M. 2017. Productive, economic and environmental effects of optimised feeding strategies in small-scale dairy farms in the Highlands of Mexico. Journal of Agricultural and Enviromental for International Development. 111: 225-243. doi:10.12895/jaeid.20171.606.

Victor, D.G., Zhou, D., Ahmed, E.H.M., Dadhich, P.K., Olivier, J.G.J., Rogner, H.H., Sheikho, K., Yamaguchi, M. 2014. Introductory Chapter. In: Climate Change 2014: Mitigation of Climate Change. Contribution of Working Group III to the Fifth Assessment Report of the Intergovernmental Panel on Climate Change. Edenhofer, O.R., Pichs-Madruga, Y., Sokona, E., Farahani, S., Kadner, K., Seyboth, A., Adler, I., Baum, S., Brunner, P., Eickemeier, B., Kriemann, J., Savolainen, S., Schlömer, C., Stechow, T.V., Zwickel T. and Minx J.C. (eds.). Cambridge University Press, Cambridge, United Kingdom and New York, NY, USA. pp 11-150.

Wildman, E.E., Jones, G.M., Wagner, P.E., Boman, R.L., Troutt, J.H.F., Lesch, T.N. 1982. A dairy cow body condition scoring system and its relationship to selected production characteristics. Journal of Dairy Science. 65: 495-501.

19

Influence of Age and Climate in the Production of Cenchrus Purpureus in the Ecuadorian Amazon Region

H.A. Uvidia-Cabadina[1]*, J.L. Ramírez-De la Rivera[2], M. de Decker[1], B. Torres[1], E.O. Samaniego-Guzmán[1], D.B. Ortega-Tenezaca[1], D.F. Reyes-Silva[1] and L.A. Uvidia Armijo[3]

[1]Universidad Estatal Amazónica. Centro de Investigación, Posgrado y Conservación de la Biodiversidad Amazónica CIPCA. Km 44 vía Puyo-Tena. Ecuador. E-mail: huvidia@uea.edu.ec
[2]Centro de Estudio de Producción Animal, Universidad de Granma, Cuba
[3]Escuela Superior Politécnica de Chimborazo. Panamericana sur km 1 ½ Riobamba, Ecuador
*Corresponding author

SUMMARY

The objective of this study was to establish the relationship between age and climatic elements and the performance indicators of *Cenchrus purpureus* vc. Maralfalfa. Sprouting ages at 30, 45, 60, 75 and 90 days were evaluated. A randomised block design was used and the variables green and dry matter yield of leaves and stems, total dry matter and dry matter production were measured. Second degree equations were established for age and the variables evaluated, as well as multiple linear curves to determine the relationship between climate factors and performance indicators; in the latter case, age was always used as an independent variable. The regression coefficients were higher than 0.96. The established regression equations explain the close relationship between age, performance, and climatic elements. These can be used to design efficient management systems of this variety. Age had a marked effect on the behaviour of the indicators evaluated, as performance increased. The established regression equations explain the close relationship between age, performance, and climatic elements. These can be used to design efficient management systems of this variety. It was concluded that age had a marked effect on the behaviour of the indicators evaluated, as performance increased.

Keywords: Age; humidity; grass; temperature; regression.

INTRODUCTION

Pastures have been proven to be an appropriate source of feed for cattle, especially in tropical countries. This is due to the high number of species that can be used, the possibility of cultivating them all year round, the ability of the ruminant to use fibrous food, the species do not compete to be used as a food for human consumption and are usually an economic source (Herrera, 2006). However, the lack of good quality forage species adapted to the environmental conditions prevailing in different livestock areas is identified as one of the most limiting problems for livestock development (Sosa, 2004).

To mitigate this situation, great efforts have been made in the introduction of new species and varieties of higher yield and quality such as *Cenchrus purpureus* vc Maralfalfa in the Ecuadorian Amazon Region. However, the influence of age and climate on performance indicators is unknown. Therefore, the objective of this study was to establish the functional relationship between age and climate and different performance indicators in *Cenchrus purpureus* vc. Maralfalfa in the Ecuadorian Amazon Region.

MATERIALS AND METHODS

Study Area

The study was carried out at the Centre of Postgraduate Research and Conservation of Amazonian Biodiversity (CIPCA) belonging to the Universidad Estatal Amazónica (Amazon State University), located at 44 kilometres along the road between Puyo and Tena, in the Carlos Julio Arosemena Tola Canton, Napo Province, at an altitude of 584masl and with the following coordinates: 01 ° 14' 4.105 " S latitude and 77 ° 53 '4.27' 'W longitude.

The average rainfall in the province during the days of the experiment was 1426mm. The temperature in the shade was 23.40°C and 23.70°C in the sun and the average relative humidity of 83.80%.

Soil was classified as inseptisol, subtype Fluvaquentic Eutrudepts. A 1m deep trial pit was made to determine the limits, where limit "A" had a thickness of 15cm. Following on from that, we clearly distinguished two different colourations every 30cm, which became yellower and yellower. The chemical composition of the soil showed a pH of 5.50, organic matter of 26.8% and nitrogen level of 1.3%. his began with selecting the terrain, (regular topography and be

free of excessive shade and watering). the removal of existing weeds and ground completely clearing was done manually. At the end of the preparation, the ground had the following characteristics: a length of 22 m and a width of 43 m, totalling 946 m². Each experimental unit had a rectangular shape of 5 m wide and 6m long with an area of 30 m². The units were separated by 1m-wide paths between each other and also between blocks.

Sowing

The cutting and transport of propagating material took place one day before sowing. It was brought from a prairie established about 3 years ago. The cuttings were selected and cut from the middle part of the Maralfalfa cane (stem) with 3 knots, with a length between 0.30 and 0.50 m and weight between 40 and 70 g.

The selected cuttings were planted or sown by placing a stake at a distance of one metre in the furrows of the plots. A disinfection of the vegetative material was carried out using Vitavax at a rate of 10 g/L of water through a fumigation pump in order to avoid the attack of possible pests or diseases.

A count of the number of cuttings was made in each experimental unit with its corresponding weight and identification by individual cards for each experimental unit. After the establishment of 150 days, a cut of uniformity was made

Sampling

The yield was measured at different sprouting ages by cutting 1 m² from each experimental unit. The total weights, leaf-stem ratio and dead material were weigh by means of a digital weighing machine.

Treatments, experimental design and experimental Measurements

A randomised block design with three replicates was used to evaluate sprouting ages at 30, 45, 60, 75 and 90 days.

The measurements observed were: green matter yield of leaves (GMYL) and of stems (GMYS), total green matter production (TGMP), dry matter yield of leaves (DMYL) and of stems (DMYS) and total production of dry matter (TPDM).

At the beginning of the evaluation in each period, a cut of uniformity was made to 20 cm of the soil (during the time period of the experiment). 25 m² plots corresponding to sprouting ages were delimited (30, 45, 60, 75 and 90) with 50 cm per side to act as borders. The area was neither irrigated nor fertilised during the experiment. The plots constituted 96 % of the pasture to be evaluated.

The yield was determined by the total cut of the plot in each treatment. The botanical composition was taken into account when expressing the yield of the variety under study. After the green weight of the total plot, the leaves and green stems were separated, weighed individually according to Herrera (2006a), dried in an air-circulating oven for 72 hours at 65 °C and this allowed us to determine their yield in DM. For this, 200 g of each sample was used with 4 replicates per treatment.

Statistical Analysis and Calculations

The data was analysed using ANOVA; the "Statistica version 8.1" system for Windows was used. To verify the normality of the data, the Kolmogorov Smirnov test and the Newman-Keuls test were used to determine differences between the means.

The regression equations (linear, quadratic, cubic, logarithmic and Gompertz) were analyzed and the descending method was used to establish the functional relationship between yield and age, as well as between yield and climate. For the selection of the best fit equation we considered a high R^2 value, a high significance, a low standard deviation of elements and estimation, a smaller mean squared error, significant contribution of equation elements and a low indetermination coefficient $(1 - R^2)$.

RESULTS AND DISCUSSION

The green matter yield increased with age (P<0.05) and a quadratic regression equation was adjusted; the highest values were reached at 90 days. Dry matter production increased with age (P<0.05), and a quadratic regression equation was adjusted; the highest value was reached at 90 days, exceeding 19 tons (Figure 1).

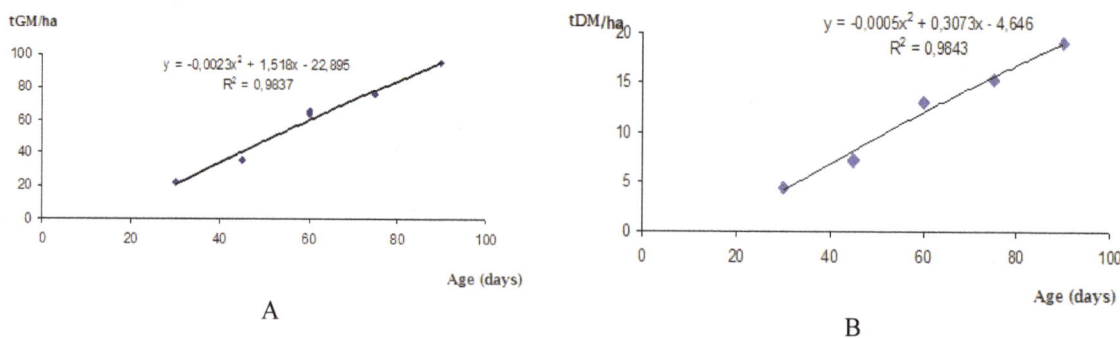

Figure 1: Effect of sprouting age on the total yield of green matter (A) and dry matter (B) (t) of *Cenchrus purpureus* vc. Maralfalfa

Recent studies by Rodríguez et al. (2014) reported linear, quadratic and cubic regression equations by establishing the functional relationship between age and dry matter yield in the *Pennisetum* species in the western region of Cuba. These results are similar to those obtained in this study, although the species are different and the experimental conditions as well. Other studies (Ramírez et al., 2010 and 2012) showed second degree equations for *Pennisetum purpureum* vc CT 169 and *Megathyrsus maximus* in rainy and less rainy seasons respectively, although it is necessary to point that the experimental conditions are different, since annual rainfall in the Ecuadorian Amazon Region exceeds 4500 mm, demonstrating the ecological plasticity of this genus.

The yield of leaves had a similar response pattern when related to age (Table 1). This is an indicator that allows one to establish the composition of the total

yield of the plant, since the leaves show: a) a high probability of increasing the photosynthetic process, b) a greater possibility of substance production for growth, c) a better accumulation of reserves for sprouting, d) a higher quantity of nutrients in the leaves than in the stems, and e) that the animal consumes more leaves than stems (Bayble et al., 2007).

The highest yield of leaves during the first weeks of sprouting can be attributed to the appearance of new offspring and the plant's need to create the substances necessary for its development (Herrera et al., 2008). However, its decline as age increases is associated with increased stem thickness and length (Romero et al., 1998). However, the study done in Mexico by Beliuchenko and Febles (1980) in *Cenchrus purpureus* v Maralfalfa differ from the aforementioned comments and the results of this research: they obtained values of yield of leaves when the cutting ages were higher than 120 days (3.90 t/ha[1]), which is similar to the behaviour of the species evaluated here, but at 75 days of sprouting.

On the other hand, the green and dry matter yield of stems increased with plant age, which amongst other aspects could be due to an increase in structural carbohydrates (cellulose and hemicellulose) as well as lignin, although they were not measured in this

experiment. However, other causes could be: water availability, development of the plant root system and time of year; these can produce morphological changes such as a decrease in foliar laminae and an increase in basal bundles (Cárdena et al., 2012). Nevertheless, it is important to note that in the Ecuadorian Amazon Region, climatic conditions are very similar throughout the year, which means a rapid plant growth, a response to what happened in this study.

Table 1. Relationship between sprouting age and phenological indicators in *Cenchrus purpureus* v Maralfalfa.

Variable	a	b	EE±	c	EE±	R^2	$1-R^2$	CMe	EE±
GMYL	-21.19	1.3	0.17	-0.01	0.001	0.81	0.19	4.28	2.07
GMYS	-1.75	0.21	0.14	0.008	0.001	0.99	0.01	2.91	1.70
DMYL	-4.24	0.26	0.03	-0.002	0.0001	0.81	0.19	0.17	0.41
DMYS	-39.9	0.04	0.02	0.002	0.0001	0.99	0.01	0.11	0.33

R^2 all at $p<0.001$. a = independent factor, b = rainfall for GMYS and DMYL, c = rainfall for the rest. Green matter yield of leaves (GMYL) and of stems (GMYS), dry matter yield of leaves (DMYL) and of stems (DMYS). CMe error square

Other studies, which evaluated the forage production of four germplasm of *Pennisetum purpureum* (Valenciaga et al, 2012) report values similar to those shown in this study. Yields were found for the maralfalfa grass (11.19 t/ha-1 stalk) maintaining an average growth rate of 113 kg/ha[1]/day[-1]. Values obtained here at the age of 75 days denote that the experimental conditions were different, which shows the adaptation of this genus to different ecosystems.

The results of the correlation analysis between the climatic variables suggest that there are correlations between the productive indicators and the climatic variables (Table 2). The relationships established between the productive indicators (GMYS, TPGM, DMYS and TPDM) and the rainfalls indicated the close dependence between these variables.

Research in Cuba conducted by Herrera and Ramos (2006), Ramirez (2010) and Ramírez et al. (2012) reported similar behaviour in Napier, king grass and *Pennisetum purpureum* v. CT 169, respectively, and reported in the first two cases that as the irrigation standard increased, the relationship between irrigation standard and yield declined. Note that in the Ecuadorian Amazon Region, annual rainfall exceeds 4500mm and can reach 6000 mm, with a monthly distribution that exceeds 300mm. This factor can contribute to the good development of this species, which by its behaviour shows the ecological plasticity that it has.

It is argued that water is the essential component of plant cells and that almost all metabolic processes depend on its presence. In addition, water is required

for the maintenance of turgor pressure and the diffusion of solutes in and between the cells; it supplies the hydrogen and oxygen that are involved in the photosynthetic process, which permits the reaffirming of the high correlations with several of the variables, both in performance and chemical composition measurements (Lösch, 1995)

The effect of rainfall on the behaviour of these morphological, biochemical and physiological processes related to the growth and quality of pasture depends on multiple factors that are closely associated with the environment, the soil and its moisture and with plant species. In this sense, literature affirmed that the growth of grasses is a function of the available moisture in the soil and that this, in turn, varies depending on the amount and distribution of precipitation, structure and slope of the soils, the radiation and temperature values, as well as the area covered by vegetation (Del Pozo, 2004). The relationship between the climatic variables of dew point and both the productive and chemical indicators reflect the relationship between humidity and rainfall, since the latter have a close correspondence with the former. By taking into account the results of the correlations, multiple linear equations were established to estimate the GMYL, GMYS, TPGM, DMYL, DMYS and TPDM based on the climatic factors. In all cases, age was used as an independent variable (Table 3). The linear multiple equations presented coefficients higher than 0.96 for all variables evaluated. The CMe values and the standard deviations of the estimation were low, providing, amongst other aspects, the reliability of the established expressions (Table 3).

Table 2. Matrix of correlation of the elements of the climate with the indicators of the performance

Variable	Total rainfall	Outdoors temperature	Temperature in the shade	Thermal sensation	Dew point	Relative humidity
GMYL	-0.61**	-0.16	0.02	-0.03	0.38	0.83**
GMYS	0.92***	-0.90***	-0.96***	-0.95***	-0.96***	-0.47*
TPGM	0.62**	-0.98***	-0.95***	-0.96***	-0.78**	-0.07
DMYL	-0.61**	-0.16	0.02	-0.03	0.37	0.83**
DMYS	0.92***	-0.90***	-0.96***	-0.95***	-0.96***	-0.47*
TPDM	0.62**	-0.98***	-0.95***	-0.96***	-0.78**	-0.07

*p<0.05   **p<0.01   ***p<0.001. Green matter yield of leaves (GMYL) and of stems (GMYS), total green matter production (TGMP), dry matter yield of leaves (DMYL) and of stems (DMYS) and total production of dry matter (TPDM).

Table 3. Simple linear models in *Cenchrus purpureus* vc Maralfalfa

Variable	a	b	EE±	c	EE±	d	EE±	R^2	$1-R^2$	CMe	EE±
GMYL	-4.92	1.44	0.08	-0.07	0.004			0.96	0.04	0.93	0.96
GMYS	1211	1.025	0.03	-14.67	2.62			0.99	0.01	3.89	1.97
TPGM	1022	1.92	0.14	-12.41	0.27	-0.047	0.008	0.99	0.01	2.87	1.69
DMYL	-0.98	0.29	0.016	-0.015	0.0009			0.96	0.04	0.03	0.18
DMYS	240.5	0.20	0.006	-2.91	0.50			0.99	0.01	0.14	0.38
TPDM	220.27	0.09	0.009	-2.65	0.14	0.006	0.0005	0.99	0.01	0.01	0.11

R^2 all at p<0,001   a = independent factor, b = rainfall for GMYS and DMYL, d = humidity for the rest, c = rainfall for the rest. Green matter yield of leaves (GMYL) and of stems (GMYS), total green matter production (TGMP), dry matter yield of leaves (DMYL) and of stems (DMYS) and total production of dry matter (TPDM). (CMe) error square

This demonstrates the reliability of the model, since the method of all possible regressors was used for its construction (Chacin, 1998). Literature reflected the use of mathematical equations to predict biomass production for tropical, temperate and some semi-arid vegetation types (Návar et al., 2001). At present, the number of studies using climate factors in mathematical expressions to explain the behaviour of production and other indicators of pasture is not extensive in the literature published in Ecuador.

Higher regression coefficients (greater than 0.93) were found in more recent studies in the species *Pennsiteum purpureum* vc CT 169 using multiple linear equations, where the yield and chemical composition and minerals P and Ca were related to climatic variables. These authors used mainly rainfall and relative humidity in their models, although the experimental conditions were different, which reaffirms the high correlation between these variables and supports the results of this study (Ramírez et al., 2014).

On the other hand, other studies, which established multiple linear equations to relate climatic variables and age, with the yield and chemical composition of the species *Megathysrus maximus*, achieved high coefficients and great adjustments when they used humidity and rainfall in their models. This is similar behaviour to that found in this work (Ramírez et al., 2011). The green matter yield increased with age (P<0.05) and a quadratic regression equation was adjusted; the highest values were reached at 90 days. Dry matter production increased with age (P <0.05), and a quadratic regression equation was adjusted; the highest value was reached at 90 days, exceeding 19 tons.

CONCLUSIONS

Age had a marked effect on the behaviour of the indicators evaluated, as performance increased. The established regression equations explain the close relationship between age, performance, and climatic elements. These can be used to design efficient management systems of this variety.

REFERENCES

Herrera, R.S. 2006. Fotosíntesis En: Pastos tropicales, contribución a la Fisiología, establecimiento, rendimiento de biomasa, producción de biomasa, producción de semillas y reciclaje de nutrientes. Ed. EDICA. ICA, La Habana, 37.

Sosa, E. Pérez, R. Ortega, RL. Zapata, B.G. 2004. Evaluación del potencial forrajero de árboles y arbustos tropicales para la alimentación de ovinos. Técnicas. Pecuarias México. 42(2):129 -144.

AOAC. 1995. Official Methods of Analysis.. 16th Ed. Ass. Off. Agric. Chem. Washington, D.C.

Rodríguez, L., Larduet R., Martínez, R., Torre, S., Verena, Herrera, Magaly, Medina, Yolaine, Noda A. 2013. Modelación de la dinámica de acumulación de biomasa en *Pennisetum purpureum* vc. king grass en el occidente de Cuba.. Revista Cubana de Ciencias Agrícolas. 47(2): 119-124.

Ramírez, J., Herrera, R., Verdecia, D., Leonard, I., Álvarez, Y. 2010. Rendimiento de materia seca y calidad nutritiva del pasto *Brachiaria brizantha* x *Brachiaria ruziziensis* vc. Mulato en un suelo fluvisol del Valle del Cauto, Cuba. Revista Cubana de Ciencias Agrícolas. 44(1): 65-72.

Ramírez, J., Herrera, R., Leonard, I., Verdecia, D., Alvarez, Y. 2012.Rendimiento y calidad de la *Brachiaria decumbens* en suelo fluvisol del Valle del Cauto, Cuba. REDVET 13(4): 1-11.

Bayble, T., Solomon, M., Prasad, N. 2007. Effects of cutting dates on nutritive value of Napier (*Pennisetum purpureum*) grass planted sole and in association with Desmodium (*Desmodium intortum*) or Lablab (*Lablab purpureus*). Livestock Research for Rural Development 19(1): 2-7.

Herrera, R., Fortes, Dayleni, García, M, Cruz, Ana, Romero Aida. 2008. Estudio de la composición mineral en variedades de *Pennisetum purpureum*. Revista Cubana de Ciencias Agrícolas. 42 (4):395-401.

Romero, C., Medina, R., Flores, R. 1998. Efecto de la fertilización nitrogenada sobre los componentes morfológicos del pasto estrella (*Cynodon plectostachyus*) en la zona de bajo Tocuyo, Estado Falcon. Revista Zootecnia Tropical. 16: 41-45.

Beliuchenko, I., Febles G. 1980. Factores que afectan la estructura de pastos puros de gramíneas. Influencia de la relación hoja-tallo y contenido químico del tallo. Revista Cubana de Ciencias Agrícolas. 14(2):167-174.

Cárdenas, R., Pinto, R., Francisco, J., Medina, F., Guevara, H., Hernández, A., Carmona, J. 2012. Producción y calidad del pasto maralfalfa. Quehacer Científico en Chapas. 1(13): 1-8

Valenciaga, Daiky, Chongo, Bertha, Herrera, R., Torres Verena, Oramas, A., Cairo, J G, Herrera Magali. 2009. Efecto de la edad de rebrote en la composición química del *Pennisetum purpureum* vc. Cuba CT 115. Revista Cubana de Ciencias Agrícolas 43(1): 73-79.

Herrera, RS, Ramos, N. 2006. Factores que influyen en la producción de biomasa y calidad. En *Pennisetum purpureum* para la ganadería tropical. EDICA, La Habana. p. 79.

Ramírez J. 2010. Rendimiento y calidad de cinco gramíneas en el Valle del Cauto. Tesis en opción al grado Científico de Doctor en Ciencias veterinarias. ICA. La Habana. 100.

Ramírez, J., Herrera, R., Leonard, I., Verdecia, D., Alvarez, Y. 2012. Rendimiento y calidad de la *Brachiaria decumbens* en suelo fluvisol del Valle del Cauto, Cuba. REDVET 13(4): 1-11.

Lösch R. Plant water relations. 1995. En: Physiology. Progress in Botany. V 56 Springer Forlag Berlin 56: 55- 96.

Del Pozo P P. 2004. Bases ecofisiológicas para el manejo de los pastos tropicales. Anuario Nuevo. Universidad Agraria de La Habana, Cuba.

Chacin F. 1998. Análisis de regresión y superficie de respuesta. Universidad Central de Venezuela. Primera Edición. Editorial Aragua. Venezuela. 274.

Ramírez, J., Leonard, I., Verdecia, D., Pérez, Yilian, Arseo, Y., Álvarez Y. 2014. Relación de dos minerales con la edad y los elementos del clima en un pasto tropical. REDVET 15(5):1-8.

Ramírez, J., Herrera, R., Leonard, I., Cisneros, M., Verdecia, D., Álvarez, Y. 2011. Relation between climatic factors, yield, and quality of *Pennisetum purpureum* cv. Cuba CT-169 in the Cauto Valley, Cuba. Cuban Journal of Agricultural Science 45(3): 293-297.

20

Genetic Characterization of South America Domestic Guinea Pig using Molecular Markers

D. F. Avilés-Esquivel[1,*], A. M. Martínez[2], V. Landi[2], L. A. Álvarez[3], A. Stemmer[4], N. Gómez-Urviola[5] and J.V. Delgado[2]

[1]Facultad de Ciencias Agropecuarias, Universidad Técnica de Ambato, Cantón Cevallos vía a Quero, sector el Tambo- La Universidad, 1801334, Tungurahua, Ecuador. Email: df.aviles@uta.edu.ec
[2]Departamento de Genética, Universidad de Córdoba, Campus Universitario de Rabanales 14071 Córdoba, España.
[3]Universidad Nacional de Colombia.
[4]Universidad Mayor de San Simón, Cochabamba, Bolivia.
[5]Universidad Nacional Micaela Bastidas de Apurímac, Perú.
*Corresponding author

SUMMARY

Twenty specific primers were used to define the genetic diversity and structure of the domestic guinea pig (*Cavia porcellus*). The samples were collected from the Andean countries (Colombia, Ecuador, Peru and Bolivia). In addition, samples from Spain were used as an out-group for topological trees. The microsatellite markers were used and showed a high polymorphic content (PIC) 0.750, and heterozygosity values indicated microsatellites are highly informative. The genetic variability in populations of guinea pigs from Andean countries was (He: 0.791; Ho: 0.710), the average number of alleles was high (8.67). A deficit of heterozygotes (F_{IS}: 0.153; p<0.05) was detected. Through the analysis of molecular variance (AMOVA) no significant differences were found among the guinea pigs of the Andean countries (F_{ST}: 2.9%); however a genetic differentiation of 16.67% between South American populations and the population from Spain was detected. A poor genetic structure was found among the Andean countries with high genetic variability. The results suggest that it is necessary to take urgent measures to prevent further genetic erosion of native guinea pigs in the Andean countries with plans for recovery and conservation of this important genetic resource in South America.

Keywords: DNA markers, *Cavia porcellus*, South America, genetic diversity.

INTRODUCTION

The guinea pig (*Cavia porcellus*) is a native animal to the Andes (Wing, 1986). This animal plays an important role in the economic income of rural families and, at the same time, it is strongly connected to cultural and religious Pre-Inca traditions (Avilés *et al.*, 2014). The guinea pig and South American camelids are a source of food due to their ability to convert poor vegetable resources to good protein (Avilés *et al.*, 2015). Guinea pig meat contains about 70% dry matter, 21.4% crude protein, 3.0% fat, 0.5% carbohydrate and 0.8% minerals, while chicken meat contains 70.2% dry matter, 18.3% crude protein, 9.3% fat, 1.2% carbohydrates and 1% minerals, which reaches commercial maturity at 3.5 months of age with an average between 800 g to 1200 g (Manjeli, 1998; Zumárraga, 2011). Since the sixteenth century, guinea pig has been introduced in Europa as a pet or scientific experimentation (Guerrini, 2003) and is now widespread in Central and South America. It even has been introduced to sub-Sahara Africa (SSA) where it has an extensive distribution and plays an important role with smallholder farmers in better nutrition and poverty reduction (Manjeli *et al.*, 1998; Matthiesen *et al.*, 2011; Maass *et al.*, 2016).

During the 1970s, in the four Andean countries (Colombia, Ecuador, Peru and Bolivia), the three commercial lines of domestic guinea pigs introduced from Peru, have been phenotypically characterized, but never using molecular marker while several studies have been conducted on guinea pig breeding with the aim to increase meat production performance (INIA, 2005). Native guinea pigs in Ecuador but also in the other Andean countires, due their lower meat production are being substituted by commercial animals without any breeding plan.

Nowadays, there exist studies on genetics of *Cavia porcellus* and its close relatives covered the phylogenetic of living lineages and domestication effects in Latin America (Spotorno *et al.*, 2004; 2006; 2007; Brust and Guenther, 2015), molecular assess of systematics, taxonomy and biogeography of the genus Cavia (Dunnum and Salazar-Bravo, 2010); and differentiation of cryptic genetics differences in wild cavies (Trillmich *et al.*, 2004); no complete genetic study has been carried out on the domestic guinea pig to understand the pattern of genetic variation in the Andean countries. Only one study has been performed with small marker panel of microsatellites in Colombia (Burgos-Paz *et al.*, 2011).

Microsatellite markers have been used, among others, for the characterization, genetic diversity and differentiation assessment, the reciprocal influence of the genetic relationships between one or more breeds' populations on each other, paternity testing and kinship studies. Currently, it is also used as a tool for genetic differentiation between domestic species (Martínez *et al.*, 2000).

In order to inquire about the genetic diversity and structure of seven domestic guinea pig populations reared in the four Andean countries; this study included one Spanish commercial population from an experimental population as an out-group. The aim of this work was to evaluate the diversity and genetic structure of guinea pig, using microsatellite markers, so as to undertake a program of genetic resources' conservation.

MATERIALS AND METHODS

Sample collection

Hair from the back part of 476 animals from four South American Andean countries were analyzed, 282 samples corresponded to the three different commercial lines: Andean (AND, 94), Inti (INTI, 94) and Peru (LPR, 94) (Chauca, 1997); these samples were obtained from ten Andean provinces from Ecuador: Carchi, Imbabura, Pichincha, Cotopaxi, Tungurahua, Bolivar, Chimborazo, Cañar, Azuay, and Loja. Samples from native Andean guinea pigs from currently guinea pig meat consuming countries were obtained: Ecuador (NTVE, 94), Colombia (COL, 17), Peru (PERU, 41) and Bolivia (BOL, 13) (Figure 1). As an out-group, 29 samples of guinea pig from Spain (SPAIN) were included, because there might be a genetic variation due to the adaptation of the environment of guinea pigs carried 500 years ago. All these native samples were obtained from the BIOCUY consortium, established within the CONBIAND Network (http://www.uco.es/conbiand/Bienvenida.html). Molecular marker analysis were carried out at the Applied Molecular Genetics laboratory from Animal Breeding Consulting Company S.L. (ABC) of the University of Cordoba, Spain.

Molecular Markers

The marker set was previously studied in Avilés *et al.* (2015). The final panel is listed in Table 1.

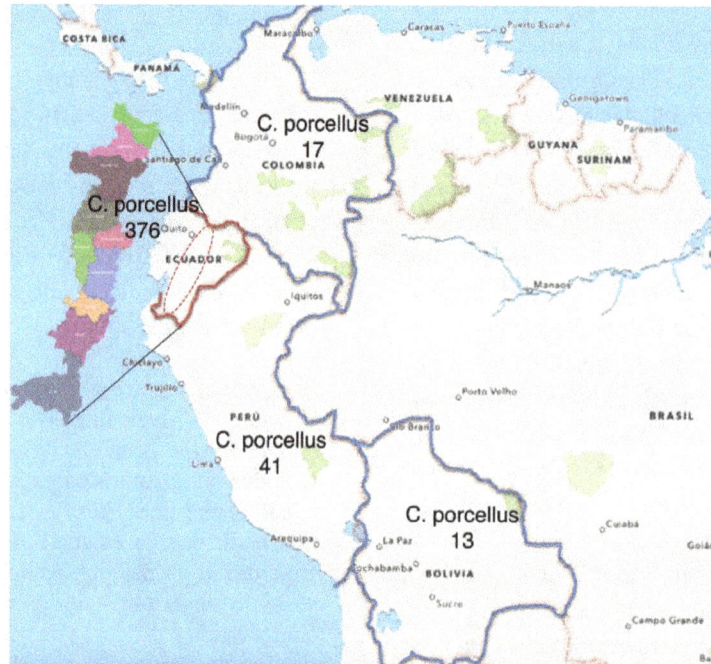

Figure 1. Geographical map of samples from Andean countries that consume guinea pig meat

Table 1. General characteristics of microsatellites

Locus	GB	RP	MX	Tm	SR	Forward	Reverse
CUY01	KP115879	GT	2	55	271-285	CTTTCAGGCAATAGGCATCC	GCAGCTTGGACTACAGAGCA
CUY02	KP115880	CA	2	55	250-262	CAAGATGCCATCAACTTTCGT	TGTTGCTGAGATGCTGCTTT
CUY03	KP115881	GT	1	55	212-252	GCAAGTCAAATTCATCCCTGA	GAGTCCTGCCAAGCAAAATC
CUY04	KP115882	GT	2	55	210-230	TCATCTCGCTTCAGCATTTG	AATGGTCAGGGGCTAGGATT
CUY05	KP115883	CA	2	55	141-163	GGCCAAAGCAGGAATGTCTA	TAGGGCAAGCATTGATGATG
CUY06	KP115884	CA	4	55	158-168	TGGCTTGCTTTCTCTTTGGT	CTGTGCTCAGCATTGCATTT
CUY07	KP115885	CA	2	55	183-197	GATGCAGTGCAGAGGAGTCA	TGTGTGGTTTTGTGTGTGAGG
CUY08	KP115886	TC	1	55	181-217	TGATTGCACCTGAGAAGTGG	CCAAGTGTTCTTGGTGCTTG
CUY09	KP115887	GT	2	55	116-130	GCTGGAATGCAAGACAAGC	TGAGTTTTCAGCTGTGATGAGT
CUY10	KP115888	GT	1	55	106-128	TTCCAAGCATTTCAGAAAACA	TGACTTCCCAACCAAGGAAA
CUY12	KP115889	AG	4	55	232-250	GGAATGGTGGCAAACTCCTA	TCTCCTCCTCCTCCTCCTTC
CUY16	KP115890	AT	3	60	223-247	TTTGAGTCAAGCCGTGAACA	GCCTGTTTTGAAACTGTTTTACTG
CUY17	KP115891	TC	4	55	152-170	TGATGGACAATATACTGGGAACC	TAGCATGCATGAAGCCCTAA
CUY18	KP115892	CA	2	55	176-214	TGTCACTTCTCACTCCACCA	TCCCAAACCTCTTGTTTGCT
CUY20	KP115893	AT	4	55	218-258	TCTTGGAAATGGCCTACATTTT	TGGTCTCTAGGGGTATCCATT
CUY22	KP115894	TC	4	55	206-232	CGAACATGCCAAGCAGATTA	ACACCAGTTCCTTGCCACAT
Cavy02*	AJ496560	AC	2	55	124-154	GGCCATTATGCCCCCCAAC	AGCTGCTCCTTGTGCTGTAG
Cavy03*	AJ496561	CT	1	55	195-225	ACAGCGATCACAATCTGCAC	GCAGTGGTAACCCAGAATGG
Cavy11*	AC192015	CT	1	55	140-180	CCGTGCTTTTCCTGTCTTTG	TGGACCCCAATCTGACATAG
Cavy12*	AC182323	AG	1	55	143-187	AGAATGCCTTTGGGACTGG	AGATCTTGCCTCTGCACTTG

GB: GenBank accession number; RP: microsatellite repeat motive; MX: polymerase chain reaction multiplex reaction where the locus amplified; Tm: annealing temperature of polymerase chain reaction; SR: size range in base pairs. * Selected Loci from Kanitz *et al.* (2009) and Asher *et al.* (2008)

Microsatellite analysis

Genomic DNA was extracted by incubating 3 hair roots in the presence of 100 µL of 5% Chelex® (Biorad, Göttingen, Germany) resin suspension at 95°C for 15 minutes, 60°C during 20 minutes and 99°C for 3 min. Twenty microsatellite loci were amplified in four multiplex PCRs divided into three electrophoresis sets (Avilés et al., 2015). The PCR products were separated through electrophoresis using a 3130Xl Genetic Analyzer® (Life Technology, Madrid, Spain), using a POP7 polymer and the internal size standard GeneScan500-Rox® (Applied Biosystems, Carlsbad, CA, USA). The allelic typification was achieved through Genescan® 3.1.2 and Genotyper® 3 software packages (Applied Biosystems, Carlsbad, CA, USA).

Statistical analyses

Mean number of alleles, observed, and unbiased expected gene diversity estimates and their standard deviations were obtained with the MS® Excel Microsatellite Toolkit software (Park, 2001) (Dublin, Ireland). The distributions of gene variability within and between breeds were studied through the analysis of F-statistics (Weir and Cockerham, 1984) as implemented in Genetix® 4.05 (Belkhir et al., 2003) (Montpellier, France). The within-breed inbreeding coefficient (F_{IS}) in each population was calculated with a 95% confidence interval. Deviations from Hardy–Weinberg equilibrium (HWE) were assessed by means of using Genepop® 3.4 software (Raymond and Rousset, 1995). To determine the structure and genetics differentiation among populations (South American and European), an analysis of molecular variance was performed (AMOVA), calculations were assessed with Arlequin® 3.01 (Excoffier and Lischer, 2010) (Lausanne, Switzerland). Genetic distances were calculated (Reynolds et al., 1983) using Populations® 1.2.28 software (Langella, 1999) (Boston, MA, USA).

A distance tree (NeighborNet) was developed from the obtained matrix D_A of Nei et al. (1983) with Splits Tree4® software (Huson & Bryant, 2006) (Tübingen, Germany) to represent the relationships between breeds graphically, as well as to depict the evidence of admixture. The version 2.3.4 of Structure® software (Pritchard et al., 2000) (Stanford, CA, USA) was used to identify the genetic structure, which identifies clusters of related individuals from multilocus genotypes and assigns individuals to identified clusters using a Bayesian algorithm based on the Markov chain Monte Carlo method. The analysis involves an admixture model with correlated allele frequencies. Eight independent runs were conducted with 50,000 interactions during the burn-in phase and 1,000,000 interactions for sampling from K=2 to K=8. The Structure results in graphic representations were obtained with the program Distruct® 1.1 (Rosenberg, 2004). The proportion of each individual genotype in each cluster or breed (q) and the probability of ancestry in other breeds were estimated.

RESULTS AND DISCUSSION

Microsatellite markers

Over the past 40 years, guinea pigs have experimented an increase in their population size from the introduction of improved lines, but the reduction of native animals might be relevant for the future sustainable utilization and conservation of this important "mini livestock" species. All the 20 microsatellite markers used in this study were successfully amplified in all the populations. A total of 216 alleles, with a mean value of 10.80 ± 3.49, were found for the 20 analysed microsatellites loci. All the microsatellites were highly polymorphic with a minimum of 6 alleles (CUY06) and a maximum of 19 (Cavy12). To evaluate the present situation, we have genetically characterized the South American guinea pig population with the efficiency of microsatellite panel has been demonstrated by the large number of alleles detected for the whole population (10.8 ± 3.40) (Avilés et al., 2015), which was higher than the values found for Ivory Coast alleles, 5.98 ± 0.37, in creole guinea pigs by Kouakuo et al. (2015); for Colombian alleles, 6.8 ± 1.64, in domestic cavies (native line and unspecified commercial lines) by Burgos-Paz et al. (2011), and 7.4 alleles were found for Brazilian wild cavies, by Kanitz et al. (2009) and 10 Uruguayan alleles by Asher et al. (2008).

Breed diversity

The mean number of alleles for all the eight populations was high (8.67±2.65), ranging from a low 4.85 (SPAIN) to a high 11.15 (INTI). Overall genetic diversity was high (H_e = 0.733 ± 0.025). F_{IS} values were significantly different from zero and ranged between 0.072 and 0.327. All the breeds showed a significant heterozygosity deficit (0.153±0.091) as shown in Table 2. The diversity ratios, represented by heterozygosity, were high in all the South American populations. The SPAIN population obtained the lowest diversity (0.504). Kouakuo et al. (2015) in Ivory Coast, and Burgos-Paz et al. (2011) in Colombia showed a lower diversity than our study. Heterozygotes deficit was found in all the populations (F_{IS} = 0.153). Kouakuo et al. (2015) and Burgos-Paz et al. (2011) showed a high heterozygotes deficit

(0.225 and 0.323) respectively. These indexes indicate high levels of genetic variability in South American population's guinea pigs.

Genetic differentiation and population structure

The values of F_{ST} (Table 3, above diagonal) corresponding to the value of genetic differentiation by pairs of breeds, ranged from 0.006 (LPR vs INTI) to 0.2829 (BOL vs SPAIN). Reynolds' pairwise genetic distance (Table 3, below diagonal) ranged from 0.0012 (LPR vs INTI) to 0.3392 (BOL vs SPAIN). Reynolds' pairwise genetic distance (Table 2, below diagonal) ranged from 0.0012 (LPR vs INTI) to 0.3392 (BOL vs SPAIN). The SPAIN population accounted for the greatest distance from all the guinea pig populations in this study. The F_{ST} value (0.029) by Wright and G_{ST} value (0,064) by Nei shows that genetic differentiation between South America populations is very small (Table 4).

Table 2. Summary of the statistics for the eight populations' genetic parameters

Pop	N	MNA	H_e	H_e Ds	H_o	H_o Ds	F_{is}	CI 95%	HWE
AND	94	10.80	0.792	0.017	0.735	0.010	0.072	0.031 - 0.107	5*
INTI	94	11.15	0.787	0.020	0.700	0.011	0.112	0.060 - 0.153	6*
LPR	94	10.90	0.789	0.019	0.709	0.011	0.103	0.056 - 0.142	8*
NTVE	94	10.90	0.797	0.019	0.697	0.011	0.127	0.072 - 0.178	9*
PERU	41	8.50	0.761	0.020	0.707	0.016	0.072	0.005 - 0.114	4*
BOL	13	5.45	0.694	0.032	0.474	0.031	0.327	0.098 - 0.457	10*
COL	17	6.80	0.736	0.020	0.556	0.027	0.250	0.057- 0.371	8*
SPAIN	29	4.85	0.504	0.051	0.424	0.021	0.162	0.035 - 0.261	4*
Mean	59.5	8.67	0.733	0.025	0.625	0.017	0.153	0.052-0.233	6.75

The following estimates were obtained through averaging across the 20 microsatellites: sample size (N), mean number of alleles (MNA), expected (He) and observed (Ho) heterozygosity, within-breed deficit in heterozygosity (F_{IS}) and the confidence interval, and the number of loci deviated from HWE proportions (HWE). Populations abbreviations: AND: commercial line Andean, INTI: commercial line Inti, LPR: commercial line Peru, NTVE: Native Ecuadorean, PERU: Native Peruvian, BOL; Native Bolivian, COL: Native Colombian, SPAIN: out-group from Spain.

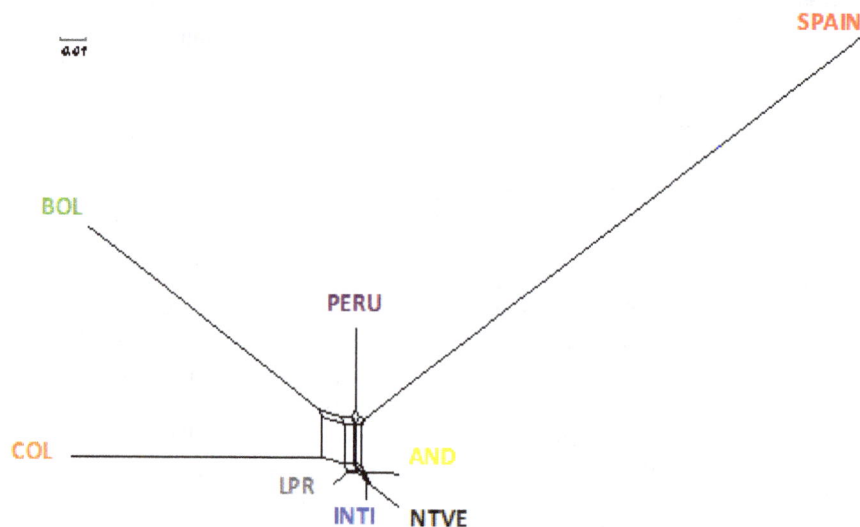

Figure 2. Neighbor-Net dendogram representing the Reynolds genetic distances between the eight studied populations.

These particular differentiations can be based on the great interchange of male and females without control and registration between Andean country markets.

This open reproduction system among the South American populations favours their migration (Figure 2).

Table 3. Estimated pairwise F_{ST} as a measure of genetic differentiation (above diagonal) and Reynolds genetic distances (below diagonal).

Pop	AND	INTI	LPR	NTVE	PERU	BOL	COL	SPAIN
AND	0	0.0028	0.0013	0.0044	0.0182	0.0409	0.0212	0.1828
INTI	0.0033	0	0.0006	0.0023	0.0191	0.03196	0.0298	0.1822
LPR	0.0019	0.0012	0	0.0047	0.0155	0.0320	0.0236	0.1839
NTVE	0.0050	0.0030	0.0054	0	0.0264	0.0476	0.0223	0.1825
PERU	0.0190	0.0202	0.0165	0.0278	0	0.0379	0.0425	0.1979
BOL	0.0440	0.0355	0.0340	0.0521	0.0420	0	0.0594	0.2829
COL	0.0236	0.0330	0.0265	0.0255	0.0465	0.0718	0	0.2530
SPAIN	0.2029	0.2026	0.2046	0.2032	0.2224	0.3392	0.2966	0

AND: commercial line Andina, INTI: commercial line Inti, PLR: commercial line Peru, NTVE: Ecuador's native, PERU: Peru's native, BOL: Bolivia's native, COL: Colombia's native and SPAIN out-group from Spain.

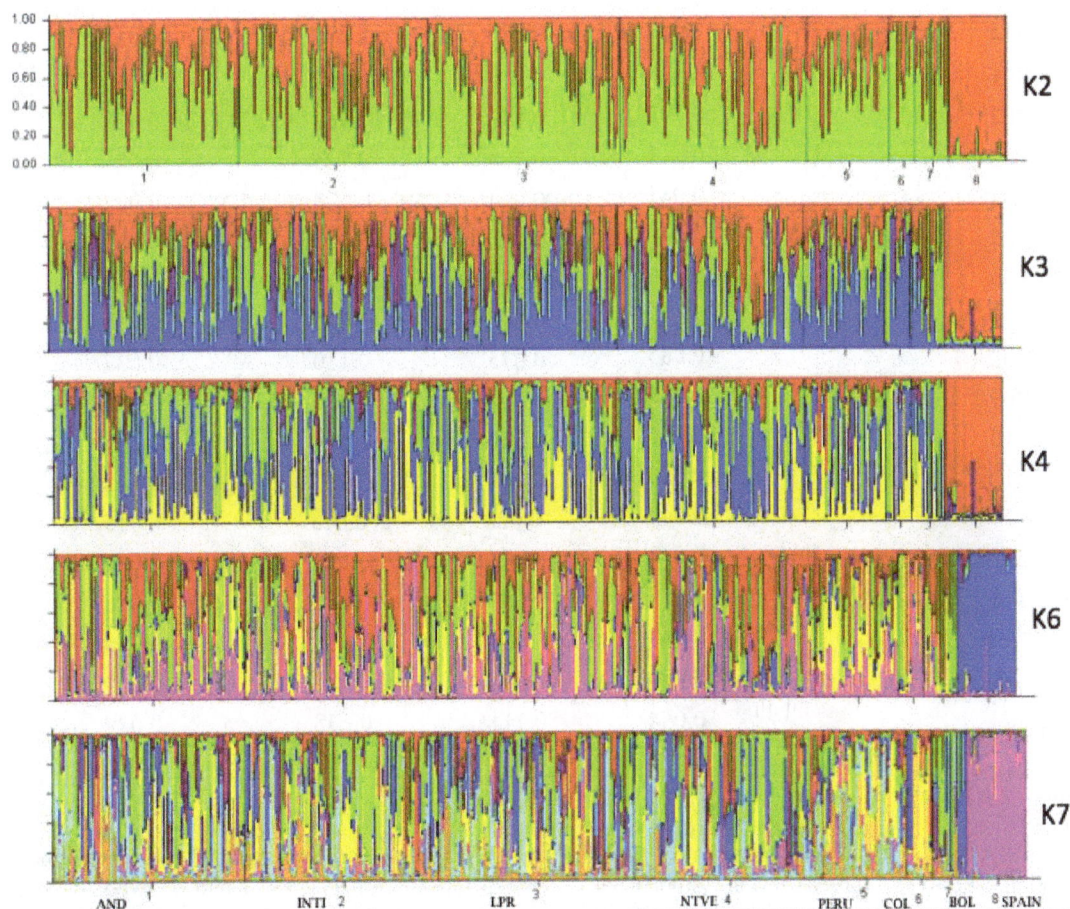

Figure 3. Graphical representation of the genetic structure of the 8 populations analysed.

The same way we can appreciate that between Andean populations we cannot find a clear population structure (Figure 3). The interesting trade started from the first settlers of South America which began to venture into new territories creating a trade route that began on the coasts of Ecuador and Peru with the trade of Spondilus or Mullu shell (*Spondilus calcyfer*) extended by the Pacific Ocean to Michoacan in Mexico to the north, crossing Central America, Colombia, Ecuador, Peru, and Bolivia (Hocquenghem, 2009), during these trips according to Sthal and Norton (1984) guinea pigs and ducks were transported to feed the crew and exchange with the pre-Columbian people. Thus, the inter-trade route, which persists to this day, was established. In the

works of Burgos- Paz *et al.* (2011) and Kouakuo *et al.* (2015) neither observed genetic structure in the populations studied

The Neighbor-Net dendogram is presented in Figure 2. The tree shows the populations from Ecuador (AND, INTI, LPR and NTVE) clustered in the same branch; COL, PERU and BOL appeared in separate clusters, while the out-group from Spain showed the greatest distance and the longer branch when comparing it to the studied South America populations. This Neighbor-Net dendogram shown the guinea pigs in the Andean population seem to have had a common origin in one single branch as showed in the studies realized by Spotorno (2004, 2006, 2007).

Table 4. Values of the coefficient of genetic variation (G_{ST}) and F statistics (F_{IS}, F_{IT} y F_{ST}).

Locus	NA	G_{ST}	F_{IS}	F_{IT}	F_{ST}
Cavy02	9	0.083	-0.002	0.040	0.041
Cavy03	13	0.066	0.171	0.196	0.031
Cavy11	17	0.032	0.103	0.116	0.014
Cavy12	18	0.030	0.328	0.336	0.013
CUY01	8	0.110	0.019	0.057	0.039
CUY02	7	0.107	0.113	0.163	0.055
CUY03	11	0.111	0.104	0.145	0.047
CUY04	9	0.081	0.094	0.128	0.037
CUY05	12	0.047	0.059	0.077	0.019
CUY06	6	0.126	0.087	0.134	0.050
CUY07	7	0.017	0.442	0.446	0.007
CUY08	17	0.085	0.075	0.110	0.037
CUY09	7	0.023	0.101	0.110	0.009
CUY10	11	0.066	0.110	0.134	0.027
CUY12	9	0.020	0.068	0.085	0.018
CUY16	11	0.040	-0.028	-0.013	0.015
CUY17	10	0.043	0.106	0.130	0.026
CUY18	10	0.063	0.163	0.185	0.026
CUY20	14	0.044	0.068	0.084	0.016
CUY22	10	0.096	0.046	0.088	0.044
Mean		0.064	0.111	0.138	0.029
Ds		0.073	0.107	0.101	0.015

AMOVA results (Table 5) indicated the differentiation between breeds was significant (16.67%), when all the South American population (or group one) and the out-group (SPAIN) (or group two) were considered. AMOVA and Structure analysis confirmed the general features observed in the Neighbor-Net dendogram. The results indicate

that, the Spain population represented the highest differentiation (16.67%) and showed the population structure. This differentiation began with the discovery of America; the colonists took guinea pigs to Europe, where they quickly became popular as exotic pets among the upper classes and royalty, including Queen Elizabeth I (Morales, 1995). In

Europe and USA, guinea pigs are considered pets, so that, Cavy clubs and associations devoted to showing and breeding guinea pigs have been established worldwide. Data Bayesian analysis through Structure program (Pritchard *et al.*, 2000) revealed no clear population structure in South American guinea pigs, while the out-group from Spain presented a clear separation as shown in Figure 3.

Table 5. Genetic variation of domestic guinea pigs between the populations of South America and Spain.

Source of variation	d.f	Sum of squares	Variance components	Percentage of variation	
Among groups	1	179.712	1.527	Va	16.67
Among populations within groups	6	110.989	0.091	Vb	0.99
Within populations	944	7117.24	7.539	Vc	82.34
Total	951	7407.941	9.157		

Fixation Indices

F_{ST}: 0.1766

F_{SC}: 0.0119

F_{CT}: 0.1667

Significance test (1023 permutations)

Vc and F_{ST}	P(rand. val < obs. val)	0.0000*
	P(rand. val = obs. val)	0.0000*
	P(rand. val <= obs. val)	0.0000+- 0.0000*
Vb and F_{SC}	P(rand. val > obs. val)	0.0000*
	P(rand. val = obs. val)	0.0000*
	P(rand. val >= obs. val)	0.0000+-0.0000*
Va and F_{CT}	P(rand. val > obs. val)	0.0000*
	P(rand. val = obs. val)	0.13294
	P(rand. val >= obs. val)	0.13294+-0.01139

*$p<0.05$. Gruop1: South America's guinea pig (AND, INTI, LPR, NTVE, COL, PERU and BOL)
Gruop2: Out-group (SPAIN)

CONCLUSION

This study has shown the guinea pigs in the Andean population seem to have had a common origin in one single branch. Over all, the results indicate the population of Latin American guinea pigs has a high genetic variability and poor population structure. The results suggest that it is necessary to take urgent measures to prevent the further genetic erosion of native guinea pigs from Andean countries. We should design and implement recovery and conservation plans for native andean guinea pigs to prevent the loss of this autochthonous genetic resource from South America. On the other hand, the comercial lines do not seem to have a clear population structure, needing to improve their marketing channel by genetically defining these popuations, without affecting the native ones.

REFERENCES

Asher, M., Lippmann, T., Epplen, J. T., Kraus, C., Trillmich, F., Sachser, N. 2008. Large males dominate: ecology, social organization, and mating system of wild cavies, the ancestors of the guinea pig. Behav Ecol Sociobiol 62: 1509-1521. Doi: 10.1007/s00265-008-0580-x

Avilés, D., Martínez, M. A., Landi, V., Delgado J. V. 2014. El cuy (*Cavia porcellus*). Un recurso andino de interés agroalimentario. Recursos Genéticos Animales 55: 87-91. Doi: https://doi.org/10.1017/S2078633614000368

Avilés, D., Landi, V., Delgado J. V., Vega-Pla, J. L., Martínez, M. A. 2015. Isolation and characterization of dinucleotide microsatellite set for a parentage and biodiversity study in domestic guinea pig (*Cavia porcellus*). Italian Journal of Animal Science. 14(3960): 615-620pp. Doi: http://dx.doi.org/10.4081/ijas.2015.3960

Belkhir, K., Borsa, P., Chikhi, L., Raufaste, N., Bonhomme, F. 2003. Genetix: 4.05 Logiciel sous WindowsTM pour la genetique des populations, In: U. d. Montpellier Ed. Montpellier, France.

Burgos-Paz, W., Ceron-Muñoz, M., Solarte-Portilla, C. 2011. Genetic Diversity and Population Structure of the Guinea pig (Cava porcellus, Rodentia, Caviidae) in Colombia. Genet. Mol. Biol. 34 (4): 711-718. Doi: 10.1590/S1415-47572011005000057

Brust V., Guenther A. 2015. Domestication effects on behavioural traits and learning performance: comparing wild cavies to guinea pigs. Animal Cognition 18(1): 99-109. Doi: 10.1007/s10071-014-0781-9

Chauca, L. 1997. Producción de cuyes (*Cavia porcellus*). FAO, Roma, Italia, Pp.77.

Dunnum, J. L., Salazar-Bravo, J. 2010. Molecular systematics, taxonomy and biogeography of the genus Cavia (Rodentia: Caviidae). J Zool Syst Evol Res 48(4): 376-388. Doi: 10.1111/j.1439-0469.2009.00561.x

Excoffier, L., Lischer, H. E. 2010. Arlequin suite ver 3.5: a new series of programs to perform population genetics analyses under Linux and Windows. Molecular Ecology Resources 10: 564-567.

Guerrini, A. 2003. Experimenting with humans and animals: from Galen to animal rights. Johns Hopkins University Press, Baltimore, MD, USA. Pp. 165.

Hocquenghem, A. M. 2009. El *Spondylus princeps* y la edad de bronce de los Andes centrales. Congreso Internacional de Americanistas. ICA.Pp. 53.

Huson, D. H., Bryant, D. 2006. Application of Phylogenetic Networks in Evolutionary Studies. Mol Biol Evol 23: 254-267. Doi: 10.1093/molbev/msj030

Instituto Nacional de Innovación Agraria. 2005. Generación de Líneas Mejoradas de Cuyes de Alta Productividad. http://www.inia.gob.pe/images/AccDirectos/publicaciones/cuyes/doc/INIA-INCAGRO2005.pdf

Kanitz, R., Trillmich, F., Bonatto, S. L. 2009. Characterization of new microsatellite loci for the South-American rodents *Cavia aperea* and *C. magna*. Conservation Genet Resour 1: 47–50. Doi: 10.1007/s12686-009-9011-1

Kouakou, P. K., Skilton, R., Apollinaire, D., Agathe, F., Beatrice, G., Clément, A. S. 2015. Genetic diversity and population structure of cavy (Cavia porcellus L) in three agro ecological zones of Côte d'Ivoire. International Journal of Agronomy and Agricultural Research 6: 27-35. http://www.innspub.net/wp-content/uploads/2015/03/IJAAR-V6No3-p27-35.pdf

Langella, O. 1999. Populations. Boston. http://bioinformatics.org/populations/.

Maass, B. L., Chauca, L., Wanjiku Ch., Sere, C. 2016. From "cuy" South Ameria to cavy in Sub-Sahara Africa: advancing development through South- South coopetarion. Conference: Tropentag" Solidarity in a competing world-fair use of resources.18-21 September, BoKU, Vienna, Austria. URL: http://www.tropentag.de/abstract.php?code=Fj4wN4Am

Manjeli, Y., Tchoumboue, J., Njwe, R. M., Teguia, A. 1998. Guinea-pig productivity under traditional management. Trop. Anim. Health Pro. 30:115-122. Doi: 10.1023/A:1005099818044

Martínez, A. M., Delgado, J. V., Rodero, A. Vega-Pla, J. L. 2000. Genetic structure of the Iberian pig breed using microsatellites. Animal Genetics. 31(5): 295-301.

Matthiesen, T., Nyamete, F., Msuya, J. M., Maass, B. L. 2011. Importance of guinea pig husbandry for the livelihood of rural people in Tanzania: a case study in Iringa region. http://www.tropen-tag.de/2011/abstracts/links/Matthiesen_llDdf2DY.pdf

Morales, E. 1995. The guinea pig: Healing, food, and ritual in the Andes. Pp. 177.

Nei, M., Tajima, F., Tateno, Y. 1983. Accuracy of estimated phylogenetic trees from molecular data. II. Gene frequency data. J Mol Evol. 19 (2):153-170

Park, S. D. E., 2001. Trypanotolerance in west african cattle and the population genetics effects of selection. Ph.D. dissertation, University of Dublin.

Pritchard, J. K., Stephens, M., Donnelly, P. 2000. Inference of Population Structure Using Multilocus Genotype Data. Genetics 155 (2):

945-959.
http://www.genetics.org/content/155/2/945

Raymond, M., Rousset, F. 1995. GENEPOP: Population genetics software for exact test and ecumenicism. J. Hered., 86, 248–249.

Reynolds, J., Weir, B. S., Cockerham, C. C. 1983. Estimation of the coancestry coefficient: Basis for a short-term genetic distance. *Genetics* 105(3): 767-769.

Rosenberg, N. A. 2004. DISTRUCT: a programme for the graphical display of population structure. Mol Ecol Notes.4:137-138. Doi: 10.1046/j.1471-8286.2003.00566.x

Spotorno, A. E., Valladares, J. P., Marín, J. C., Zeballos, H. 2004. Molecular diversity among domestic guinea-pigs (*Cavia porcellus*) and their close phylogenetic relationship with the Andean wild species Cavia tschudii. Rev Chil Hist Nat 77: 243–250. Doi: http://dx.doi.org/10.4067/S0716-078X2004000200004

Spotorno, A. E., Marín, J. C., Manríquez, G., Valladares, J. P., Rico, E., Rivas C. 2006. Ancient and modern steps during the domestication of guinea pigs (*Cavia porcellus L.*). J Zool. 270: 57–62. Doi: 10.1111/j.1469-7998.2006.00117.x

Spotorno, A. E., Manriquez, G., Fernandez, L. A., Marín, J. C., Gonzalez, F., Wheeler, J. 2007. Domestication of guinea pigs from a southern Peru-northern Chile wild species and their middle Pre-Columbian mummies. In: Kelt DA, Lessa EP, Salazar-Bravo J, Patton JL. Ed. The Quintessential Naturalist: Honoring the Life and Legacy of Oliver P. P. Pearson. Univ Cal Pubs Zool, Berkeley, Los Angeles, and London. Pp. 1-981. Doi: http://dx.doi.org/10.1525/california/9780520098596.003.0014

Stahl, P., Norton, P. 1984. Animales domésticos y las implicaciones del intercambio precolombino desde Salamago, Ecuador. Miscelánea Antropológica Ecuatoriana 4: 83-92.

Trillmich F., Kraus C., Künkele J., Asher M., Clara M, Dekomien G., Epplen J. T., Saralegui A., Sachser N. 2004. Species-level differentiation of two cryptic species pairs of wild cavies, genera *Cavia* and *Galea*, with a discussion of the relationships between social systems and phylogeny in the Caviinae. Canadian Journal of Zoology 82(3): 516-524. https://doi.org/10.1139/z04-010

Weir, B. S., Cockerham, C. C. 1984. Estimating F-statistics for the analysis of population structure. Evolution 38: 1358-1370. Doi: 10.2307/2408641

Wing, E. S. 1986. D9omestication of Andean mammals. En: F. Vuilleumier & M. Monasterio. Eds., High altitude tropical biogeography, pp. 246-264. Oxford University Press y American Museum of Natural History, Oxford

Zumárraga, S. 2011. Invovaciones Gastronómicas del cuy. Imbabura desde: http://repositorio.utn.edu.ec/handle/123456789/1139

Permissions

All chapters in this book were first published in TSA, by Universidad Autónoma de Yucatá¡n; hereby published with permission under the Creative Commons Attribution License or equivalent. Every chapter published in this book has been scrutinized by our experts. Their significance has been extensively debated. The topics covered herein carry significant findings which will fuel the growth of the discipline. They may even be implemented as practical applications or may be referred to as a beginning point for another development.

The contributors of this book come from diverse backgrounds, making this book a truly international effort. This book will bring forth new frontiers with its revolutionizing research information and detailed analysis of the nascent developments around the world.

We would like to thank all the contributing authors for lending their expertise to make the book truly unique. They have played a crucial role in the development of this book. Without their invaluable contributions this book wouldn't have been possible. They have made vital efforts to compile up to date information on the varied aspects of this subject to make this book a valuable addition to the collection of many professionals and students.

This book was conceptualized with the vision of imparting up-to-date information and advanced data in this field. To ensure the same, a matchless editorial board was set up. Every individual on the board went through rigorous rounds of assessment to prove their worth. After which they invested a large part of their time researching and compiling the most relevant data for our readers.

The editorial board has been involved in producing this book since its inception. They have spent rigorous hours researching and exploring the diverse topics which have resulted in the successful publishing of this book. They have passed on their knowledge of decades through this book. To expedite this challenging task, the publisher supported the team at every step. A small team of assistant editors was also appointed to further simplify the editing procedure and attain best results for the readers.

Apart from the editorial board, the designing team has also invested a significant amount of their time in understanding the subject and creating the most relevant covers. They scrutinized every image to scout for the most suitable representation of the subject and create an appropriate cover for the book.

The publishing team has been an ardent support to the editorial, designing and production team. Their endless efforts to recruit the best for this project, has resulted in the accomplishment of this book. They are a veteran in the field of academics and their pool of knowledge is as vast as their experience in printing. Their expertise and guidance has proved useful at every step. Their uncompromising quality standards have made this book an exceptional effort. Their encouragement from time to time has been an inspiration for everyone.

The publisher and the editorial board hope that this book will prove to be a valuable piece of knowledge for researchers, students, practitioners and scholars across the globe.

List of Contributors

Hugo Antonio Ruiz-Piña and Francisco Javier Escobedo-Ortegon
Laboratorio de Zoonosis y otras Enfermedades Transmitidas por Vector, Centro de Investigaciones Regionales "Dr. Hideyo Noguchi", Universidad Autónoma de Yucatán

Edwin Gutierrez-Ruiz, Roger Ivan Rodriguez-Vivas, Manuel Bolio-Gonzalez and Dianelly Ucan-Leal
Departamento de Salud Animal y Medicina Preventiva, Facultad de Medicina Veterinaria y Zootecnia, Universidad Autónoma de Yucatán

J. C. Mazetto Júnior
Institute of Agricultural Sciences at the Federal University of Uberlandia. Amazonas Av., Umuarama. Uberlândia-MG, Brazil

R. C. A. Sene, R. T. Thuler, J. L. R. Torres and E. F. A. Moreira
Federal Institute of Triangulo Mineiro (IFTM-Uberaba), 4000 João Batista Ribeiro St., Uberaba-MG, Brazil

F. H. Iost Filho
University of Sao Paulo, College of Agriculture "Luiz de Queiroz" (ESALQ-USP), 11 Padua Dias Av., Piracicaba-SP, Brazil

Jesús Jarillo-Rodríguez, Epigmenio Castillo-Gallegos, Braulio Valles de la Mora and Eliazar Ocaña-Zavaleta
Centro de Enseñanza, Investigación y Extensión en Ganadería Tropical de la FMVZ-UNAM. Facultad de Medicina Veterinaria y Zootecnia, Apartado Postal136, Martínez de la Torre C.P. 93600, Veracruz, México

Luis Ramírez y Avilés
Facultad de Medicina Veterinaria y Zootecnia de la Universidad Autónoma de Yucatán. Mérida, Yucatán, Km 15.5 Carretera Mérida-Xmatkuil. Mérida, México

C.Ogoshi
Agricultural Research and Rural Extension Enterprise of Santa Catarina (EPAGRI), Experimental Station of Caçador, Caçador, SC, Brazil

F. S. Carlos, C. R. C. Bittencourt and R. D. Almeida
Rio Grandense Rice Institute (IRGA), Rice Experiment Station, Cachoeirinha, RS, Brazil

A. R. Ulguim and A. J. Zanon
Federal University of Santa Maria (UFSM), Santa Maria, RS, Brazil

Márcia Maria Mauli, Lúcia Helena Pereira Nóbrega, Danielle Medina Rosa and Michelle Tonini
Universidade Estadual do Oeste do Paraná- UNIOESTE – Cascavel – PR, Brazil

Antonio Pedro Souza Filho
EMBRAPA Amazônia Oriental – Belém – PA, Brazil

Adriana Maria Meneghetti
Universidade Tecnologica Federal do Paraná – UTFPR – Santa Helena – PR, Brazil

José C. Segura-Correa, Juan G. Magaña-Monforte and Jesús R. Aké-López
Facultad de Medicina Veterinaria y Zootecnia, Universidad Autónoma de Yucatán, Km. 15.5 carretera Mérida-Xmatkuil, A.P.4-116, Itzimná, Mérida, Yucatán, México

Victor M. Segura-Correa
Centro de Investigación Regional del Sureste, INIFAP km 25 carretera Mérida-Motul, C.P. 97454, Mocochá, Yucatán, México

V. Herrera-Yunga
Carrera de Zootecnia. Facultad de Ciencias Pecuarias, Escuela Superior Politécnica de Chimborazo (ESPOCH). Avenida Panamericana km ½ Riobamba, Ecuador

J. Labanda and G. Escudero-Sanchez
Carrera de Medicina Veterinaria y Zootecnia, Universidad Nacional de Loja (UNL). La Argelia, 110150 Loja – Ecuador

F. Castillo and A. Torres
Centro de Biotecnología, Universidad Nacional de Loja (UNL). La Argelia, 110150 Loja – Ecuador

M. Capa-Morocho
Carrera de Ingeniría Agronómica, Universidad Nacional de Loja (UNL). La Argelia, 110150 Loja – Ecuador
Facultad de Ciencias Agropecuarias, Universidad Técnica de Ambato, Carretera Cevallos-Quero, 180350 Cevallos, Tungurahua, Ecuador

R. Abad-Guamán
Carrera de Medicina Veterinaria y Zootecnia, Universidad Nacional de Loja (UNL). La Argelia, 110150 Loja – Ecuador
Facultad de Ciencias Agropecuarias, Universidad Técnica de Ambato, Carretera Cevallos-Quero, 180350 Cevallos, Tungurahua, Ecuador

Kolawole E. Law-Ogbomo and Agbonsalo U. Osaigbovo
Department of Crop Science, Faculty of Agriculture, University of Benin, PMB 1154, Benin City 300001

Oscar José Smiderle, Manoel Luiz Silva Neto, Jerri Edson Zilli and Krisle Silva
Empresa Brasileira de Pesquisa Agropecuária Roraima, Depto. de Pesquisa de Sementes, Caixa Postal 133, 69301-970, Boa Vista, RR, Brasil

Aline das Graças Souza
Universidade Federal de Pelotas, Instituto de Biologia, Depto de Botânica, Campus Universitário s/n. Capão do Leão, Pelotas, RS Brasil

D. Guevara-Freire, L. Valle-Velástegui, M. Barros-Rodríguez, Carlos Vásquez, H. Zurita-Vásquez, J. Dobronski-Arcos and P. Pomboza-Tamaquiza
Facultad de Ciencias Agropecuarias, Universidad Técnica de Ambato, Carretera Cevallos-Quero, 180350 Cevallos, Tungurahua, Ecuador

Hernán Zurita-Vásquez, Luciano Valle, Marcia Buenaño, Deysi Guevara, Gonzalo Mena and Carlos Vásquez
Facultad de Ciencias Agropecuarias, Universidad Técnica de Ambato, Carretera Cevallos-Quero, 180350 Cevallos, Tungurahua, Ecuador

P. T. Tabot, S. N. Mbega and N. F. J. Tchapga
Department of Agriculture, Higher Technical Teachers' Training College Kumba, University of Buea. Kumba, Cameroon

Montiel Nola
Cátedra Libre de Autismo, Vicerrectorado Académico, Universidad del Zulia, 4001, Maracaibo, Venezuela

Yenddy Carrero
Laboratorio de Biología Molecular y Celular. Facultad de Ciencias de la Salud. Universidad Técnica de Ambato. Ambato-Ecuador

Kendy Eduardo Urdaneta
Research Division, Autism Immunology Unit of Maracaibo, 4001, Maracaibo, Venezuela
Cátedra Libre de Autismo, Vicerrectorado Académico, Universidad del Zulia, 4001, Maracaibo, Venezuela
Departamento de Biología, Facultad Experimental de Ciencias, Universidad del Zulia, Maracaibo, Venezuela

María Andrea Castillo
Research Division, Autism Immunology Unit of Maracaibo, 4001, Maracaibo, Venezuela

Cátedra Libre de Autismo, Vicerrectorado Académico, Universidad del Zulia, 4001, Maracaibo, Venezuela

Mathieu Renouf and Charly Lubin
Laboratorio de Biología Molecular y Celular. Facultad de Ciencias de la Salud. Universidad Técnica de Ambato. Ambato-Ecuador
Faculté de Médecine et de Pharmacie. Université de Poitiers. Poitiers-France

Neomar Semprún-Hernández
Research Division, Autism Immunology Unit of Maracaibo, 4001, Maracaibo, Venezuela
Cátedra Libre de Autismo, Vicerrectorado Académico, Universidad del Zulia, 4001, Maracaibo, Venezuela
Laboratorio de Biología Molecular y Celular. Facultad de Ciencias de la Salud. Universidad Técnica de Ambato. Ambato-Ecuador

Miguel A. Domínguez-Muñoz and Fernando Martínez-Cordero
Department of Reproduction, College of Veterinary Medicine and Animal Science, Autonomous University of Tamaulipas, Mexico

Miguel Mellado
Department of Animal Nutrition, Universidad Autónoma Agraria Antonio Narro, Saltillo, México

Cecilia C. Zapata-Campos and Jaime Salinas-Chavira
Department of Nutrition, College of Veterinary Medicine and Animal Science, Autonomous University of Tamaulipas, Mexico

Claudio Arzola-Alvarez
Department of Animal Nutrition, College of Animal Science and Ecology, Universidad Autónoma de Chihuahua, Chihuahua, México

G.N. Karuku
Department of Land Resources and Agricultural Technology, University of Nairobi

M. Maobe
Department of Plant Science and Crop Protection, University of Nairobi

Yuneisy Milagro Agüero-Fernández, Luis Guillermo Hernández-Montiel, Bernardo Murillo-Amador, Alejandra Nieto-Garibay, Enrique Troyo-Diéguez and Carlos Michel Ojeda-Silvera
Centro de Investigaciones Biológicas del Noroeste S.C., La Paz, Baja California Sur, México

Ramón Zulueta-Rodríguez
Facultad de Ciencias Agrícolas, Universidad Veracruzana, Campus Xalapa, Xalapa, Veracruz, México Ramón

M. Valle-Parra, P. Pomboza-Tamaquiza, M. Buenaño-Sánchez, D. Guevara-Freire, C. Vásquez and M. Pérez-Salinas
Facultad de Ciencias Agropecuarias, Universidad Técnica de Ambato, Cantón Cevallos vía a Quero, sector el Tambo- La Universidad, 1801334, Tungurahua, Ecuador

P. Chasi-Vizuete
Universidad Técnica de Cotopaxi. Faculty of Agricultural Sciences and Natural Resources. Av. San Felipe s/n Latacunga, Ecuador

D.F. Avilés-Esquivel, M.A. Montero, H. Zurita-Vásquez and M. Barros-Rodríguez
Facultad de Ciencias Agropecuarias, Universidad Técnica de Ambato, Carretera Cevallos-Quero, 180350 Cevallos, Tungurahua, Ecuador

Maria Mitsi Nallely Becerril-Gil, Felipe López-Gonzalez, Julieta Gertrudis Estrada-Flores and Carlos Manuel Arriaga-Jordán
Instituto de Ciencias Agropecuarias y Rurales (ICAR), Universidad Autónoma del Estado de México. Campus UAEM El Cerrillo, El Cerrillo Piedras Blancas, Toluca, Estado de México. C.P. 50090

H.A. Uvidia-Cabadina, M. de Decker, B. Torres, E. O. Samaniego-Guzmán, D. B. Ortega-Tenezaca and D. F. Reyes-Silva
Universidad Estatal Amazónica. Centro de Investigación, Posgrado y Conservación de la Biodiversidad Amazónica CIPCA. Km 44 vía Puyo-Tena. Ecuador

J. L. Ramírez-De la Rivera
Centro de Estudio de Producción Animal, Universidad de Granma, Cuba

L. A. Uvidia Armijo
Escuela Superior Politécnica de Chimborazo. Panamericana sur km 1 ½ Riobamba, Ecuador

D. F. Avilés-Esquivel
Facultad de Ciencias Agropecuarias, Universidad Técnica de Ambato, Cantón Cevallos vía a Quero, sector el Tambo- La Universidad, 1801334, Tungurahua, Ecuador

A. M. Martínez, V. Landi and J. V. Delgado
Departamento de Genética, Universidad de Córdoba, Campus Universitario de Rabanales 14071 Córdoba, España

L. A. Álvarez
Universidad Nacional de Colombia

A. Stemmer
Universidad Mayor de San Simón, Cochabamba, Bolivia

N. Gómez-Urviola
Universidad Nacional Micaela Bastidas de Apurímac, Perú

Index

www.ingramcontent.com/pod-product-compliance
Lightning Source LLC
Chambersburg PA
CBHW050456200326
41458CB00014B/5201